APPLIED PHYSICAL CHEMISTRY
WITH
MULTIDISCIPLINARY APPROACHES

Innovations in Physical Chemistry: Monograph Series

APPLIED PHYSICAL CHEMISTRY WITH MULTIDISCIPLINARY APPROACHES

Edited by

A. K. Haghi, PhD
Devrim Balköse, PhD
Sabu Thomas, PhD

Apple Academic Press Inc.
3333 Mistwell Crescent
Oakville, ON L6L 0A2 Canada

Apple Academic Press Inc.
9 Spinnaker Way
Waretown, NJ 08758 USA

© 2018 by Apple Academic Press, Inc.
First issued in paperback 2021
No claim to original U.S. Government works
ISBN 13: 978-1-77463-638-1 (pbk)
ISBN 13: 978-1-77188-606-2 (hbk)

Library and Archives Canada Cataloguing in Publication

Applied physical chemistry with multidisciplinary approaches / edited by A.K. Haghi, PhD, Devrim Balköse, PhD, Sabu Thomas, PhD.
(Innovations in physical chemistry: monograph series)
Includes bibliographical references and index.
Issued in print and electronic formats.
ISBN 978-1-77188-606-2 (hardcover).--ISBN 978-1-315-16941-5 (PDF)
1. Chemistry, Physical and theoretical. 2. Chemistry, Technical. I. Thomas, Sabu, editor II. Balköse, Devrim, editor III. Haghi, A. K., editor IV. Series: Innovations in physical chemistry. Monograph series

QD453.3.A67 2018   541   C2018-900170-4   C2018-900171-2

Library of Congress Cataloging-in-Publication Data

Names: Haghi, A. K., editor. | Balköse, Devrim, editor. | Thomas, Sabu, editor.
Title: Applied physical chemistry with multidisciplinary approaches / editors, A.K. Haghi, PhD, Devrim Balköse, PhD, Sabu Thomas, PhD.
Description: Toronto : Apple Academic Press, 2018. | Series: Innovations in physical chemistry. Monograph series | Includes bibliographical references and index.
Identifiers: LCCN 2018000021 (print) | LCCN 2018000709 (ebook) | ISBN 9781315169415 (ebook) | ISBN 9781771886062 (hardcover : alk. paper)
Subjects: LCSH: Chemistry, Physical and theoretical. | Chemistry, Technical.
Classification: LCC QD453.3 (ebook) | LCC QD453.3 .A67 2018 (print) | DDC 541--dc23
LC record available at https://lccn.loc.gov/2018000021

Apple Academic Press also publishes its books in a variety of electronic formats. Some content that appears in print may not be available in electronic format. For information about Apple Academic Press products, visit our website at **www.appleacademicpress.com** and the CRC Press website at **www.crcpress.com**

ABOUT THE EDITORS

A. K. Haghi, PhD

A. K. Haghi, PhD, is the author and editor of 165 books, as well as 1000 published papers in various journals and conference proceedings. Dr. Haghi has received several grants, consulted for a number of major corporations, and is a frequent speaker to national and international audiences. Since 1983, he served as professor at several universities. He is currently Editor-in-Chief of the *International Journal of Chemoinformatics and Chemical Engineering* and *Polymers Research Journal* and on the editorial boards of many international journals. He is also a member of the Canadian Research and Development Center of Sciences and Cultures (CRDCSC), Montreal, Quebec, Canada. He holds a BSc in urban and environmental engineering from the University of North Carolina (USA), an MSc in mechanical engineering from North Carolina A&T State University (USA), a DEA in applied mechanics, acoustics and materials from the Université de Technologie de Compiègne (France), and a PhD in engineering sciences from the Université de Franche-Comté (France).

Devrim Balköse, PhD

Devrim Balköse, PhD, is currently a retired faculty member in the Chemical Engineering Department at Izmir Institute of Technology, Izmir, Turkey. She graduated from the Middle East Technical University in Ankara, Turkey, with a degree in chemical engineering. She received her MS and PhD degrees from Ege University, Izmir, Turkey, in 1974 and 1977, respectively. She became Associate Professor in macromolecular chemistry in 1983 and Professor in process and reactor engineering in 1990. She worked as a research assistant, assistant professor, associate professor, and professor between 1970 and 2000 at Ege University. She was the Head of the Chemical Engineering Department at Izmir Institute of Technology, Izmir, Turkey, between 2000 and 2009. She is now a retired faculty member in the same department. Her research interests are in polymer reaction engineering, polymer foams and films, adsorbent development, and moisture sorption. Her research projects are on nanosized zinc borate production, ZnO polymer composites, zinc borate lubricants, antistatic additives, and metal soaps.

Sabu Thomas, PhD

Sabu Thomas, PhD, is a Professor of Polymer Science and Engineering at the School of Chemical Sciences and Director of the International and Inter University Centre for Nanoscience and Nanotechnology at Mahatma Gandhi University, Kottayam, Kerala, India. The research activities of Professor Thomas include surfaces and interfaces in multiphase polymer blend and composite systems; phase separation in polymer blends; compatibilization of immiscible polymer blends; thermoplastic elastomers; phase transitions in polymers; nanostructured polymer blends; macro-, micro- and nano-composites; polymer rheology; recycling; reactive extrusion; processing–morphology–property relationships in multiphase polymer systems; double networking of elastomers; natural fibers and green composites; rubber vulcanization; interpenetrating polymer networks; diffusion and transport; and polymer scaffolds for tissue engineering. He has supervised 68 PhD theses, 40 MPhil theses, and 45 Masters theses. He has three patents to his credit. He also received the coveted Sukumar Maithy Award for the best polymer researcher in the country for the year 2008. Very recently, Professor Thomas received the MRSI and CRSI medals for his excellent work. With over 600 publications to his credit and over 23,683 citations, with an h-index of 75, Dr. Thomas has been ranked fifth in India as one of the most productive scientists.

INNOVATIONS IN PHYSICAL CHEMISTRY: MONOGRAPH SERIES

This new book series, Innovations in Physical Chemistry: Monograph Series, offers a comprehensive collection of books on physical principles and mathematical techniques for majors, non-majors, and chemical engineers. Because there are many exciting new areas of research involving computational chemistry, nanomaterials, smart materials, high-performance materials, and applications of the recently discovered graphene, there can be no doubt that physical chemistry is a vitally important field. Physical chemistry is considered a daunting branch of chemistry— it is grounded in physics and mathematics and draws on quantum mechanics, thermodynamics, and statistical thermodynamics.

Innovations in Physical Chemistry has been carefully developed to help readers increase their confidence when using physics and mathematics to answer fundamental questions about the structure of molecules, how chemical reactions take place, and why materials behave the way they do. Modern research is featured throughout also, along with new developments in the field.

Editors-in-Chief

A. K. Haghi, PhD
Editor-in-Chief, International Journal of Chemoinformatics and Chemical Engineering and Polymers Research Journal; Member, Canadian Research and Development Center of Sciences and Cultures (CRDCSC), Montreal, Quebec, Canada
Email: AKHaghi@Yahoo.com

Lionello Pogliani, PhD
University of Valencia-Burjassot, Spain
Email: lionello.pogliani@uv.es

Ana Cristina Faria Ribeiro, PhD
Researcher, Department of Chemistry, University of Coimbra, Portugal
Email: anacfrib@ci.uc.pt

BOOKS IN THE SERIES

- High-Performance Materials and Engineered Chemistry
- Applied Physical Chemistry with Multidisciplinary Approaches
- Methodologies and Applications for Analytical and Physical Chemistry
- Physical Chemistry for Engineering and Applied Sciences: Theoretical and Methodological Implication
- Theoretical Models and Experimental Approaches in Physical Chemistry: Research Methodology and Practical Methods
- Engineering Technology and Industrial Chemistry with Applications
- Modern Physical Chemistry: Engineering Models, Materials, and Methods with Applications
- Engineering Technologies for Renewable and Recyclable Materials: Physical-Chemical Properties and Functional Aspects
- Physical Chemistry for Chemists and Chemical Engineers: Multidisciplinary Research Perspectives
- Chemical Technology and Informatics in Chemistry with Applications

CONTENTS

LIST OF CONTRIBUTORS

Burcu Alp
Department of Chemical Engineering, Izmir Institute of Technology, Gulbahce Urla, İzmir, Turkey

Devrim Balköse
Department of Chemical Engineering, Izmir Institute of Technology, Gulbahce Urla, İzmir, Turkey.
E-mail: devrimbalkose@gmail.com

Carla Baptista
Serviço Endocrinologia, Diabetes e Metabolismo, Centro Hospitalar Universitário Coimbra, Coimbra, Portugal

Rajashekhar Bhajantri
Department of Physics, Karnatak University, Dharwad, Karnataka, India

Ana M. T. D. P. V. Cabral
Department of Chemistry, Coimbra Chemistry Centre, University of Coimbra, 3004-535 Coimbra, Portugal
Faculty of Pharmacy, University of Coimbra, Azinhaga Sta. Comba, 3000-548 Coimbra, Portugal.
E-mail: acabral@ff.uc.pt

Francisco Carrilho
Serviço Endocrinologia, Diabetes e Metabolismo, Centro Hospitalar Universitário Coimbra, Coimbra, Portugal

Gloria Castellano
Facultad de Veterinaria y Ciencias Experimentales, Departamento de Ciencias Experimentales y Matemáticas, Universidad Católica de Valencia San Vicente Mártir, Guillem de Castro-94, E-46001 València, Spain

Ching Kang Chen
School of Business and Economics, University of Brunei, Bandar Seri Begawan, Brunei

Sara Pires de Oliveira
Science and Technology Faculty, University of Coimbra, Coimbra, Portugal

Miguel A. Esteso
U.D. Química Física, Facultad de Farmacia, Universidad de Alcalá, 28871 Alcalá de Henares, Madrid, Spain
Department of Chemistry, Coimbra Chemistry Centre, University of Coimbra, 3004-535 Coimbra, Portugal. E-mail: miguel.esteso@uah.es

Pedro Furtado
Science and Technology Faculty, University of Coimbra, Coimbra, Portugal

Arunkumar Jayakumar
Auckland University of Technology, Auckland, New Zealand

Ajith James Jose
Post Graduate and Research Department of Chemistry, St. Berchmans College (Autonomous), Changanassery, India. E-mail: ajithjamesjose@gmail.com

Sam John
Post Graduate and Research Department of Chemistry, St. Berchmans College (Autonomous),
Changanassery, India

Vinita Kapoor
Department of Chemistry, Sri Venkateswara College, Delhi University, Delhi, India. E-mail: vinita-arora85@gmail.com

Raghvendra Kumar Mishra
International and Inter University Centre for Nanoscience and Nanotechnology, Mahatma Gandhi
University, Priyadarshini Hills P.O., Kottayam, Kerala 686560, India. E-mail: raghvendramishra4489@
gmail.com

Raquel Monteiro
Science and Technology Faculty, University of Coimbra, Coimbra, Portugal

Ishwar Naik
Government Arts and Science College, Karwar, Karnataka, India. E-mail: iknaik@rediffmail.com

Jagadish Naik
Department of Physics, Mangalore University, Mangalore, Karnataka, India

Sukanchan Palit
Department of Chemical Engineering, University of Petroleum and Energy Studies, P. O. Bidholi via
Premnagar, Dehradun 248007, Uttarakhand, India. E-mail: sukanchan68@gmail.com, sukanchan92@
gmail.com

Meha J. Prajapati
Department of Chemistry, Sardar Patel University, Vallabh Vidyanagar 388120, Anand, Gujarat, India

Mekha Susan Rajan
Post Graduate and Research Department of Chemistry, St. Berchmans College (Autonomous),
Changanassery, India

Ana C. F. Ribeiro
Department of Chemistry, Coimbra Chemistry Centre, University of Coimbra, 3004-535 Coimbra,
Portugal. E-mail: anacfrib@ci.uc.pt

Anjalipriya S.
Post Graduate and Research Department of Chemistry, St. Berchmans College (Autonomous),
Changanassery, India

Sevdiye Atakul Savrik
Department of Chemical Engineering, İzmir Institute of Technology, Gulbahce Urla, Izmir, Turkey.
E-mail: sevdiyeatakul@gmail.com

Anamika Singh
Department of Botany, Maitreyi College, University of Delhi, Delhi, India

Rajeev Singh
Department of Environment Studies, Satyawati College, University of Delhi, Delhi, India.
E-mail: 10rsingh@gmail.com

Cemal İhsan Sofuoğlu
Department of Computer Engineering, İzmir Institute of Technology, Gulbahce Urla, İzmir, Turkey

Kiran R. Surati
Department of Chemistry, Sardar Patel University, Vallabh Vidyanagar 388120, Anand, Gujarat, India.
E-mail: kiransurati@yahoo.co.in

Heru Susanto
Department of Computer Science and Information Management, Tunghai University, Taichung, Taiwan
Computational Science, Indonesian Institute of Sciences, Serpong, Indonesia

Sabu Thomas
International and Inter University Centre for Nanoscience and Nanotechnology, Mahatma Gandhi University, Priyadarshini Hills P.O., Kottayam, Kerala 686560, India. E-mail: sabuchathukulam@yahoo.co.uk

Francisco Torrens
Institut Universitari de Ciència Molecular, Edifici d'Instituts de Paterna, Universitat de València, P. O. Box 22085, E-46071 València, Spain

Carolina Travassos
Science and Technology Faculty, University of Coimbra, Coimbra, Portugal

R. R. Usmanova
Ufa State Technical University of Aviation, Ufa 450000, Bashkortostan, Russia. E-mail: Usmanovarr@mail.ru

Fatma Üstün
Department of Chemical Engineering, Izmir Institute of Technology, Gulbahce Urla, İzmir, Turkey. E-mail: fmutlus@gmail.com

Francisco J. B. Veiga
Faculty of Pharmacy, University of Coimbra, Azinhaga Sta. Comba, 3000-548 Coimbra, Portugal
Faculty of Pharmacy, REQUIMTE/LAQV, Pharmaceutical Technology, University of Coimbra, Coimbra, Portugal. E-mail: fveiga@ci.uc.pt

Luis M. P. Verissimo
U.D. Química Física, Facultad de Farmacia, Universidad de Alcalá, 28871 Alcalá de Henares, Madrid, Spain
Department of Chemistry, Coimbra Chemistry Centre, University of Coimbra, 3004-535 Coimbra, Portugal. E-mail: luis.verissimo@uc.pt

G. E. Zaikov
N. M. Emanuel Institute of Biochemical Physics, Russian Academy of Sciences, Moscow 119991, Russia. E-mail: chembio@chph.ras.ru

LIST OF ABBREVIATIONS

AAc	acrylic acid
ADCIS	Advanced Concepts in Imaging Software
ADCY	adenylate cyclase
AFC	alkaline fuel cell
AFM	atomic force microscopy
AI	artificial intelligence
AMIA	American Medical Informatics Association
ANN	artificial neural network
AOPs	advanced oxidation processes
AUT	Auckland University of Technology
BDT	benzodithiophene
BHJ	bulk heterojunction
BIKB	BioInfoKnowledgeBase
BN	boron nitride
BT	benzothiadiazole
CA	cluster analysis
CCG	catalyst-coated gas
CCM	catalyst-coated membrane
CL	catalyst layer
CLAHE	contrast-limited adaptive histogram equalization
CNS	central nervous system
CO	crystal orbital
COR	catalyst-to-oil ratio
CPT	cyclopentadithiophene
DAMC	direct alcohol fuel cell
DBPs	disinfection by-products
DFBT	difluorobenzothiadiazole
DMFC	direct methanol fuel cell
DoE	Department of Energy
DOS	density of states
DPP	diketopyrrolopyrolle
DR	diabetic retinopathy
DTC	dithienocarbazole
DTP	dithienopyrrole

EA	electron affinity
EBI	European Bioinformatics Institute
EC	electrocoagulation
ECPs	electrically conducting polymers
EDG	electron-donating group
EDS	energy dispersion spectroscopy
EDX	energy-dispersive X-ray
EMO	evolutionary multiobjective optimization
EOs	essential oils
EWG	electron-withdrawing group
FCC	fluid catalytic cracking
FCCU	fluid catalytic cracking unit
FDA	Food and Drug Administration
FDFCC	flexible dual-riser fluid catalytic cracking
FET	field-effect transistors
FF	fill factor
FI-HG-AAS	flow injection hydride generation atomic absorption spectrometry
FLD	Fisher linear discriminant
FTIR	Fourier transform infrared
GA	genetic algorithm
GDLs	gas diffusion layers
HDTMOS	hexadecyltrimethoxysilane
HGO	heavy gas oil
HGP	human genome project
HHV	higher heating value
HIV	human immunodeficiency virus
HOMO	highest occupied molecular orbit
HPLC	high-performance liquid chromatography
HT-SOFC	high-temperature solid oxide fuel cell
IC	interconnecting layer
IDT	indacenodithiophene
IIM	inverse iteration methods
IP	ionization potential
IPN	inverse participation number
IT	information technology
IT-SOFC	intermediate-temperature solid oxide fuel cell
LCST	lower critical solution temperature
LP	liquid paraffin
LUMO	lowest unoccupied molecular orbit

MCFC	molten carbonate fuel cell
MEA	membrane and electrode assembly
MOEAs	multiobjective evolution algorithms
MOSA	multiobjective simulated annealing
MTBE	methyl tert-butyl ether
NCBI	National Center for Biotechnology Information
NDMA	N-nitrosodimethylamine
NDs	neurodegenerative diseases
NFC	negative factor counting
NGS	next-generation sequencing
NPDR	nonproliferative diabetic retinopathy
NSGA	nondominated sorting genetic algorithm
OCV	open circuit voltage
OFETs	organic field-effect transistors
OLEDs	organic light-emitting diodes
OPV	organic photovoltaic
PA	polyacetylene
PAA	polyacrylic acid
PAc	polyacene
PAcA	polyacenacene
PAFC	phosphoric acid fuel cell
PANI	polyaniline
PCA	principal components analysis
PCC	Pearson's correlation coefficient
PCDDs	polychlorinated dibenzodioxins
PCE	power conversion efficiency
PCP	pentachlorophenol
PCR	polymerase chain reaction
PDB	Protein Data Bank
PDFA	polydifluoroacetylene
PDR	proliferative diabetic retinopathy
PEI	polyethylenimine
PEM	proton-exchange membrane
PEMFC	proton-exchange membrane fuel cell
PFSA	perfluorosulfonic acid
PFu	polyfuran
PL	polylysine
PMAA	polymethacrylic acid
PMFA	polymonofluoroacetylene
PO	Pareto-optimal

PPh	polyphenanthrene
PPN	polyperinaphthalene
PPV	polyphenylene vinylene
PPy	polypyrrole
PSS	polystyrene sulfonate
PTFE	polytetrafluoroethylene
PTh	polythiophene
PV	photovoltaic
PVDF	polyvinylidene fluoride
RFC	reversible fuel cell
RFLPs	restriction fragment length polymorphisms
SA	simulated annealing
SARS	severe acquired respiratory syndrome
SBML	systems biology markup language
SEM	scanning electron microscope
SERB	Science and Engineering Research Board
SiNPs	silica nanoparticles
SNP	single-nucleotide polymorphism
SOFC	solid oxide fuel cell
SPCs	special pseudo components
SPION	super-paramagnetic iron oxide nanoparticles
SVM	support vector machines
TD-DFT	time-dependent density functional theory
TEM	transmission electron microscopy
TPD	thienopyrrolodione
TT	thienothiophene
UCST	upper critical solution temperature
UV	ultraviolet
VGO	vaccum gas oil
VOCs	volatile organic compounds
WAO	wet air oxidation
WAXD	wide-angle X-ray diffraction
XRD	X-ray diffraction
ZB UFPs	zinc borate ultrafine powders

PREFACE

Physical chemistry emphasizes the intersection of chemistry, mathematics, physics, and the resulting applications across many disciplines of science.

Physical chemistry applies physics and mathematics to problems that interest chemists, biologists, and engineers. Physical chemists use theoretical constructs and mathematical computations to understand chemical properties and describe the behavior of molecular and condensed matter. Their work involves manipulations of data as well as materials. Physical chemistry entails extensive work with sophisticated instrumentation and equipment as well as state-of-the-art computers.

With the development of a variety of exciting new areas of research involving computational chemistry, nano-, and smart materials, there can be no doubt that physical chemistry is a vitally important field. It is also perceived as the most daunting branch of chemistry, being necessarily grounded in physics and mathematics. Physical chemistry need not appear as a large assortment of different, disconnected, and sometimes intimidating topics. Instead, physical chemistry provides a coherent framework for chemical knowledge, from the molecular to the macroscopic level.

The applications from these multidisciplinary fields illustrate methods that can be used to model physical processes, design new products, and find solutions to challenging problems.

Highlighting illustrative case studies, technological applications, and theoretical and foundational concepts, this book is a crucial reference source for graduate students interested in the key concepts for modern technologies and optimization of new processes.

This volume combines up-to-date research findings and relevant theoretical frameworks on applied chemistry, materials, and chemical engineering.

Multidisciplinary approaches present an up-to-date review of modern materials and chemistry concepts, issues, and recent advances in the field. Distinguished scientists and engineers from key institutions worldwide have contributed chapters that provide a deep analysis of their particular subjects. At the same time, each topic is framed within the context of a broader more multidisciplinary approach, demonstrating its relationship and interconnectedness to other areas. The premise of this book, therefore, is to offer both a comprehensive understanding of applied science and engineering as a whole and a thorough knowledge of individual subjects. This approach appropriately conveys the basic fundamentals, state-of-the-art technology, and applications of the involved disciplines, and further encourages scientific collaboration among researchers.

This volume emphasizes the intersection of chemistry, mathematics, physics, and the resulting applications across many disciplines of science and explores applied physical chemistry principles in specific areas, including the life chemistry, environmental sciences, geosciences, and material sciences.

PART I
Applications of Polymers

CHAPTER 1

SMART POLYMERS: A SMART APPROACH TO LIFE

AJITH JAMES JOSE*, MEKHA SUSAN RAJAN, ANJALIPRIYA S., and SAM JOHN

Post Graduate & Research Department of Chemistry, St. Berchmans College (Autonomous), Changanassery, India

Corresponding author. E-mail: ajithjamesjose@gmail.com

CONTENTS

ABSTRACT

Smart polymers, also called stimuli-responsive or environment-sensitive polymers, are intelligent polymers that respond in a dramatic way to slight changes in the environment such as pH, temperature, dual stimuli, light, and phase transition. These systems are able to recover their initial state when the sign or stimuli end. They undergo fast and reversible changes in the microstructure from a hydrophilic to a hydrophobic state that are triggered by small stimuli in the environment. Due to their unique characteristics, such as versatility and tunable sensitivity, they found a variety of applications in biomedical and nanotechnology fields. The exploitation of smart polymeric systems for delivering bioactive agents, including peptide and protein drugs was one of the major breakthrough discoveries. Interest in stimuli-responsive polymers is steadily gaining attention especially in the fields of controlled and self-regulated drug delivery. Smart polymers are representing promising means for targeted drug delivery, tissue engineering, cell culture, textile engineering, radioactive wastage, gene carriers, glucose sensors, and protein purification. The present chapter aims to highlight various kinds of smart polymers and their applications.

1.1 INTRODUCTION

Smart polymer is one that responds to an external stimulus in a controlled, reproducible, and reversible manner. They can respond to a single stimulus or multiple stimuli, for example, temperature, pH, electric or magnetic field, light intensity, biological molecules that induces macroscopic responses in the material, such as swelling/collapse or solution-to-gel transition depending on the physical state of the chains.[1] The term "smart polymers" was coined due to the similarity of stimuli-responsive polymers with biopolymers. There is a strong belief that nature has always been struggling for smart solution in creating life. The dream for scientists is not only to mimic biological processes, and therefore understand them better, but also to create novel species and invent new processes.[2]

Scientists studying the natural polymers found in living organisms (proteins, carbohydrates, and nucleic acids) have examined how they behave in biological systems as they perform their structural and physiological roles. That information is being put to use to develop similar man-made polymeric substances with definite properties and the ability to respond to changes in their environment. Smart polymers are becoming increasingly more common, as scientists learn about the chemistry and trigger that foster conformational changes in polymer structures and create ways to take advantage of them, and eventually control them. New polymeric materials are being chemically devised that recognize specific environmental changes in biological systems, and adjust in a predictable manner, making them useful tools for drug delivery or other metabolic control mechanisms. These polymers are potentially very useful for a variety of applications including some related to biotechnology and biomedicine. Smart polymer nanocarriers for drug delivery applications play a crucial role in the development of highly active and selective treatments, as they permit a controlled delivery of drugs in the right place at the right time. Better knowledge of the molecular biology and synthesis of new polymers with stimulus-sensitive moieties gave rise to more effective, specifically localized action and personalized therapies. Smart polymers sensitive to the presence of some biomarkers are extensively used in targeting specific disease conditions (e.g., smart polymers sensitive to folate receptor can be used to deliver anticancer agents to tumor cells). Polymeric smart coatings have been developed that are capable of both detecting and removing hazardous nuclear contaminants. Polymer-bound smart catalysts are useful in waste minimization, catalyst recovery, and catalyst reuse. Such applications of smart materials involving catalysis chemistry, sensor chemistry, and chemistry relevant to decontamination methodology are especially significant to environmental problems.[3] Smart polymers are biocompatible, strong, resilient, flexible, and easy to sharpen and color.

1.2   CLASSIFICATION OF SMART POLYMERS

1.2.1   pH-SENSITIVE SMART POLYMERS

The pH-sensitive smart polymers are polyelectrolytes containing weak acid or basic groups which can either accept or release protons in response to pH changes in surrounding environment. Polymers with a large number of ionizable groups are known as polyelectrolytes. The ionizable group acts as hydrophilic or hydrophobic part of the polymer. Thus, the transition between soluble and insoluble state is created by decreasing net charge of the polymer molecule. The polymers' electric charge can be decreased by decreasing its pH, neutralizing the electric charge, and reducing the hydrophilicity or increasing hydrophobicity of the macromolecule.

In the human body, we can see remarkable changes of pH that can be used to direct therapeutic agents to a specific body area, tissue, or cell compartment. These conditions make the pH-sensitive polymers ideal for pharmaceutical systems for the specific delivery of therapeutic agents.[4] Polyacids or polyanions (such as carboxylic acid or sulfonic acid) having greater number of ionizable acid groups can accept protons at low pH values and release protons at high pH values. Due to the electrostatic repulsion of the negatively charged groups, the polymer swells as the pH increases. Most of the anionic pH-sensitive smart polymers are based on polyacrylic acid (PAA) (Carbopol) or its derivatives, polymethacrylic acid (PMAA), poly (ethylenimines) (PEI), poly (L-lysine), and poly (N,N-dimethylaminoethyl methacrylamide). The pH-sensitive polymers containing sulfonamide group are another example of polyacid polymers. These polymers have pKa values in the range of 3–11 and the hydrogen atom of the amide nitrogen is readily ionized to form polyacids. Narrow pH range and good sensitivity are the major advantages of these polymers over carboxylic acid-based polymers. Examples of cationic polyelectrolytes are poly (N,N-dialkyl aminoethyl methacrylates), poly (lysine) (PL), PEI, and chitosan. Polybases or polycations are protonated at high pH values and positively ionized at neutral or low pH.

1.2.2   THERMORESPONSIVE POLYMERS

Thermoresponsive polymers undergo abrupt change in their solubility in response to small change in temperature. By maintaining physiochemical stability and biological activity thermoresponsive polymers control the rate of release of incorporated drug and its aqueous solution exhibits temperature-dependent and reversible sol–gel transitions near body temperature. A common characteristic feature of thermoresponsive polymers is the presence of hydrophobic group such as ethyl, methyl, and propyl groups.[5] There are two main types of thermoresponsive polymers; the first presents a lower critical solution temperature (LCST) while the second presents an upper critical solution temperature (UCST). LCST and UCST are the respective critical temperature points below and above which the polymer and solvent are completely miscible. Thus, for example, a polymer solution below the LCST is a clear, homogeneous solution while a polymer solution above the LCST appears cloudy (LCST also known as cloud point). This happens because it is energetically more favorable. In particular, considering the free energy of the system using the Gibbs equation $\Delta G = \Delta H - T\Delta S$ (G: Gibbs free energy, H: enthalpy, and S: entropy), the phase separation is more favorable when the temperature is increased mostly due to the entropy of the system. Specifically, the main driving force is the entropy of the water, that when the polymer is not in solution, the water is less ordered and has higher entropy. This is also termed as "hydrophobic effect."[6–8] Below the LSCT, the enthalpy term, related to the hydrogen bonding between the polymer and the water molecules, is responsible for the polymer dissolution. When raising the temperature above the LCST, the entropy term (hydrophobic interactions) dominates leading to polymer precipitation.[9] A well-known temperature-sensitive polymer is poly (N-isopropylacrylamide) (PNIPAAm). PNIPAAm has a LCST of around 32°C, a very useful temperature for biomedical applications since it is close to the body temperature (37°C). Adjustment of the LCST of PNIPAAm has been achieved by copolymerizing with hydrophilic or hydrophobic monomers rendering the overall hydrophilicity of the polymer higher or lower, respectively.[10–13]

Examples of polymers that show temperature-sensitive character are PNIPAAM, poly (ethylene oxide)–poly (ethylene oxide) triblock copolymers (PEO–PPO–PEO), and poly (ethylene glycol)–poly (lactic acid)–poly (ethylene glycol) triblocks (PEG–PLA–PEG). The most widely used temperature-sensitive polymers include poly (N-alkyl substituted acrylamides) and PNIPAAm with transition temperature of 32°C and poly (N-vinyl alkyl amides) such as poly (N-vinyl iso-butyramide) with transition temperature of 39°C.[14]

LCST system is preferred over UCST systems for drug delivery technologies as high temperature is needed for the latter, which is unfavorable for heat-labile drugs and molecules. According to the phase response to the temperature change, polymers are subdivided into negatively thermosensitive, positively thermosensitive, and thermoreversible types. The major advantages of thermosensitive polymeric systems are the avoidance of toxic organic solvents, the ability to deliver both hydrophilic and lipophilic drugs, reduced systemic side effects, site specific drug delivery, and sustained release properties.[15]

1.2.3 POLYMERS WITH DUAL STIMULI RESPONSIVENESS

These are the polymeric structures sensitive to both temperature and pH, obtained by the simple combination of ionization and hydrophobic (inverse thermosensitive) functional groups. This approach is mainly achieved by the copolymerization of monomers bearing these functional groups, combining temperature-sensitive polymers with polyelectrolytes (semi-interpenetrating polymer networks, SIPN and interpenetrating polymer networks, IPN) or by the development of new monomers that respond simultaneously to both stimuli.[16–20] The benefit of dual-stimuli-responsive polymeric system is that one stimulus can be used to load the carrier with the drug while a second one can be used to trigger the release. To obtain a temperature and pH-sensitive polymer, it is only necessary to combine temperature-sensitive monomers [e.g., poly (N-isopropylacrylamide-co-methacrylic acid and PNIPAm)] with pH-sensitive monomers (e.g., acrylic acid, AA and methacrylic acid, MAA).[21] Dual-stimuli-responsive polymers are obtained by combining two different polymers resulting in the formation of a block copolymer that has two dissimilar LCSTs. These materials show a well-defined core–shell structure consisting of temperature-sensitive cores with pH-sensitive shells. A pH- and temperature-responsive copolymer of NIPAAm and acrylic acid (AA) was developed for use as an oral matrix drug delivery system.

These dual-responsive pH-/temperature-sensitive copolymers overcome the limitations of both temperature- and pH-sensitive polymers, such as clogging of needle during injection and reduction in the development of acidic conditions due to degradation of polymer chains.

1.2.4 PHASE-SENSITIVE SMART POLYMERS

Phase-sensitive smart polymers are largely used to prepare biocompatible formulations of proteins for controlled delivery in biologically active and conformationally stable form. Compared to other systems, phase-sensitive polymeric systems have advantages such as low cost, ease of manufacture, high protein and peptides loading capacity, less traumatic manufacturing conditions, and the major advantage is the lysozymal enzyme release. In this approach, a water-insoluble biodegradable polymer such as poly (D,L-lactide) and poly (D,L-lactide-co-glycoide) dissolves in pharmaceutically accepted solvent to which a drug is added forming a solution or suspension. After injecting the formulation into the body, the water-miscible organic solvent dissipates and water penetrates into the organic phase. This causes the phase separation and precipitation of the polymer forming a depot at the site of injection.[22,23] Organic solvents used include hydrophobic solvents (such as triacetin, ethyl acetate, and benzylbenzoate) and hydrophobic solvents (such as N-methyl-2-pyrrolidone and tetraglycol). Major applications of phase-sensitive smart polymer lie in lysozyme release, controlled release of several proteins and use of emulsifying agents in phase-sensitive formulations to increase the stability of drug.[24]

1.2.5   FIELD-RESPONSIVE POLYMERS

Field-responsive polymers respond to the application of electric, magnetic, sonic, or electromagnetic fields. Its fast response time and controlled drug release rate make it more special than other stimuli-responsive polymers.

1.2.5.1   LIGHT-SENSITIVE POLYMERS

Since the stimulus of light can be imposed instantaneously and can be delivered in specific amounts with high accuracy, it renders light-responsive hydrogels highly advantageous over others. Also, the capacity for instantaneous delivery of the sol–gel stimulus renders light-responsive polymers potentially applicable for the development of optical switches, display units, and ophthalmic drug delivery systems. Encapsulated drug can be released or active after irradiation with a light source from outside the body. Light-sensitive drug carriers are fabricated from polymers that contain different photosensitizers such as azobenzene, stilbene, and triphenylmethane.[25–27] These systems are biocompatible, biodegradable, polymerizable, and at least partially water-soluble macromers.[28] The macromers include at least one water-soluble region, a region which is biodegradable, and two free radical-polymerizable regions. Many of the problems of two-phase systems, such as they cannot be recycled, result in increasingly expensive bioproducts, purification processes, and environmental pollution, have been overcome by the use of light-sensitive smart polymers.[29] On the basis of wavelength of light that triggers the phase transition, polymer systems can be UV-sensitive or visible-sensitive. Visible light-sensitive polymers are preferred over UV-sensitive polymers because of their ready availability, safety, and ease of use.

Limitations of light-sensitive polymers include:

- Inconsistent response due to the leaching of noncovalently bound chromophores during swelling or contraction of the system;
- Slow response of hydrogel toward the stimulus;
- Dark toxicity.

1.2.5.2 ELECTRIC FIELD-SENSITIVE POLYMERS

An electric field in the form of an external stimulus offers numerous advantages. They have wide application in the field of controlled drug delivery, artificial muscle actuations, energy transductions, and sound dampening as they transform electrical energy into mechanical energy. These polymers are pH-responsive as the delivery systems exploiting this external stimulus are obtained from polyelectrolytes, which contain a relatively large amount of ionizable groups along the backbone chain. Application of electric field for drug-releasing process causes a change in pH and results in degradation and bending of polymer chains by disrupting the hydrogen bonds. Naturally occurring polymers such as chitosan, alginate, and hyaluronic acid are commonly employed for the preparation of electroresponsive polymers.

1.2.5.3 MAGNETIC-SENSITIVE POLYMERS

Magnetic targeting is based on the attraction of magnetic micro- and nanoparticles to an external magnetic field source. In the presence of a magnetic field gradient, a translational force will be experienced on the particle/drug complex to effectively trap the complex in the field at the target site and is later pulled toward the magnet.[30] In the presence of an external magnetic field, super paramagnetic iron oxide nanoparticles (SPION) are capable of delivering drug particles and fixing them at the intended site while the drug is being released. This is termed magnetic drug targeting. Magnetic nanoparticles can also be encapsulated within liposomes.[31–33]

Magnetic drug delivery systems possess three main advantages:

- Visualization of drug delivery vehicles;
- Ability to control and guide the movement of drug carriers through magnetic fields;
- Thermal heating which has been used to control drug release or produce tissue ablation.

1.2.6 MULTISTIMULI-RESPONSIVE POLYMERS

Compared with conventional single- or dual-stimuli-based polymer materials, multiple stimuli-responsive polymer would be more intriguing since more functions and finer modulations can be achieved through more parameters.[34] They offer a wide range of applications and their properties can be tuned through several mechanisms. Multistimuli-responsive polymeric materials can be obtained by the incorporation of different functional groups, which are responding to different stimuli. Combinations of light and temperature, temperature and pH, and light and electric field have been reported.

1.3 APPLICATIONS OF SMART POLYMERS

1.3.1 SMART DRUG DELIVERY SYSTEMS

Drug delivery refers to approaches, formulations, technologies, and systems for administering a pharmaceutical compound (drug) to achieve a therapeutic effect in humans or animals. The application of smart polymers shows great promise due to modulated or pulsating drug release pattern to mimic the biological demand. These carriers allow delivery of drug at right time and concentration by only releasing the drug in response to an external stimulus. The mechanisms of drug release include ejection of the drug from the gel as the fluid phase synergizes out, drug diffusion along a concentration gradient, electrophoresis of charged drugs toward an oppositely charged electrode, and liberation of the entrapped drug as the gel or micelle complex erodes. In a temperature-sensitive polymer, a dilute solution (1–3%) of the polymer is watery liquid, while on warming to body temperature, the solution gels becoming viscous and clinging to surface in a "bioadhesive" form. These smart polymers become viscous and cling to the surface in a "bioadhesive" form therefore providing an effective way to administer drugs, either topically or mucosae, over longer timescales, by dissolving them in the solution, which also contains hydrophobic regions. Thermosensitive polymers, as carriers for DNA delivery, act as an intelligent mode of gene transfection. Drug release from thermoresponsive self-assembled polymeric micelles composed of cholic acid and PNIPAAm was studied as model system to examine the rate of drug release based on temperature changes.[3]

In case of oral drug delivery, poly (acrylic acid) has been formulated into drug-loaded particles, where they retain their therapeutic cargo in the acidic environment of the stomach and release the encapsulated drug in the alkaline environment of the small intestine due to the ionization of the carboxylic acid groups and swelling of the polymer matrix. Chitosan/glycerophosphate pharmaceutical system is a good alternative to pharmaceutical implants due to its porosity and greater ability to promote the controlled delivery of macromolecules and drugs with low solubility in water. The incorporation of paclitaxel (a hydrophobic anticancer drug) inhibits the growth of the cancer cells by its sustained delivery on target site.[35,36]

1.3.2 REVERSIBLE BIOCATALYST

Smart polymers can be used to design reversible soluble/insoluble biocatalyst. Reversible biocatalyst catalyzes enzyme reaction in their soluble state and thus can be used in reactions with insoluble/poorly soluble substrates. Reversible soluble biocatalyst is formed by the phase separation of smart polymers in aqueous solutions following a slight change in the external conditions, when the enzyme molecule is bound covalently to the polymer. As the reaction is complete, the conditions are changed to cause the catalyst to precipitate so that it can be separated from the product and used in the next cycle, after redissolution. Stimuli that are used for the development of reversibly soluble biocatalysts include pH, temperature, ionic strength, and addition of chemical species such as calcium.

For example, trypsin immobilized on a pH-responsive copolymer of methyl methacrylate and MAA is used for repeated hydrolysis of casein. A biocatalyst sensitive to magnetic field is produced by immobilizing invertase and γ-Fe_2O_3 in poly (N-isopropylacrylamide-co-acrylamide) gel. The heat generated by exposure of γ-Fe_2O_3 to a magnetic field causes the gel to collapse, which is followed by a sharp decrease in the rate of sucrose hydrolysis. Polymeric smart coatings capable of detecting and removing hazardous nuclear wastes are of great importance in sensor and catalyst chemistry. Such catalyst having advantages of both homogeneous and heterogeneous catalyst can serve as convenient "chemical switches" sensitive to light changes in external conditions.[3]

1.3.3 GLUCOSE SENSORS

The pH-sensitive polymers have played a notable role in the fabrication of insulin delivery systems for the treatment of diabetic patients. Delivering insulin is different from delivering other drugs, since insulin has to be delivered in an exact amount at the exact time of need. Many devices have been developed for this purpose and all of them have a glucose sensor built into the system. In a glucose-rich environment, such as the bloodstream after a meal, the oxidation of glucose to gluconic acid catalyzed by glucose oxidase

(GluOx) can lower the pH to approximately 5.8. This enzyme is probably the most widely used enzyme in glucose sensing, and makes possible to apply different types of pH-sensitive hydrogels for modulated insulin delivery.[37]

1.3.4 TISSUE ENGINEERING

Developing stimulus-responsive biomaterials with easy to tailor properties is a highly desired goal of tissue engineering community. Tissue engineering aims to restore, maintain, or improve tissue functions that are defective or have been lost by different pathological conditions, either by developing biological substitutes or by reconstructing tissues with the use of scaffolds. Smart hydrogels constitute promising materials for such scaffolds for two reasons. First, their interior environment is aqueous. Second, they can release the cells at the appropriate place in response to a suitable stimulus. A potential application is in the repair of damaged cartilage sites as in rheumatoid arthritis. The change of surface properties from hydrophobic above the critical temperature of the polymer grafted to hydrophilic below it has been successfully used for detachment of mammalian cells. PNiPAAm-grafted surfaces which are hydrophobic at 37°C are normally used for mammalian cell cultivation. The temperature is above the critical temperature for the grafted polymer and cells are growing well on it. Decrease in temperature results in surface transition to hydrophilic state and the cells are easily detached from the solid substrate without any damage. This technology has been significantly developed for the cultivation of cell sheets with designed shape for tissue engineering. PNiPAAm was covalently grafted onto tissue culture polystyrene dish surfaces by electron beam irradiation with mass patterns. Epithelial and mesenchymal cells of lung or bovine aortic endothelial cells can also be cultivated. Microglia or human monocytes and monocyte-derived macrophages were also successfully cultivated on PNiPAAm-grafted substrates and released by decreasing temperature. At present, the low-temperature liftoff of cell sheets from surfaces grafted with smart polymer presents a mature technique, which constitutes an important step on the way to the cultivation of functional 3-D tissues.[38,39]

1.3.5 PROTEIN FOLDING

The major challenge in the ongoing biochemical research lies in attaining the native structure and functions of proteins, which can be crossed using the refolding process. The use of smart polymer reduces the hydrophobicity of surfactants which facilitates or hinders the conformational transition of unfolded protein, depending upon the magnitude of unfolded protein. Refolding of bovine carbonic anhydrase was examined in presence of PPO–Ph–PEG at various temperatures. The refolding yield of carbonic anhydrase was strongly enhanced and aggregated formation of PPO–Ph–PEG at specific temperature of 50–55°C. Eudragit S-100, a pH-sensitive smart polymer, is supposed to increase the rate of refolding and refolding percentage of denatured protein. This was found to assist refolding of α-chymotrypsin, which is known to bind to the polymer rather than nonspecifically.[40,41]

1.3.6   OIL RECOVERY

Water in a well can be blocked by the use of smart polymeric materials that inhibit the water influxes. Fracturing fluids are used to fill the artificial fractures of oil layers. This artificial system has a high permeability with respect to the oil comparison with the rocks.[42]

1.3.7   MOLECULAR GATES AND SWITCHES

The latest thrilling breakthrough achieved by the group of Stayton and Hoffman, at the University of Washington, USA, involves the use of smart polymers that provide size-selective switches to turn proteins on and off. The carefully controlled placement of the polymer ensures that when a stimulus is applied, the collapsing/swelling of a gel causes the active site of the protein to be blocked/unblocked. In one of the early examples, PNIPAAm was linked to streptavidin at a site located just above its biotin-binding site. When the temperature is raised above the LCST of the hydrogel, it collapses covering the active site. Biotin can no longer bind to streptavidin, thus the polymer effectively acts as "molecular gate." The concept of physical blocking of recognition sites by a collapsed form has also been utilized in design of photoswitches for ligand association which might be useful in bioprocessing, biosensors, etc.

1.3.8   IN BIOTECHNOLOGY AND MEDICINE

Smart polymers can be physically mixed with or chemically conjugated to biomolecules to yield a large family of polymer–biomolecule system that can respond to biological, physical as well as chemical stimuli. Proteins and oligopeptides, sugars, polysaccharides, single- and double-stranded oligonucleotides, DNA plasmids, simple lipids, phospholipids, and synthetic drug molecules fall in the category of biomolecules that can be polymer conjugated. These polymer–biomolecule complexes are referred to as affinity smart biomaterials or intelligent bioconjugates. Also, such polymers have been used in developing smart surfaces and smart hydrogels that can respond to external stimuli. Such polymeric biomaterials have shown a range of different applications in the field of biotechnology and medicine.

1.3.9 AUTONOMOUS FLOW CONTROL IN MICROFLUIDICS

The concept of "lab in a chip" has evolved out of efforts to miniaturize analytical instruments. By using photolithography on a chip, one can create microchannels and work with very small volumes. Smart materials show considerable promise in designing microactuators for autonomous flow control inside these microfluidic channels. Saitoh et al.[43] have explored the use of glass capillaries coated with PNIPAAm for creating an on/off valve for the liquid flow. Below LCST, the PNIPAAm-coated capillary allowed the flow of water, and above LCST, the flow was blocked as the coating was now hydrophobic. Beebe et al.,[44] on the other hand, used a pH-sensitive methacrylate to control the flow inside the microchannels. The hydrogel-based microfluid valve opened or closed depending upon the pH of the solution. The design has the potential of being self-regulating/autonomous since the valve can be controlled by feedback of H^+ produced or consumed in the reaction. Undoubtedly, we will see many other innovative designs for such applications in coming years.

1.3.10 IN PROTEIN PURIFICATION

The uses of smart polymers include the concentration of protein solutions along with the isolation as well as purification of biomolecules. Recombinant thermostable lactate dehydrogenase from the thermophile *Bacillus stearothermophilus* was purified by affinity partitioning in an aqueous two-phase polymer system formed by dextran and a copolymer of N-vinyl caprolactam and 1-vinyl imidazole. The enzyme is partitioned preferentially into the copolymer phase in presence of Cu ions. The enzyme lactate dehydrogenase from porcine muscle has better access to the ligands and binds to the column. With the decrease in temperature, the polymer molecules undergo transition to a more expanded coil conformation. Finally, the bound enzyme is replaced by the expanded polymer chains. This system was used for lactate dehydrogenase purification.[45]

1.3.11   IN GENE THERAPY

The aim of gene therapy includes curing genetic diseases and viral infections, slowing down tumor growth, and stopping neurodegenerative diseases.[46] Most promising applications of smart polymer are as nonviral gene carriers. Polyelectrolytes have high potential as biomaterials in developing oppositely charged molecules. Naked DNA is very difficult to incorporate into the cells because it is negatively charged and it has a very large size at physiological conditions. Liposomes and polycations form the two major classes of nonviral gene delivery methods to condense DNA in charge-balanced nanoparticles that can be carried into cell compartments.

Both have to be cationic in nature in order to form complexes with anionic DNA, and the complex has to have net positive charge to interact with the anionic cell membrane and undergo endocytosis. The design has to confirm two contradictory requirements during endocytosis. While attaching to the coil and forming endosome, the binding between the carrier and the DNA has to be quite high. On the other hand, for DNA to move into nucleus to initiate transcription, the complex should be easy to dissociate. It is here that the stimuli-sensitive polymers are uniquely suited to fulfill the dual requirement, as the stimulus can control the binding to DNA. PEI is a highly cationic synthetic polymer that condenses DNA in solution, forming complexes that are readily endocytosed by many cell types. Chitosan, a biocompatible and reabsorbable cationic aminopolysaccharide, has also extensively been used as DNA carrier. Hoffman's group has dedicated great efforts to obtain new delivery systems to introduce efficiently biomolecules to intracellular targets.[47-49] They mimicked the molecular machinery of some viruses and pathogens that are able to sense the lowered pH gradient of the endosomal compartment and become activated to destabilize the endosomal membrane. This mechanism enhances protein or DNA transport to the cytoplasm from intracellular compartments such as endosome. They demonstrated the utility of poly (2-propylacrylic acid) (PPAA) to enhance protein and DNA intracellular delivery.

1.3.12   SMART POLYMER REDUCES RADIOACTIVE WASTE

Scientists reported the development of a new polymer that reduces the amount of radioactive waste produced during routine operation of nuclear reactor. In the study, the researchers created an absorbent material that, unlike unconventional ion exchange resins, has the unique ability of disregarding iron-based ions. The polymer's high selectivity increases its appeal.[50]

1.3.13   STIMULI-RESPONSIVE SURFACES

The change in surface properties of a thermoresponsive polymer with respect to temperature has been of wide interest and used in tissue culture applications, that is, hydrophobic nature above the critical temperature and hydrophilic below CST. Mammalian cells are cultivated on a hydrophobic solid culture dishes and are usually detached from it by protease treatment which also causes damage to the cells. This is rather an inefficient way in which some detached cells are able to adhere onto new dishes because the rest are damaged. At temperature of 370°C, a substrate surface coated with grafted PNIPAAm is hydrophobic because this temperature is above the critical temperature of the polymer and the cells grow well. However, the surface became hydrophilic, when temperature is decreased to 200°C, so that the cells can be easily detached without any damage. These cells can be used for further culturing. The cells are detached maintaining the cell–cell junction. This enables the collection of cultured cells as a single sheet. Cell sheet is highly effective when transplanted to patients due to tight communication between cells and cells. This technology has recently been commercialized. Temperature-sensitive size-exclusion chromatography is used for high protein resolution with low nonspecific interactions. The nonlinear response of smart polymers is what makes them so unique. A significant change in structure and properties can be induced by a very small stimulus. Once that change occurs, a predictable all-or-nothing response occurs with complete uniformity throughout the polymer.[51]

1.3.14   SMART POLYMERS IN TEXTILE ENGINEERING

Textiles have also experienced great improvements through the incorporation of various kinds of smart polymers for their formation. A series of polymer fibers with a shape-memory effect were developed. First, a set of shape-memory polyurethanes with very hard segment content were synthesized and their solutions were spun into fibers through wet spinning. It was found that the fibers showed less shape fixity but more shape recovery compared with thin films. Further studies revealed that the recovery stress of fibers was higher than that of thin films. The smart fibers may exert the recovery forces of the shape-memory polymers to an extreme extent in the direction of the fiber axis and therefore provide a possibility for producing high-performance acutators.[52]

1.4 FUTURE SCOPE

The smart and genetically produced biomaterials and drug delivery systems have made great progress in recent years. The intelligent response of a smart polymer gel can be used in diverse novel applications such as sensors, actuators, chemical and biological separation techniques, membranes, controlled drug delivery methods, and even some consumer products. New smart polymers with transition temperatures and pH levels in the range at which certain biomolecules are most stable (4–5°C and pH 5–8) will be developed and introduced commercially.[53] Many delivery systems based on polymers have progressed to the clinical and in some cases to the commercial production. Although these delivery systems encounter many challenges, they possess a potential approach toward the controlled release of bioactive reagents. One of the revolutionary applications is the idea of smart toilettes that analyze urine and help identify health problems. In environmental biotechnology, smart irrigation systems have also been proposed. It would be supremely useful to have a system that turns on and off, and controls fertilizer concentrations, based on soil moisture, pH, and nutrient levels. Many innovative approaches to targeted drug delivery systems that self-regulate based on their unique cellular surroundings are also under investigation. The use of smart polymers in biomedicine faces many problems such as the possibility of toxicity or incompatibility of artificial substances in the body, including degradation products and by-products. However, smart polymers have amazing applications in biotechnology and biomedical applications if these obstacles can be overcome.

1.5 CONCLUSIONS

Biopolymers (such as proteins, carbohydrates, and nucleic acid) are basic components of living organic systems that are responsible for the construction and operation of the complicated machinery of cells. Stimuli-responsive polymers represent a smart approach to mimic the behavior of these biopolymers, which undergo drastic conformational change at a critical point while remaining stable over a wide range of environmental conditions. Conjugation of synthetic smart polymers with protein has great applications in nanotechnology and biotechnology. They possess biomedical applications in numerous structures from diagnosis (e.g., detecting and imaging) to treatment (e.g., tissue engineering). Smart polymers sensitive to the presence of some biomarkers could be helpful in targeting specific disease conditions. All these applications show very promising results in scientific world, opening everyday a new window of development. Over the next couple of years, smart polymers will conquer all fields of science with its attracting features.

KEYWORDS

- smart polymer
- stimuli-responsive polymer
- drug delivery
- tissue engineering
- textile engineering

REFERENCES

1. Aguilar, M. R.; Roman, S. J. Introduction to Smart Polymers and Their Application. In *Smart Polymers and Their Applications*; Aguilar, M. R., Roman, S. J., Eds.; Woodhead Publishing: Cambridge, 2014; pp 1–7.
2. Kumar A. Smart Polymeric Biomaterials: Where Chemistry and Biology Can Merge. Available at http://www.iitk.ac.in/directions/dirnet 7/PP~ASHOK~FFF-pdf
3. Grey, H. N.; Bergbreiter, D. E. Application of Polymeric Smart Materials to Environmental Problems. *Environ. Health. Perspect.* **1997,** *105,* 55–63.
4. Grainger, S. J.; El-Sayed, M. E. H. Stimuli-sensitive Particles for Drug Delivery. In *Biologically-responsive Hybrid Biomaterials: A Reference for Material Scientists and Bioengineers;* World Scientific Publishing Co. Pte. Ltd., **2010**; pp 171–189.
5. Qiu, Y.; Park, K. Environment Sensitive Hydrogels for Drug Delivery. *Adv. Drug. Deliv. Rev.* **2001,** *53,* 321–339.
6. Klouda, L.; Mikos, A. G. Thermoresponsive Hydrogels in Biomedical Applications. *Eur. J. Pharm. Biopharm.* **2008,** *68,* 34–45.

7. Lutz, J. F. Polymerization of Oligo (Ethylene Glycol) (Meth) Acrylates: Toward New Generations of Smart Biocompatible Materials. *J. Polym. Sci Part A* **2008**, *46*, 3459–3470.

8. Southall, N. T.; Dill, K. A.; Haymet, A. D. J. A View of the Hydrophobic Effect. *J. Phys. Chem. B* **2002**, *106*, 521–533.

9. Macewan, S. R.; Callahan, D. J.; Chilkoti, A. Stimulus-responsive Macromolecules and Nanoparticles for Cancer Drug Delivery. *Nanomedicine* **2010**, *5*(5), 793–806.

10. Doorty, K. B.; Golubeva, T. A.; Gorelov, A. V.; Rochev, Y. A.; Allen, L. T.; Dawson, K. A.; Gallagher, W. M.; Keenan, A. K. Poly (N-isopropylacrylamide) Co-polymer Films as Potential Vehicles for Delivery of an Antimitotic Agent to Vascular Smooth Muscle Cells. *Cardiovasc. Pathol.* **2003**, *12*, 105–110.

11. Hacker, M. C.; Klouda, L.; Ma, B. B.; Kretlow, J. D.; Mikos, A. G. Synthesis and Characterisation of Injectable, Thermally and Chemically Gelable, Amphiphilic Poly (N-isopropylacrylamide)-based Macromers. *Biomacromolecules.* **2008**, *9*, 1558–1570.

12. Feil, H.; Bae, Y. H.; Feijen, J.; Kim, S. W. Effect of Comonomer Hydrophilicity and Ionization on the Lower Critical Solution Temperature of N-isopropylacrylamide Copolymers. *Macromolecules* **1993**, *26*, 2496–2500.

13. Yin, X.; Hoffman, A. S.; Stayton, P. S. Poly (N-isopropylacrylamide-co-propylacrylic acid) Copolymers that Respond Sharply to Temperature and pH. *Biomacromolecules* **2006**, *7*, 1381–1385.

14. Cao, Y. L.; Ibarra, C.; Vacanti, C. Preparation and Use of Thermoresponsive Polymers. In *Tissue Engineering Methods and Protocols;* Morgan, J. R.; Yarmush, M. L., Eds.; Humana Press: Totowa, 2006; pp 75–84.

15. Shidhaye, S.; Badshah, F.; Prabhu, N.; Parikh, P. Smart Polymers: A Smart Approach to Drug Delivery. *World J. Pharm. Res.* **2014**, *3*, 159–172.

16. Aguilar, M. R.; Elvira, C.; Gallardo, A.; Vazquez, B.; Roman, J. S. Smart Polymers and Their Applications as Biomaterials. In *Topics in Tissue Engineering,* 2007; Vol. 3, pp 1–27.

17. Bulmus, V.; Ding, Z.; Long, C. J.; Stayton, P. S.; Hoffman, A. S. Site-specific Polymer—Streptavidin Bioconjugate for PH-controlled Binding and Triggered Release of Biotin. *Bioconjug. Chem.* **2000**, *11*, 78–83.

18. Brazel, C. S.; Peppas, N. A. Synthesis and Characterization of Thermo- and Chemo-mechanically Responsive Poly (N-isoproylacrylamide-co-methacrylic acid) Hydrogels. *Macromolecules* **1995**, *28*, 8016–8020.

19. Kuckling, D.; Alder, H.-J. P.; Ardnt, K. F.; Ling, L.; Habicher, W. D. Temperature and pH-dependent Solubility of Novel Poly (N-isoropylacrylamide)-copolymers. *Macromol. Chem. Phys.* **2000**, *201*, 273–280.

20. Zareie, H. M.; Volga Bulmus, E.; Piskin, E.; Gunning, A. P.; Morris, V. J.; Hoffman, A. S. Investigation of a Stimuli-responsive Copolymer by Atomic Force Microscopy. *Polymer* **2000**, *41*, 6723–6727.

21. You, J.; Almeda, D.; Ye, G. J. C.; Auguste, D. T. Bioresponsive Matrices in Drug Delivery. *J. Biol. Eng.* **2010**, *4*, 15.

22. Ravivarapu, H. B.; Moyer, K. L.; Dunn, R. L. Sustained Activity and Release of Leuprolide Acetate from an In Situ Forming Polymeric Implant. *AAPS Pharm. Sci. Tech.* **2000**, *1*, 1–8.

23. Eliza, P. E.; Kost, J. Characterization of a Polymeric PLGA Injectable Implant Delivery Systems for the Controlled Release of Proteins. *J. Biomed. Mater. Res.* **2000**, *50*, 388–396.

24. Higuchi, T. Mechanism of Sustained—Action Mechanism. *Pharm. Sci. J.* **1963**, *52*, 1145–1149.
25. Budhall, B. M.; Marquez, M.; Velev, O. D. Microwave, Photo- and Thermally Responsive PNIPAm-gold Nanoparticle Microgels. *Langmuir* **2008**, *24*, 11959–11966.
26. Angelatos, A. S.; Radt, B.; Caruso, F. Light-responsive Polyelectrolyte/Gold Nanoparticle Microcapsules. *J. Phys. Chem. B* **2005**, *109*, 3071–3076.
27. Alvarez-Lorenzo, C.; Bromberg, L.; Concheiro, A. Light-sensitive Intelligent Drug Delivery Systems. *Photochem. Photobiol.* **2009**, *85*, 848–860.
28. Gan, L. H.; Gan, Y. Y.; Roshan Dean, G. Poly (N-acrylol-N'-propylpiperazine): A New Stimuli Responsive Polymer. *Macromolecules* **2000**, *33*, 7893–7897.
29. Ghizal, R.; Fatima, G. R.; Sreevastava, S. Smart Polymers and Their Applications. *Int. J. Eng. Tech. Manag. Appl. Sci.* **2014**, *2*, 104–115.
30. Grief, A. D.; Richardson, G. Mathematical Modeling of Magnetically Targeted Drug Delivery. *J. Magn. Magn. Mater.* **2005**, *293*, 455–463.
31. Pankhurt, Q. A.; Connody, J.; Jones, S. K.; Dobson, J. Application of Magnetic Nanoparticles in Biomedicine. *J. Phys. D: Appl. Phys.* **2003**, *36*, 167–181.
32. Mizogami, S.; Mizutani, M.; Fukuda, M.; Kavabata, K. Abnormal Ferromagnetic Behavior for Pyrolytic Carbon Under Low Temperature Growth by CVD Method. *Synth. Met.* **1991**, *43*, 3271–3272.
33. Bawa, P.; Pillay, V.; Choonara, Y. E.; duToit, L. C. Stimuli-responsive Polymers and Their Applications in Drug Delivery. *Biomed. Mater.* **2009**.
34. Cao, Z. Q.; Wang, G. J. Multi-stimuli-responsive Polymer Material: Particles, Films, and Bulk Gels. *Chem. Rec.* **2016**, *16*(3), 1398–1435.
35. Gil, E. S.; Hudson, S. M. Stimuli-responsive Polymers and Their Bioconjugates. *Prog. Polym. Sci.* **2004**, *29*(12), 1173–1222.
36. Ruel-Gariepy, E.; Leroux, J. In Situ-forming Hydrogels—Review of Temperature-sensitive Systems. *Eur. J. Pharm. Biopharm.* **2004**, *58*(2), 409–426.
37. Podual, K.; Doyle III, F. J.; Peppas, N. A. Preparation for Dynamic Response of Cationic Copolymer Hydrogels Containing Glucose Oxidase. *Polymer* **2000**, *41*, 3975–3983.
38. Kikuchi, A.; Okano, T. Nanostructured Designs of Biomedical Materials: Application of Cell Sheet Engineering to Functional Regenerative Tissues and Organs. *J. Control Release* **2005**, *101*, 69–84.
39. Shimizu, T.; Yamato, M.; Kikuchin, A.; Okano, T. Cell Sheet Engineering for Myocardial Tissue Reconstruction. *Biomaterials* **2003**, *24*, 2309–2316.
40. Kukoi, R.; Morita, S.; Ota, H.; Umakoshi, H. Protein Refolding Using Stimuli-responsive Polymer-modified Aqueous Two-phase Systems. *J. Chromatogr. B Biomed. Sci. Appl.* **2000**, *743*, 215–223.
41. Chen, Y. J.; Huang, L. W.; Chin, H. C.; Lin, S. C. Temperature-responsive Polymer-Assisted Protein Refolding. *Enzyme Microb. Technol.* **2003**, *32*, 120–130.
42. Shashkina, Yu. A.; Zaroslov, Yu. D.; Smirnov, V. A.; Philippova, O. E.; Kholehlov, A. R.; Priyakhina, T. A.; Churochkina, N. A. Hydrophobic Aggregation in Aqueous Solutions of Hydrophobically Modified Polyacrylamide in the Vicinity of Overlap Concentration. *Polymer* **2003**, *44*(8), 2289–2293.
43. Saltoh, T.; Suzuki, Y.; Hiralde, M. Preparation of Poly (N-isoproplyacrylamide)-modified Glass Surface for Flow Control in Microfluidics. *Anal. Sci.* **2002**, *18*, 203–205.
44. Beebe, D. J.; Moore, J. S.; Bauer, J. M.; Bauer, J. M.; Yu, Q.; Liu, R. H.; Devadoss, C.; Jo, B. Functional Hydrogel Structures for Autonomous Flow Control Inside Microfluidic Channel. *Nature* **2000**, *404*, 588–590.

45. Matttiasson, B.; Dainyak, M. B.; Galaev, I. Y. Smart Polymers and Protein Purification; *Polymer-Plast. Technol. Eng.* **1998,** *37,* 303–308.
46. Verma, I.; Sonia, M. Gene Therapy-promise, Problems, and Prospects. *Nature* **1997,** *389,* 239–242.
47. Stayton, P. S.; Hoffman, A. S.; Murty, N., et al. Molecular Engineering of Proteins and Polymers for Targeting and Intercellular Delivery of Therapeutics. *J. Control. Release.* **2000,** *65,* 203–220.
48. Stayton, P. S.; El-Sayed, M. E.; Murthy, N., et al. Smart Delivery Systems for Biomolecular Therapeutics. *Orthod. Craniofac. Res.* **2005,** *8,* 219–225.
49. Stayon, P. S.; El-Sayed.; Hoffman, A. S. Smart Polymeric Carriers for Enhanced Intercellular Delivery of Therapeutic Macromolecules. *Expert Opin. Biol. Ther.* **2005,** *5,* 23–32.
50. Bhaskarapillai, A. Synthesis and Characterization of Imprinted Polymers for Radioactive Waste Reduction. *Ind. Eng. Chem. Res.* **2009,** *48*(8), 3730.
51. Ruan, C., Zeng, K.; Grimes, C. A. A Mass-sensitive pH Sensor Based on a Stimuli-responsive Polymer. *Anal. Chim. Acta.* **2003,** *497,* 123–131.
52. Ji, F. L.; Zhu, Y.; Hu, J. L.; Liu, Y.; Yeung, L. Y.; Ye, G. D. Smart Polymer Fibers with Shape Memory Effect. *Smart Mater. Struct.* **2006,** *15,* 1547–1554.
53. Patil, N. V. Smart Polymers Are in the Biotech Future. *Bioprocess Int.* **2006,** 42–46.

CHAPTER 2

AN ASSESSMENT ON POLYMER ELECTROLYTE MEMBRANE FUEL CELL STACK COMPONENTS

ARUNKUMAR JAYAKUMAR

Mechanical Engineering Department, Auckland University of Technology, Auckland, New Zealand

E-mail: ajayakum@aut.ac.nz

CONTENTS

ABSTRACT

Polymer electrolyte membrane (PEM) fuel cells can be considered as a clean and reliable system for at least the next generation of power generation if we can circumvent the technical issues and minimize the related cost of stack components. Understanding the functional characteristics of the integral stack components of (PEM) fuel cell stack can be a significant contribution to fuel cell development and commercialization. The present chapter aims to review critically the integral stack components of PEM fuel cells. The chapter also emphasizes the limitations encountered with the current electrode materials, gas diffusion material, and also the durability of these stack materials. A snapshot in the advancement of electrode materials and the economics of the stack components is also envisaged.

2.1 GENERAL

Energy is an indispensable constituent and a key driver for the sustainable growth of a nation and the existence of life cannot be imagined without it. Conventional energy systems result in poor air quality and cause billions of dollars each year to the governments of various countries. In addition, these energy systems also have poor operating efficiency. Finding a clean and efficient energy system will be one of the tactical challenges to circumvent these issues and fuel cell systems can be a comprehensive solution because they curb emissions such as carbon monoxide, carbon dioxide, sulfur, nitrogen oxides, etc. These toxic gases not only lead to several lung diseases but also lead to the lethal greenhouse effect and global warming. In addition, fuel cells also have economic implications by reducing the country's dependence on imported petroleum in the transportation sector. The operating efficiency of fuel cells is as high as 85% for combined power and heat.[1] Their competence is credited due to the fact that chemical energy is directly converted to electrical energy, unlike traditional generators where chemical energy is initially converted to mechanical energy [by internal combustion (IC) engine] and then to electrical energy (by generator) as illustrated in Figure 2.1.

In fuel cells, there is neither combustion nor emission and as a consequence, there is a high electrical efficiency than in conventional generators of parallel size and rating.[2] In addition, fuel cells are the only systems which can deliver power from micro- to megawatt applications. However, the types of fuel cell systems to be integrated into a particular application will depend on the compatibility constraint and economic considerations.

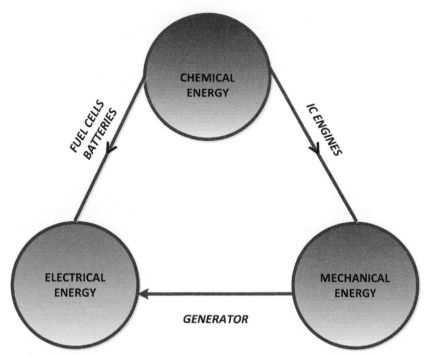

FIGURE 2.1   Energy conversion path of conventional power systems.

2.2   FUEL CELL AND ITS PERFORMANCE

The basic concept of the fuel cell (irrespective of the type) is that it produces electricity with hydrogen and oxygen as fuel in the process, producing water and heat as the by-products. The anode, cathode, and net reaction of a PEM fuel cell is as follows:

$$H_2 \rightarrow 2H^+ + 2e^- \qquad\qquad E_a^o = 0V$$

$$2H^+ + 2e^- + {}_2 \rightarrow H_2O \qquad\qquad E_c^o = 1.229V$$

$$H_2 + 1/2O_2 \rightarrow H_2O \qquad\qquad E^o = E_c^o - E_a^o = 1.229V$$

The ideal standard potential (Ev) at 298 K for a fuel cell in which H_2 and O_2 react is 1.229 V with a liquid water product or 1.18 V with a gaseous water product.[3] The impact of temperature on the ideal voltage, E, for the oxidation of hydrogen is also shown in Table 2.1 for the various types of fuel cells.[3] Each case assumes gaseous products as its basis.

TABLE 2.1   Ideal Voltage as a Function of Cell Temperature.

Temperature	Polymer electrolyte membrane fuel cell (PEMFC)		Alkaline fuel cell (AFC)	Phosphoric acid fuel cell (PAFC)	Molten carbonate fuel cell (MCFC)	Intermediate temperature solid oxide fuel cell (IT-SOFC)	High-temperature solid oxide fuel cell (HT-SOFC)
	25°C	80°C					
Ideal voltage	1.18	1.17	1.16	1.14	1.03	0.99	0.91

The open circuit voltage (OCV) of a fuel cell is also strongly influenced by the reactant concentrations and the maximum ideal potential occurs when the reactants at the anode and cathode are pure. In an air-fed system or if the feed to the anode is other than pure dry hydrogen, the cell potential will be reduced. Similarly, the concentration of reactants at the exit of the cell will be lower than at the entrance. This reduction in partial pressure leads to a Nernst correction that reduces the OCV locally, often by as much as 250 mV in higher temperature cells.[3] Because the electrodes should be highly conductive and the electrode within one cell consequently has close to uniform voltage, low OCV affects the operation of the entire cell. This significantly impacts on the achievable cell operating voltage and consequently the system efficiency of especially the higher temperature fuel cells.[3]

2.2.1   FUEL CELL EFFICIENCY

For a fuel cell, the useful energy output is the electrical energy produced and energy input is the enthalpy of hydrogen. Assuming that all of the Gibbs free energy can be converted into electrical energy, the maximum possible (theoretical) efficiency of a fuel cell at 25°C by using the hydrogen higher heating value (HHV) is:[4]

$$\eta = \frac{\Delta G_f^o}{\Delta H^o} = \frac{237.1 \text{ kJ/mol}}{286 \text{ kJ/mol}} = 83\%$$

If both ΔG_f^o and ΔH^o are divided by nF, the fuel cell efficiency is expressed as the ratio of two potentials: F—Faraday's constant (96485.3415 C mol^{-1}); n—2.

$$\eta = \frac{\Delta G_f^o / nF}{\Delta H^o / nF} = \frac{1.23}{1.48} = 83\% \ (25°C, 1 \text{ atm})$$

where 1.23 V is the theoretical cell potential and 1.48 V is the potential corresponding to hydrogen's HHV or the thermoneutral potential.[4] The theoretical efficiency is sometimes also known as the thermodynamic efficiency or the maximum efficiency limit. However, in fuel cells, the fuel is typically not completely converted, as is common in most types of heat engines.

The thermal efficiency of a hydrogen/oxygen (air) fuel cell can then be written in terms of the actual cell voltage:

$$= \frac{\text{Useful Energy}}{\Delta H} = \frac{\text{Useful Power}}{\left(\dfrac{\Delta G}{0.83}\right)} = \frac{\text{Volts}_{\text{actual}} \times \text{Current}}{\text{Volts}_{\text{ideal}} \times \text{Current} / 0.83} = \frac{(0.83)(V_{\text{actual}})}{E_{\text{ideal}}}$$

As mentioned previously, the ideal voltage of a cell operating reversibly on pure hydrogen and oxygen at 1 atm pressure and 25°C is 1.229 V. Thus, the thermal efficiency of an actual fuel cell operating at a voltage of V_{cell} based on the HHV of hydrogen is given by:

$$\eta = 0.83 \times V_{\text{cell}} / E_{\text{ideal}} = 0.83 \times V_{\text{cell}} / 1.229 = 0.675 \times V_{\text{cell}}.$$

To attain the total stack efficiency of fuel cell, the ideal thermodynamic and voltage efficiency should be multiplied by utilization factor (μ), which refers to the fraction of the total fuel or oxidant introduced into a fuel cell that reacts electrochemically, that is, hydrogen/oxygen (air) consumed during the course of the reaction.

$$\text{Fuel Cell Stack Efficiency} (\eta_{FC}) = \eta \times \mu.$$

And as a consequence, the practical fuel cell stack efficiency is always lower than the theoretical efficiency.

2.3 CONVENTIONAL FUEL CELLS TYPES

Fuel cells are basically classified by the chemistry of electrolyte that it incorporates. Table 2.2 classifies the list and characteristics of widely used fuel cells, namely alkaline fuel cell (AFC), phosphoric acid fuel cell (PAFC), molten carbonate fuel cell (MCFC), solid oxide fuel cell (SOFC), and polymer electrolyte membrane fuel cell (PEMFC).

TABLE 2.2 Classification of Conventional Fuel Cells.

Fuel cell type	Electrolyte/charge carrier	System efficiency/ operating temperature	Advantages
Alkaline fuel cell (AFC)	Potassium hydroxide, soaked in a matrix KOH (aq) **OH⁻**	60% 100–110°C	Aqueous electrolyte promotes fast cathode reaction and high performance.
Phosphoric acid fuel cell (PAFC)	Phosphoric acid soaked in a matrix **H⁺**	40–45% 150–200°C	Can use impure H_2 as fuel and can tolerate up to 1.5% CO at operating temp.
Molten carbonate fuel cell (MCFC)	Carbonate solution CO_3^-	50–60% 650°C	High operating temperature, therefore no expensive catalysts and can operate on cheap fuels.
Solid oxide fuel cell (SOFC)	Yttria-stabilized zirconia, or more recently, lanthanide-doped ceria **O –**	50–60% 800–1000°C	High operating temperature, therefore no noble metal catalysts and can operate on cheap fuels.
Polymer electrolyte membrane fuel cell (PEMFC)	Perfluoro-sulfonic acid (Nafion) **H⁺**	60% 60–80°C	High power density, low temperature and quick start-up (excellent for vehicle applications).

2.4 NONCONVENTIONAL FUEL CELL TYPES

2.4.1 DIRECT METHANOL OR ALCOHOL FUEL CELL

Direct methanol or alcohol fuel cell (DMFC or DAFC) is basically a subset of PEM fuel cells that incorporate polymer membrane as the electrolyte. In these systems, the anode catalyst draws the hydrogen from liquid fuels (hydrogen carrier), eliminating the need for a fuel reformer.

The empirical equation that governs the oxidation reactions of alcohols is given by:[5]

$$C_nH_{2n+1}OH + (2n-1)\,H_2O \rightarrow nCO_2 + 6nH^+ + 6ne^-.$$

However, traces of carbon dioxide are produced in these kinds of fuel cells depending on the type and concentration of alcohol. Fuel cells utilizing

2-propanol and other alcohols as a hydrogen carrier will be a versatile power source to energize mobile and other portable electronic kits.[5] The limitation in hydrogen storage low operating temperature and high-energy density makes these kind of fuel cell systems an appropriate candidate for low power applications that do not require uninterrupted operation.

2.4.2 REGENERATIVE FUEL CELL

The regenerative fuel fell (RFC) is capable of operating on both the fuel cell and electrolyzer mode. While operating on the fuel cell mode, it produces electricity and in the reverse mode acts as an electrolyzer for hydrogen generation. The greatest advantage of RFC is that it can provide power on demand but during times of high power production from other technologies (such as when high winds speed to an excess of available wind power), RFCs can convert the surplus electricity into hydrogen (energy vector) and store it, which is much economical to convert the power and store it in the expensive battery banks.

The polarity of the cell during the electrolyzer mode is opposite to that for the fuel cell mode and the following reactions describe the chemical process in the hydrogen generation mode:

At cathode: $H_2O + 2e^- \rightarrow H_2 + O^{2-}$.

At anode: $O^{2-} \rightarrow 1/2O_2 + 2e^-$.

Overall: $H_2O \rightarrow 1/2O_2 + H_2$.

The pertinent equations for the fuel cell mode are already elaborated in Section 2.2.

2.5 LIMITATION OF VARIOUS FUEL CELLS (EXCLUDING PEM FUEL CELL)

There are numerous types of fuel cells which are elaborated in Sections 2.3 and 2.4. The AFC was the first fuel cell technology to be put into service and make the generation of electricity from hydrogen feasible.[6] NASA has used it since the mid-1960s in Apollo-series missions and on the space shuttle. The limitations with AFC are the need of liquid electrolyte which occupies a large space and limits the power density. In addition, they have issues pertaining to electrolyte poisoning.[6] PAFC is a type of fuel cell that uses liquid phosphoric acid as an electrolyte. PAFC demonstrated excellent performance for most of the distributed power generation applications in terms of power, efficiency, and low emissions. However, the challenge that this technology needs to overcome for becoming a mass product is the cost. Moreover, they have a lower power density compared to PEM fuel cell. MCFC and SOFC belong to the family of high-temperature fuel cells which operate around 1000°C. For the MCFC, the problems are brought by the highly corrosive electrolyte in terms of electrode stability, electrolyte containment in the cell matrix, and gas sealing of the stack. In the SOFC, the thermal mismatch between the ceramic cell and stack components induces material degradation.[7] In addition, these fuel cells (MCFC and SOFC) have a very poor power density and selection of appropriate electrodes material for optimal operation is the greatest challenge of these fuel cell systems.

2.6   ADVANTAGES OF PEM FUEL CELLS

Among the fuel cells extensively studied, most efforts have been made for the development of PEMFCs.[8] They are a promising alternative for clean power source for automotive application because they are compact systems and their start-up process is very fast. The sealing of PEMFC electrode is easier than other types of fuel cells. Material specifications of PEMFC electrodes are solid and substantial.[9] The following reasons are attributed to the supremacy of PEMFCs compared to other fuel cells.

- High stack energy efficiency of 65% and high power density of 2 kW/kg.[10]
- Have a low operating temperature (~80°C) which allows rapid start-up and cold start-up (−20°C).[10]
- PEM fuel cell can vary the output to match the demands.

The aforementioned features are further enhanced when integrated with their modular design, making PEM fuel cells a potential system for deployment in portable power, transportation, and off-grid power applications.[11] In addition, unlike the IC engines where the efficiency is maximum with the highest loads, the PEM fuel cell efficiency is high with partial loads. This is advantageous because in typical driving conditions such as urban and suburban scenarios, most of the time the vehicle is demanding a small fraction of the nominal FC power.[12]

2.7 PEM FUEL CELL PRINCIPLE OF OPERATION

A schematic representation of PEM fuel cell revealing the vital components and ion transport is illustrated in Figure 2.2.[13]

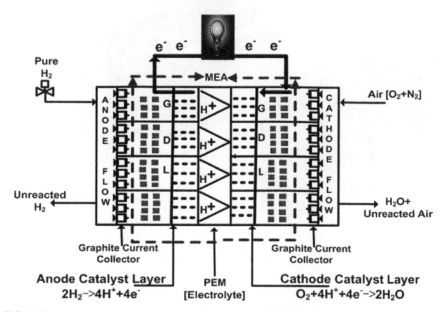

FIGURE 2.2 A plan of PEM fuel cell stack component and ion transfer.

2.8 PEM FUEL CELL STACK COMPONENTS

Understanding the PEM fuel cell stack components is a prerequisite for a successful commercialization of PEM fuel cell stack and as a consequence, in the present assessments, we intended to provide details about key components of PEM fuel cell stack, namely the

- membrane and electrode assembly,
- gas diffusion layer (GDL),
- bipolar plate (graphite plate),
- current collectors and end plate,
- gasket.

The integral PEM fuel cell stack components are structured and illustrated in Figure 2.3.

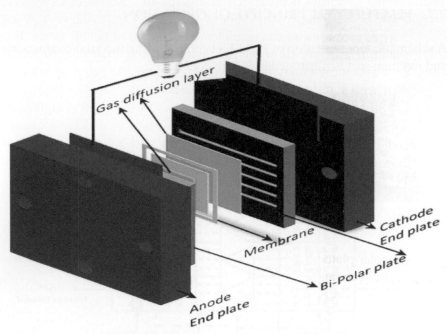

FIGURE 2.3   An exploded view of a PEM fuel cell stack components.

2.8.1 BIPOLAR OR SEPARATOR PLATE

The function of the bipolar plate is to transfer electrons between adjacent fuel cells. There are channels or grooves cut into the anode and cathode sides of the bipolar plates that provide a low-resistance path for oxidizer and fuel to reach the catalytic zone. The bipolar plate may be made of either metal (stainless steel, titanium, aluminum, and several metal alloys) or a conductive carbon polymer (high-density graphite made of polymer resin-impregnated graphite). Bipolar plates have been subjected to different coatings to improve their corrosion resistance. These conductive plates act as an anode for one half-cell and as a cathode for the adjacent half-cell. As to the geometrical configurations of the gas flow fields, a variety of different designs are known and the conventional designs typically comprise either pin, integrated cooling flow fields, interdigitated flow field parallel, or serpentine designs of flow-field channels.[14] An optimal design is required for the bipolar plates based on the application and the operating conditions under which the fuel cell stack is utilized. Figure 2.4 represents the bipolar plate with a serpentine flow.

FIGURE 2.4 Bipolar plate with serpentine flow.

The bipolar plate usually incorporates flow conduits for the fluid feeds and efficient heat transfer. In addition, bipolar plates act as a physical barrier to the fuel and oxidizer streams of adjacent cells in a stack.

2.8.1.1 REQUIREMENTS FOR BIPOLAR PLATES

The prerequisites for bipolar plates are to evenly distribute the reactant gases over the respective active electrode surface to minimize the concentration over potential; high values of electronic conductivity for current collection; high mechanical strength for stack integrity; impermeability to reactant gases for safe operation; resistance to corrosion in severe cell environment for long lifetime; and cheap materials, easy and automated fabrication for low cost. The bipolar plates also provide a structural support and path for the products of the reaction to be evacuated from the cell.[14] In addition, these plates also assist in an effective thermal management. To achieve these characteristics of bipolar plates, a wide combination of physical and chemical properties that are often contradictory has been proposed and investigated over the years.[15] As a result, a set of targets and requirements have been established to develop suitable material for the fabrication of bipolar plates. A summary of such requirements and targets is presented below.[16]

- Bulk electrical conductivity (in-plane): >100 S cm^{-1};
- Hydrogen permeability: <2 × 10^{-6} cm^3 cm^{-2} s^{-1};
- Corrosion rate: <16 × 10^{-6} A cm^{-2};
- Interfacial contact resistance: <20 mΩ cm^2 at 150 N cm^{-2};

- Tensile strength: >41 MPa;
- Flexural strength: >59 MPa;
- Thermal conductivity: >10 W m^{-1} K^{-1};
- Thermal stability: up to 120°C;
- Chemical and electrochemical stability in acidic environments;
- Low thermal expansion;
- Acceptable hydrophobicity (or hydrophilicity).

2.8.2 MEMBRANE AND ELECTRODE ASSEMBLY

In the case of a PEM fuel cell, the membrane and electrode assembly (MEA) is the structure consisting of an electrolyte (PEM) with surfaces coated with catalyst (CL)/carbon/binder layers and sandwiched by GDLs and current collectors. A typical MEA primarily includes membrane, the CLs, and GDLs, and their functions are elaborated in Table 2.3.

TABLE 2.3   Integral Membrane and Electrode Components and Their Function.

S. No.	Components	Function
1	Polymer electrolyte membrane (PEM)	Conducts protons and impedes the flow of electrons.
	For example, Nafion	Must be thin (to reduce ohmic loss) and should have high protonic conductivity (to conduct protons).
2	Catalyst layers (CLs)	Active site, where the reaction takes place.
	For example, Pt, Ru	A good catalyst must have optimal mass transport characteristics.
3	Gas diffusion layers (GDLs)	Facilitates diffusion of reactant gases across the catalyst layer.
	For example, C-paper and cloth	Aids the transportation of electricity from individual catalyst site in MEA to the current collectors.
		Prevents the water flooding in the electrode by consistently removing the product water and also ironically maintains water on the catalyst layer surface to enhance ionic conductivity.
		Allows for efficient heat transfer.

The integration of CLs and GDLs is termed as gas diffusion electrodes. The performance of a PEM fuel cell is substantially influenced by the surface morphology of these electrodes.

2.8.2.1 MEMBRANE

The polymer electrolyte membranes are materials that permit the transport of hydrogen ions from the anode to the cathode where they interact with oxygen (cathode) and electrons to form water. Perfluorinated polymer membranes Nafion® (introduced by Dupont in 1966) has been a standard of extensive study for a long time due to their high protonic conductivity and relatively low equivalent weight. The membrane must be well hydrated to remain conductive and must be resistant to the reducing environment at the cathode as well as the harsh oxidative environment at the anode. In general, these membranes rely on appropriate humidification to efficiently transport the protons. In addition, conventional membranes are not recommended to be used for temperatures above 80°C, since the membrane would dry out and deteriorate. Figure 2.5 represents a typical catalyst-coated membrane (CCM).

FIGURE 2.5 Catalyst-coated membrane.

Identifying an appropriate protonic membrane that possesses the required combination of high ionic conductivity, mechanical strength, dehydration resistance, chemical stability, and gas permeability along with reasonable low cost will be a key to the success of PEM fuel cells.

2.8.2.2 MEMBRANE REQUIREMENTS

The following membrane properties need to be optimized for fuel cell applications: (1) rapid ion (proton) transport; (2) good mechanical, chemical, and thermal properties requiring the selection of a suitable polymer backbone, and possibly membrane reinforcement; (3) low gas permeability; and (4) low levels of swelling.[17] Advanced self-humidifying membranes have been developed by Rowshanzamir et al.[18] which eliminate the need for humidification and thereby significantly reduce the complexity and cost of the fuel cell system, as there is no need of a humidification system for improving the proton conductivity of the membrane. Generation of water molecules on the Pt particles embedded in the membrane by the recombination of permeated hydrogen and oxygen was attributed to the higher performance of a single cell prepared using the self-humidifying membrane.

2.8.2.3   ELECTROCATALYSTS LAYER

The CL is the critical component in PEM fuel cell stack used to accelerate the electrochemical reaction of hydrogen and oxygen in the cell. Normally, expensive noble metals are required for the electrochemical oxidation and reduction of hydrogen and oxygen, respectively. Platinum nanoparticles supported on carbon can be a promising material for both the anode and cathode. Platinum is by far the most effective catalysts element used for PEM fuel cells, and nearly all current PEM fuel cells use Pt on porous carbon catalysts with smaller platinum particles having higher active surface area. Physically the catalyst layer is in direct contact with the membrane and the GDL. Pt-/Ru-based alloys may instead be used in the anode of fuel cells using "impure" hydrogen, that is, hydrogen obtained by steam reforming of hydrocarbon fuels.[19]

2.8.2.4   ELECTROCATALYST LAYER REQUIREMENTS

In both anode and cathode, the CL is the location of the half-cell reaction in a PEM fuel cell.[20] The CLs need to be designed so as to generate high rates of the desired reactions and minimize the amount of catalyst necessary for reaching the required levels of power output. To meet the goal, the following requirements need to be considered: (1) large three-phase interface in the CL, (2) efficient transport of protons, (3) easy transport of reactant and product gases and removal of condensed water, and (4) continuous electronic current passage between the reaction sites and the current collector.

The CL is either applied to the membrane (CCM) or as an independent layer. However, the objective is to place the catalyst particles, platinum or platinum alloys, within close proximity of the membrane.

The catalyst ink usually includes catalyst, carbon powder, binder, and solvent. Sometimes, other additives are added to improve the dispersion of the components and stabilize the catalyst ink. The catalyst either covers the surface of the GDL or directly coats the surface of the membrane (CCM). The CL usually consists of: (1) an ionic conductor such as perfluorosulfonic acid (PFSA) ionomer to provide a passage for protons to be transported in or out, (2) metal catalysts supported on a conducting matrix such as carbon to provide a means for electron conduction, and (3) a water-repelling agent such as polytetrafluoroethylene (PTFE) to provide sufficient porosity for the gaseous reactants to be transferred to catalyzed sites. Every individual factor must be optimized to provide the preeminent performance of a CL.

A mixture of carbon-supported platinum catalyst along with Nafion ionomer is used to prepare the CL. Hydrophobicity is introduced into the electrode for better nonwetting characteristics by the addition of requisite amount of Teflon.

One method of increasing the performance of platinum catalysts is to optimize the size and shape of the platinum particles. Decreasing the particles' size alone increases the total surface area of catalyst available to participate in reactions per volume of platinum used but recent studies have demonstrated additional ways to make further improvements in catalytic performance.

Normally, the platinum nanoparticles are spread on a larger surface of carbon powder, thus acting as a supporting medium. While smaller particles are generally desirable, the final performance of the catalyst in the operating environment is a consequence of many factors. Several studies have shown that the oxygen reduction reaction exhibits a particle-size effect in the range of 1–5 nm. Uniform catalyst dispersion, observable by transmission electron microscopy (TEM), is essential for good platinum utilization and is shown in Figure 2.6.[21]

The first generation of polymer electrolyte membrane fuel cells (PEMFC) used PTFE-bound Pt black electrocatalysts that exhibited reasonable durability, however at a high cost (expensive platinum loadings of 4 mg/cm^2).[22] This is commonly achieved by developing methods to increase the utilization of the platinum that is deposited. Recently, platinum loadings as low as 0.04 mg/cm^2 have been reported using advanced fabrication techniques.[23] In addition to catalyst loading, there are a number of CL properties that have to be carefully optimized to achieve high utilization of the catalyst material, reactant diffusivity, ionic and electrical conductivity, and the level of hydrophobicity all have to be carefully balanced. Usually, the CL is reinforced by a thick porous electrode support layer termed as GDL.

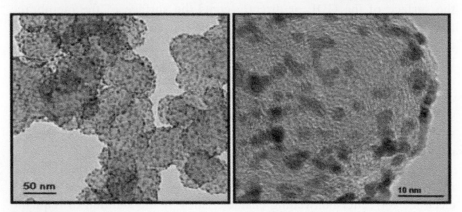

FIGURE 2.6 Transmission electron microscopy (TEM) image of DURA-lystTM-1140 (40% Pt/C catalyst).

2.8.2.5 GAS DIFFUSION LAYER (GDL)

The GDL electrically interconnects the catalyst and current collector. Often, an intermediate porous layer is added between the GDL and CL to ease the transitions between the large pores in the GDL and small porosity in the CL. The GDL mainly acts as a pathway to the CL from the flow channels and consists of two layers, a hydrophobic agent (polytetrafluoroethylene) and a macroporous layer made of Toray carbon paper or a carbon cloth that is covered with a microporous layer. GDL also aids to remove the product water or else flooding can occur where water blocks the GDL site and this limits the reactants' ability to access the catalyst and significantly decreases the performance. Teflon can be coated onto the GDL to limit the possibility of flooding. Several GDL microscopic variables such as porosity, tortuosity, and permeability impact the behavior of the fuel cells.[24] The functions of an ideal GDL are already elaborated in Table 2.1. Altering the composition of the diffusion layer can lead to substantial improvements in the performance of the cell; the scanning electron microscope (SEM) image of carbon cloth and carbon paper is shown in Figure 2.7.

FIGURE 2.7 SEM image of carbon cloth and carbon paper.

2.8.2.6 GDL REQUIREMENTS

GDL must be porous, electrically conductive, and thin. The reactants must be able to reach the catalyst but conductivity and porosity can act as opposing forces. Optimally, the GDL should be composed of about one-third Nafion or 15% PTFE. The carbon particles used in the GDL can be larger than those employed in the catalyst because surface area is not the most important variable in this layer. GDL should be around 15–35 μm thick to balance the required porosity with mechanical strength.[20] There are generally two types of GDL that are commercially used in the PEM fuel cell. One is the use of nonwoven carbon paper while the other one is a woven carbon cloth. The characteristics of the carbon paper and the carbon cloths are elaborated in Table 2.4.

TABLE 2.4 Difference in General Characteristics of Carbon Cloth and Paper.

Nonwoven carbon paper	Woven carbon cloth
Very brittle	Very flexible
Hardly any compressibility	Good compressibility
Less thickness hence less ohmic loss	More thickness hence more ohmic loss
Optimal for nonhumid operation	Optimal for humid operation

The thickness of the gas diffusion medium is vital because the thinner the material, the lesser the resistance and this would lead to less ohmic loss. The platinum which is used to catalyze the hydrogen catalyzes the carbon present in the carbon paper or cloth thus contributing to the degradation of the GDL. Noncarbon support structures offer higher stability to corrosion. Metallic parts are remarkable candidates for their excellent mechanical and thermal characteristics.[25] The applicability of titanium sinter as gas diffusion backing in a PEMFC was studied by Hottinen et al.[26] and its results revealed that the contact resistance between the titanium sinter and MEA is quite large. This can be reduced quite efficiently with platinum coating on the surface of the sinter. However, titanium sinters may be advantageous in applications where high current densities are not necessary. This is especially the case in applications where a simplified fuel cell structure is favorable. In addition, metal-based GDLs are not prone to voltage oxidation unlike carbon-based GDL.

2.8.3   GASKETS

The MEA, which is considered as the heart of the PEM fuel cell stack, is sandwiched between two bipolar plates and gaskets further around the edges of the MEA so as to make it sealed. These gaskets are typically made of a polymer and their thickness has to be appropriately selected, considering potential creep of the materials. Hard gaskets support well-defined gaps; however, they may be compromised in their sealing properties, do not compensate tolerances very well, and may cause mechanical distortion on the bipolar plates, which in turn can cause cracks after a long time. On the other hand, with soft gaskets, it is more difficult to control the compression ratio but they are superior with respect to tolerances and mechanical stabilization of the stack.[27]

2.9   DEGRADATION MECHANISM OF CONVENTIONAL PEM FUEL CELL STACK COMPONENT

Similar to any other power system, PEM fuel cell stack performance degrades over operating time. Chemical and mechanical degradation can occur in the electrode materials during its operation. The degradation is dependent on materials, fabrication, and operating conditions. The changes in temperature and relative humidity associated with transitions between low and high power can have adverse effects on the component properties and the integrity of the fuel cell system. For example, transient automotive operating conditions, specifically power (or voltage) cycling, can exacerbate a fuel cell's degradation and reduce its components' durability/reliability as well.[28] Active research is moving ahead worldwide to develop cheap, advanced, and durable material that degrades at a lower rate and is capable of operating better in the wider range of operating condition (voltage, temperature, and relative humidity). Korean researchers[29] designed a new MEA with a double-layered cathode comprising an inner CL prepared by the conventional decal transfer method and an outer CL directly coated on gas diffusion media. In summary, the major failure modes for PEM fuel cell components, including the membrane, catalyst/CL, GDL, bipolar plate, and sealing material, are listed in detail in Table 2.5.[28]

TABLE 2.5 Major Failure Modes of Components in PEM Fuel Cells.

Component	Failure modes	Causes
Membrane	Mechanical degradation	Mechanical stress due to nonuniform press pressure, inadequate humidification, or penetration of the catalyst and seal material traces
	Thermal degradation	Thermal stress, thermal cycles
	Chemical/electrochemical degradation	Contamination, radical attack
Catalyst layer	Loss of activation	Sintering or dealloying of electrocatalyst
	Conductivity loss	Corrosion of electrocatalyst support
	Decrease in mass transport rate of reactants	Mechanical stress
	Loss of reformate tolerance	Contamination
	Decrease in water management ability	Change in hydrophobicity of materials due to Nafion or PTFE dissolution
GDL	Decrease in mass transport	Degradation of backing material
	Decrease in water management ability	Mechanical stress, change in the hydrophobicity of materials
	Conductivity loss	Corrosion
Bipolar plate	Conductivity loss	Corrosion, oxidation
	Fracture/deformation	Mechanical stress
Sealing gasket	Mechanical failure	Corrosion, mechanical stress

The Department of Energy (DoE, USA) has set up targets for fuel cells for various applications and is listed as follows: for stationary applications, it is assumed that a fuel cell system should be capable of a minimum lifetime of 40,000 h (with 8000 h of uninterrupted service at >80% rated power). Automotive applications have a lifetime target of 5000 h at a degradation rate of less than 1% per 1000 h, with up to 20,000 operational hours required for buses.[30]

Degradation of fuel cell performance originates from various mecha-nisms; it may involve all of the main components of the stack (i.e., elec-trolyte, electrodes, and bipolar plates) and is often specific to the type of fuel cell technology. These degradation mechanisms may be reversible (e.g., water management in a PEMFC), partially reversible (e.g., poisoning by fuel impurities), or irreversible (e.g., sintering of electrodes or electrolyte rupture), and may be exacerbated or remedied by excursions from steady-state operation such as the cycling of temperature, electrical load, humidifi-cation level, electrode oxidation state, etc.

A recently published book by Büchi et al. details the extent to which each component of the fuel cell can contribute to performance degradation and pays testament to the importance and level of activity in this area.[31]

2.10   DEVELOPMENT OF ADVANCED MATERIAL FOR PEM FUEL CELL STACK

The degradation mechanisms contribute to the performance decline of fuel cell stack components, and, as a consequence, development of advanced and durable material resistant to corrosion and degradation. Durability of the PEMFC depends on several factors, and among them, corrosive nature of the supported carbons is one of the most significant one.[32] From the perspective of thermodynamics, carbon is prone to oxidation under the operating conditions of the cells, especially the PTFE and carbon composite of the GDLs are susceptible to chemical attack (i.e., OH^- radicals, as elec-trochemical by-products) and electrochemical (voltage) oxidation.[33] Oxida-tion of support carbon could also lead to changes in surface hydrophobicity which decreases the gas transport. In order to reduce the support carbon corrosion from the electrode surface and to improve the performance and durability of the cells concurrently, various approaches have been made by many researchers.[33,34]

The heart of the PEM fuel cell stack is the MEA and the performance and durability of a PEMFC intensely depend on the structure and fabri-cation process of the MEA.[20,35-40] The widely incorporated techniques are "catalyst-coated gas (CCG) diffusion layer" in which a catalyst is coated onto GDL. In the CCG technique, catalyst particles can penetrate into the highly porous GDL, and the subsequent hot pressing can improve interfacial contact between the electrodes and the membrane but destroy the pore struc-tures of CLs and GDLs. The alternate technique being the "CCM," where the catalyst is coated onto the membrane. In this technique, the GDL retains its intrinsic properties such as pore structures, and catalyst particles are not lost while the interfacial resistance is relatively high and the membrane is swollen during catalyst coating.[39–41] Sundar et al.[42] have aimed to reduce the rate of carbon support corrosion by incorporating nonequilibrium impregnation technique and Figure 2.8 shows the surface view of the nanocatalyzed electrodes.

FIGURE 2.8 Surface view of the nanocatalyzed electrode.

The longtime stability of platinum-nanocatalyzed MEA was compared to be superior to the conventional one. Development of durable self-humidifying MEAs is also a critical role in keeping the membrane hydrated, especially in the high current density range. On the contrary, the effective removal of product water from the cathode is necessary to prevent the well-known flooding phenomenon,[43,44] which impedes the transport of reactant gases to the catalytic active sites and sharply decreases the cell performance.[43–45] Thus, a MEA must optimally manage both the humidification as well as flooding issues. Lina Fukuharaa et al.[46] prepared a polymer electrolyte membrane through a novel technique of sulfonation of natural rubber grafted with polystyrene. The membrane was found to form nanomatrix channel consisting of natural rubber particle of about 1 μm in average diameter.

Lu et al.[44] incorporated a new technique by adding a microporous layer to the top surface of the gas diffusion layer (GDL) to take advantage of the capillary force to remove product water. Qi et al.[45] and Eom et al.[43] improved cell performance by adding hydrophobic polytetrafluoroethylene to the surface of the gas diffusion layers. Kongkanand et al.[47] achieved high-power PEM fuel cell performance even with an ultralow Pt loading. Compared to the conventional Pt/C, their catalyst material exhibited superior performance particularly at high current density. These advancements in the PEM fuel cell components authenticate that PEM fuel cells can be soon commercialized efficaciously.

2.11   ECONOMICS OF PEM FUEL CELL STACK COMPONENTS

The United States DoE has specified long-term targets for PEM fuel cells for vehicular applications and targeted by the year 2015; fuel cells for these applications were expected to be 60% efficient and cost US $30/kW.[48] Catalysts are a major cost factor for PEM fuel cells, accounting for 77% of the stack costs.[49] In addition, they are also a major contributing factor for fuel cell durability, as sintering, particle growth, dissolution, agglomeration, etc. lead to performance losses in the fuel cell stack. Considerable effort has been directed at decreasing catalyst cost by increasing catalyst activity and decreasing the platinum and noble metal content of the catalysts. Significant improvements have been obtained with Pt alloy catalysts.[50] Kongkanand et al.[47] showed that Pt loading can be reduced to a level as low as 0.025 mg cm^{-2} without noticeable transport-related losses. This suggests considerable potential for further fuel cell cost reduction. Researchers are also trying to use Pt-free catalysts for the oxygen reduction reaction with these catalysts. The total cost of a car with the PEM fuel cell system is $500–600/kW which is 10-fold higher compared to cars using IC engine. Total cost of the PEMFCs includes the costs of assembly process, bipolar plate, platinum electrode, membrane, and peripherals.[9]

2.12   DISCUSSION AND INTERPRETATION

Integral PEM fuel cell stack components, namely membrane and GDL exhibit nonlinear properties. For instance, membrane should neither be hydrophilic nor hydrophobic as well the too thin GDLs cannot provide good electrical contact between the current collecting plates and the CL and too thick exhibit high electrical resistance. And as a consequence, the membrane's water absorption as well as the thickness of GDL need to be optimized.

More information relating to efficiency and cell parameters is required to correlate cost and performance of the PEM fuel cell stack components. For example, if we can operate the fuel cell stack less than the rated power (less current contributes to less I^{2}R loss), it might reduce the degradation of the stack components. However, such an act depends on the end user's requirement and the customers cannot be influenced to operate a PEM fuel cell at a lower rated power. The economic sustainability of the PEM fuel cell stack depends on the capital cost as well as the durability of its components.

PEM fuel cells with the current platinum loadings and membrane (Nafion) are strictly not practicable for commercialization. One main goal of the catalyst design is to increase the catalytic activity of platinum by decreasing the loading. Alternative cheap catalyst materials can be a promising solution to supersede this issue. Similar trends should be followed in the development of advanced membrane materials. Nafion membranes are expensive and not environmentally friendly with regards to recycling and disposal of fluorinated polymers.[8] Researches on alternative non-Pt-based catalysts and non-Nafion membranes operating at higher temperature (ca. 120°C) might also play a significant role in the development of cheap PEM fuel cell stack. From an engineering perspective, controlling the operating parameters can also enhance the performance as well the durability of the PEM fuel cell stack. A fuzzy logic model designed by Arunkumar et al.[51] to pragmatically control the operating parameters is one such example.

It should be noted that there has been a steady decrease in the cost reduction and in 2009, the actual cost of a fuel cell stack and balance of plant was US $61/kW,[52] still short of US $30/kW (DoE, USA) target for 2017. It is very essential to understand the components involved in PEM fuel cell fabrication in order to make PEM fuel cells competitive against battery technologies, as well against the IC engine technologies for automotive applications.

2.13 CONCLUSION

The developments of PEM fuel cell stack components have received more attention recently and this chapter provides a comprehensive outlook on the key functional components and its economic implications. The proposed review paper is one of its kinds where the divergent factors on both technical and nontechnical constraints are summarized. The factors affecting the durability and degradation mechanisms of each component are also explicitly conversed. Development of advanced functional and cost-effective materials with desired electrical, chemical, mechanical, and thermal property would pave significant way for the PEM fuel cell stack development.

PEM fuel cell a 19th-century invention, a 20th-century technology development, will be a ubiquitous power system of the 21st century if the key stack components such as bipolar plate, membrane, CL, and gas diffusion backings can be developed efficiently and economically.

ACKNOWLEDGMENTS

This work was supported by the IBTec, Auckland University of Technology (AUT) grant by the Doctoral Fellowship and the author would like to acknowledge their support. The authors are also grateful to Ms. Jo Stone for effective proofreading.

KEYWORDS

- **PEM fuel cell**
- **membrane**
- **catalyst layer**
- **bipolar plate**
- **gas diffusion layer**
- **degradation**
- **scanning electron microscope**
- **transmission electron microscopy**

REFERENCES

1. Kordesch, K. V.; Simader, G. R. Fuel Cell Systems: Sections 4.6–4.8. In *Fuel Cells and Their Applications*; pp 133–179.
2. Pethaiah, S. S.; Arunkumar, J.; Ramos, M.; Al-Jumaily, A.; Manivannan, N. The Impact of Anode Design on Fuel Crossover of Direct Ethanol Fuel Cell. *Bull. Mater. Sci.* **2016,** *39*(1), 273–278.
3. EG & G Technical Services. *Fuel Cell Handbook*; EG & G Technical Services Inc.: Albuquerque, NM, DOE/NETL-2004/1206, 2004.
4. Zhang, J. *PEM Fuel Cell Electrocatalysts and Catalyst Layers: Fundamentals and Applications*; Springer-Verlag London Ltd, 2008.
5. Kumar, J. A.; Kalyani, P.; Saravanan, R. Studies on PEM Fuel Cells Using Various Alcohols for Low Power Applications. *Int. J. Electrochem. Sci.* **2008,** *3*(8), 961–969.
6. McLean, G.; Niet, T.; Prince-Richard, S.; Djilali, N. An Assessment of Alkaline Fuel Cell Technology. *Int. J. Hydrogen Energy* **2002,** *27*(5), 507–526.
7. Li, X.; Fields, L.; Way, G. Principles of Fuel Cells. *Platin. Metals Rev.* **2006,** *50*(4), 200–201.
8. Qiu, B.; Lin, B.; Qiu, L.; Yan, F. Alkaline Imidazolium- and Quaternary Ammonium-functionalized Anion Exchange Membranes for Alkaline Fuel Cell Applications. *J. Mater. Chem.* **2012,** *22*(3), 1040–1045.

9. Taner, T. Alternative Energy of the Future: A Technical Note of PEM Fuel Cell Water Management. *J. Fundam. Renew. Energy Appl.* **2015,** *5*(3), 1–4

10. Drive, U. *Fuel Cell Technical Team Roadmap*; US Drive Partnership, 2013.

11. Devanathan, R.; Idupulapati, N.; Baer, M. D.; Mundy, C. J.; Dupuis, M. Ab Initio Molecular Dynamics Simulation of Proton Hopping in a Model Polymer Membrane. *J. Phys. Chem. B* **2013,** *117*(51), 16522–16529.

12. Friedman, D.; Moore, R. PEM Fuel Cell System Optimization. *Proc. Electrochem. Soc.* **1998,** *27*, 407–423.

13. Jayakumar, A.; Sethu, S. P.; Ramos, M.; Robertson, J.; Al-Jumaily, A. A Technical Review on Gas Diffusion, Mechanism and Medium of PEM Fuel Cell. *Ionics* **2015,** *21*(1), 1–18.

14. Li, X.; Sabir, I. Review of Bipolar Plates in PEM Fuel Cells: Flow-field Designs. *Int. J. Hydrogen Energy* **2005,** *30*(4), 359–371.

15. Karimi, S.; Fraser, N.; Roberts, B.; Foulkes, F. R. A Review of Metallic Bipolar Plates for Proton Exchange Membrane Fuel Cells: Materials and Fabrication Methods. *Adv. Mater. Sci. Eng.* **2012,** *2012*.

16. Cunningham, B. D.; Huang, J.; Baird, D. G. Development of Bipolar Plates for Fuel Cells from Graphite Filled Wet-lay Material and a Thermoplastic Laminate Skin Layer. *J. Power Sources* **2007,** *165*(2), 764–773.

17. Brandon, N. P.; Skinner, S.; Steele, B. C. Recent Advances in Materials for Fuel Cells. *Annu. Rev. Mater. Res.* **2003,** *33*(1), 183–213.

18. Rowshanzamir, S.; Peighambardoust, S. J.; Parnian, M. J.; Amirkhanlou, G. R.; Rahnavard, A. Effect of Pt-Cs$_{2.5}$H$_{0.5}$PW$_{12}$O$_{40}$ Catalyst Addition on Durability of Self-humidifying Nanocomposite Membranes Based on Sulfonated Poly (Ether Ether Ketone) for Proton Exchange Membrane Fuel Cell Applications. *Int. J. Hydrogen Energy* **2015,** *40*(1), 549–560.

19. Haile, S. M. Fuel Cell Materials and Components. *Acta Mater.* **2003,** *51*(19), 5981–6000.

20. Litster, S.; McLean, G. PEM Fuel Cell Electrodes. *J. Power Sources* **2004,** *130*(1), 61–76.

21. He, C.; Desai, S.; Brown, G.; Bollepalli, S. PEM Fuel Cell Catalysts: Cost, Performance, and Durability. *Interface-Electrochem. Soc.* **2005,** *14*(3), 41–46.

22. Wilson, M. S.; Valerio, J. A.; Gottesfeld, S. Low Platinum Loading Electrodes for Polymer Electrolyte Fuel Cells Fabricated Using Thermoplastic Ionomers. *Electrochim. Acta* **1995,** *40*(3), 355–363.

23. O'Hayre, R.; Lee, S.-J.; Cha, S.-W.; Prinz, F. B. A Sharp Peak in the Performance of Sputtered Platinum Fuel Cells at Ultra-low Platinum Loading. *J. Power Sources* **2002,** *109*(2), 483–493.

24. Espinoza, M.; Andersson, M.; Yuan, J.; Sundén, B. Compress Effects on Porosity, Gasphase Tortuosity, and Gas Permeability in a Simulated PEM Gas Diffusion Layer. *Int. J. Energy Res.* **2015,** *39*(11), 1528–1536.

25. Tolochko, N. K.; Mozzharov, S. E.; Yadroitsev, I. A.; Laoui, T.; Froyen, L.; Titov, V. I.; Ignatiev, M. B. Balling Processes During Selective Laser Treatment of Powders. *Rapid Prototy. J.* **2004,** *10*(2), 78–87.

26. Hottinen, T.; Mikkola, M.; Mennola, T.; Lund, P. Titanium Sinter as Gas Diffusion Backing in PEMFC. *J. Power Sources* **2003,** *118*(1), 183–188.

27. Kundler, I.; Hickmann, T. Bipolar Plates and Gaskets: Different Materials and Processing Methods. In *High Temperature Polymer Electrolyte Membrane Fuel Cells*; Springer: Cham, 2016; pp 425–440.

28. Silva, R.; Paiva, T.; Rangel, C. MEA Degradation and Failure Modes in PEM Fuel Cells. **2009**.
29. Kim, G.; Eom, K.; Kim, M.; Yoo, S. J.; Jang, J. H.; Kim, H.-J.; Cho, E. Design of an Advanced Membrane Electrode Assembly Employing a Double-Layered Cathode for a PEM Fuel Cell. *ACS Appl. Mater. Interfaces* **2015**, *7*(50), 27581–27585.
30. Fowler, M.; Mann, R.; Amphlett, J.; Peppley, B.; Roberge, P. Reliability Issues and Voltage Degradation. *Handbook of Fuel Cells;* Wiley Online Library, 2005.
31. Büchi, F. N.; Inaba, M.; Schmidt, T. J. *Polymer Electrolyte Fuel Cell Durability*; Springer: New York, 2009; pp 29–53
32. Maass, S.; Finsterwalder, F.; Frank, G.; Hartmann, R.; Merten, C. Carbon Support Oxidation in PEM Fuel Cell Cathodes. *J. Power Sources* **2008**, *176*(2), 444–451.
33. Kou, Z.; Cheng, K.; Wu, H.; Sun, R.; Guo, B.; Mu, S. Observable Electrochemical Oxidation of Carbon Promoted by Platinum Nanoparticles. *ACS Appl. Mater. Interfaces* **2016**.*8*(6), 3940–3947
34. Wang, X.; Li, W.; Chen, Z.; Waje, M.; Yan, Y. Durability Investigation of Carbon Nanotube as Catalyst Support for Proton Exchange Membrane Fuel Cell. *J. Power Sources* **2006**, *158*(1), 154–159.
35. Frey, T.; Linardi, M. Effects of Membrane Electrode Assembly Preparation on the Polymer Electrolyte Membrane Fuel Cell Performance. *Electrochim. Acta* **2004**, *50*(1), 99–105.
36. Artyushkova, K.; Atanassov, P.; Dutta, M.; Wessel, S.; Colbow, V. Structural Correlations: Design Levers for Performance and Durability of Catalyst Layers. *J. Power Sources* **2015**, *284*, 631–641.
37. Sun, L.; Ran, R.; Shao, Z. Fabrication and Evolution of Catalyst-coated Membranes by Direct Spray Deposition of Catalyst Ink onto Nafion Membrane at High Temperature. *Int. J. Hydrogen Energy* **2010**, *35*(7), 2921–2925.
38. Thanasilp, S.; Hunsom, M. Effect of Pt:Pd Atomic Ratio in Pt–Pd/C Electrocatalyst-coated Membrane on the Electrocatalytic Activity of ORR in PEM Fuel Cells. *Renew. Energy* **2011**, *36*(6), 1795–1801.
39. Thanasilp, S.; Hunsom, M. Effect of MEA Fabrication Techniques on the Cell Performance of Pt–Pd/C Electrocatalyst for Oxygen Reduction in PEM Fuel Cell. *Fuel* **2010**, *89*(12), 3847–3852.
40. Rajalakshmi, N.; Dhathathreyan, K. Catalyst Layer in PEMFC Electrodes—Fabrication, Characterisation and Analysis. *Chem. Eng. J.* **2007**, *129*(1), 31–40.
41. Tang, H.; Wang, S.; Jiang, S. P.; Pan, M. A Comparative Study of CCM and Hot-pressed MEAs for PEM Fuel Cells. *J. Power Sources* **2007**, *170*(1), 140–144.
42. Pethaiah, S. S.; Kalaignan, G. P.; Ulaganathan, M.; Arunkumar, J. Preparation of Durable Nanocatalyzed MEA for PEM Fuel Cell Applications. *Ionics* **2011**, *17*(4), 361–366.
43. Eom, K.; Cho, E.; Jang, J.; Kim, H.-J.; Lim, T.-H.; Hong, B. K.; Lee, J. H. Optimization of GDLs for High-performance PEMFC Employing Stainless Steel Bipolar Plates. *Int. J. Hydrogen Energy* **2013**, *38*(14), 6249–6260.
44. Lu, Z.; Daino, M. M.; Rath, C.; Kandlikar, S. G. Water Management Studies in PEM Fuel Cells, Part III: Dynamic Breakthrough and Intermittent Drainage Characteristics from GDLs with and Without MPLs. *Int. J. Hydrogen Energy* **2010**, *35*(9), 4222–4233.
45. Qi, Z.; Kaufman, A. Improvement of Water Management by a Microporous Sublayer for PEM Fuel Cells. *J. Power Sources* **2002**, *109*(1), 38–46.
46. Fukuhara, L.; Kado, N.; Kosugi, K.; Suksawad, P.; Yamamoto, Y.; Ishii, H.; Kawahara, S. Preparation of Polymer Electrolyte Membrane with Nanomatrix Channel Through

Sulfonation of Natural Rubber Grafted with Polystyrene. *Solid State Ion.* **2014,** *268,* 191–197.

47. Kongkanand, A.; Subramanian, N. P.; Yu, Y.; Liu, Z.; Igarashi, H.; Muller, D. A. Achieving High-Power PEM Fuel Cell Performance with an Ultralow-Pt-Content Core–Shell Catalyst. *ACS Catal.* **2016,** *6*(3), 1578–1583.

48. Moore, R.; Hauer, K.; Ramaswamy, S.; Cunningham, J. Energy Utilization and Efficiency Analysis for Hydrogen Fuel Cell Vehicles. *J. Power Sources* **2006,** *159*(2), 1214–1230.

49. Carlson, E.; Kopf, P.; Sinha, J.; Sriramulu, S.; Yang, Y. *Cost Analysis of PEM Fuel Cell*; National Renewable Energy Laboratory NREL, Systems for Transportation, Report No. NREL/SR-560-39104, 2005.

50. Marcinkoski, J.; Kopasz, J. P.; Benjamin, T. G. Progress in the US DOE Fuel Cell Subprogram Efforts in Polymer Electrolyte Fuel Cells. *Int. J. Hydrogen Energy* **2008,** *33*(14), 3894–3902.

51. Jayakumar, A.; Ramos, M.; Al-Jumaily, A. In *A Novel Fuzzy Schema to Control the Temperature and Humidification of PEM Fuel Cell System,* ASME 2015 13th International Conference on Fuel Cell Science, Engineering and Technology Collocated with the ASME 2015 Power Conference, the ASME 2015 9th International Conference on Energy Sustainability, and the ASME 2015 Nuclear Forum, American Society of Mechanical Engineers, 2015; pp V001T06A005-V001T06A005.

52. Spendelow, J.; Marcinkoski, J. *Fuel Cell System Cost-2009*; US Department of Energy, DOE Hydrogen Program Record 2009, pp 9012.

CHAPTER 3

THEORETICAL STUDIES ON MOLECULAR DESIGNING OF NOVEL ELECTRICALLY CONDUCTING POLYMERS USING GENETIC ALGORITHM

VINITA KAPOOR

Department of Chemistry, Sri Venkateswara College, Delhi University, Delhi, India

E-mail: vinitaarora85@gmail.com

CONTENTS

ABSTRACT

Electrically conducting polymers have attracted a great deal of attention ever since their discovery. There have been enormous experimental and theoretical efforts in the area because of the versatile properties of these materials which combine the high conductivity of pure metals while being corrosion resistant, light in weight, and easy to process. These features make them one of the most widely applicable classes of materials. Their applications include super capacitors, electronic devices, organic light-emitting diodes, optical devices, smart windows, sensors, batteries, solar cells, artificial muscles, memory devices, nanoswitches, and many more. The chemical structures of these organic conjugated polymers can be manipulated in several ways to obtain properties as desired. In order to "tailor-make" these polymers and obtain desired properties out of them, it is very important to first understand the relationship between the chemical structure of the polymer and its properties. Efforts are thus being made to make these materials intrinsically conducting and more processable.

When it comes to designing a polymer, there can be enormous ways in which the homopolymer units may be arranged to form a polymeric chain. As a result, the task of synthesizing a polymeric chain with desired properties becomes very cumbersome. To simplify this problem, we have endeavored to the use of a computational method of optimization for designing of low bandgap polymers, known as the genetic algorithm (GA). GA has proved to be a very versatile technique for theoretical designing of low bandgap polymers which possess high level of delocalization.

3.1   MOLECULAR DESIGNING

Molecular designing includes manufacturing as well as designing of new molecules which may not exist in nature or be stable beyond a very narrow range of conditions. It is an extremely challenging task requiring manual manipulation of molecules theoretically through simulations, or experimentally via synthetic techniques. The methodology adapted for manufacturing may vary with the applications for which these materials are desired. The work presented in the present chapter is related to the theoretical designing of novel low bandgap organic conjugated polymers. There has been significant expansion in research and tremendous market growth of molecular electronic devices, organic field-effect transistors (FET), light-emitting devices, biosensors, and solar cells during the past decade. This represents the chief driving force for the designing of low bandgap polymers with tailored electronic properties. In order to successfully design and synthesize novel materials, it is necessary to have a fundamental understanding of the relationship between the chemical structure of polymer and its electronic and conduction properties.

3.1.1 STRUCTURE–PROPERTY RELATIONSHIP

The structure of the molecules comprising any material determines the properties that the material possesses. "Structure determines properties" is a powerful concept in chemistry and in all fields in which chemistry is important including environmental science, biology, biochemistry, polymer science, medicine, engineering, and nutrition.

A great deal of information about molecular properties can be obtained merely from the way in which the atoms of molecules are connected, without resort to more elaborate aspects of molecular structure. With the development of new and more sophisticated indices and methods, this field can be expected to be of even greater importance in the future.

3.1.2 ELECTRICALLY CONDUCTING POLYMERS

One very flourishing outcome of the structure–property relationship is the field of electrically conducting polymers (ECPs). There has been significant expansion in research and tremendous market growth of ECPs in the form of molecular electronic devices, light-emitting devices, biosensors, and solar cells during the past two decades. This represents the chief driving force for the synthesis[1–9] of π-conjugated systems with tailored electronic properties. Since the above applications require active materials with a specific blend of properties, the search for the low bandgap polymers has formed a separate branch of study known as *bandgap engineering*.

However, one of the fundamental challenges in the field of conducting polymers is to design low bandgap intrinsically conducting polymers so that there is no need to dope them. This is because the process of doping of ECPs is often the source of chemical instability in them. Another problem often associated with doped polymers is their poor processibility, which is restricted to a great extent because of the insolubility and infusibility of these polymers.

In order to successfully design and synthesize such novel materials, it is necessary to have a fundamental *understanding of the relationship between the chemical structure of polymer and its electronic and conduction properties.* The magnitude of the bandgap (E_g) and the energy positions of the HOMO [highest occupied molecular orbital or ionization potential (IP)] and LUMO [lowest unoccupied molecular orbital or electron affinity (EA)] energy levels are the most important characteristics for determining the optical and electrical properties of a given polymer. Through numerous manipulations and modifications of the chemical structures of conjugated polymers, bandgaps as small as 0.5–1.0 eV have been achieved using various synthetic routes.[10–13]

Several methods of theoretical investigations are also being used by researchers to explore the electronic structure and hence the conduction properties of ECPs. Theoreticians have been successful in achieving very low bandgap ECPs through structural modifications in the conjugated carbon backbone chain, using efficient molecular modeling techniques available as a result of fast-growing computational competence.

3.1.3 SOME APPLICATIONS OF ECPS

ECPs have fascinated mostly because of their numerous applications in solar cells, lightweight batteries, sensors, optical data storage devices, molecular electronic devices, etc. Several conducting polymers such as polyacetylene (PA), polythiophene (PTh), polypyrrole (PPy), polyaniline (PANI), etc. have been reported as excellent electrode materials for rechargeable batteries,[14] as well as good candidates for electrochromic displays and thermal smart windows.[15] Scientists have used PPy films in a neurotransmitter as a drug release system into the brain.[16] ECPs have also been used to fabricate diodes, capacitors, resistors,[17] FET, and printed circuit boards. PANI has turned out to be one of the most extensively commercialized electronic polymers, with diverse applications ranging from electromagnetic shielding to an anti-rust agent to nanoscience. Some essential applications of ECPs are discussed below.

3.1.3.1 BATTERIES AND PHOTOVOLTAICS

Conducting polymers are promising materials for this application due to the fact that they may be electrochemically switched and possess useful mechanical properties in terms of tensile strength and flexibility. In polymers where both *p*- and *n*-doping processes are feasible, they are bought to use as both positive and negative electrodes in the same battery system. This concept was first explored by Mac Diarmid and Heeger in the early 1980s in case of PA.[18,19] The battery developed by them had PA. This battery when compared to a lead acid battery is not only lighter in weight but also has both higher energy and power densities. The polymer electrodes of these batteries have a longer shelf life because the ions involved in the delivery and storage of charge come from the solution rather than from the electrodes themselves, and therefore, these electrodes are saved from the mechanical wear brought about by the dissolution and redeposition of electrode materials during charge–discharge cycles in ordinary batteries. Another advantage of polymer electrode batteries is the absence of toxic materials in them and therefore the disposal problems are minimized. Afterwards, several efficient conducting polymer-based batteries have been reported.[20-27] Currently, there exists an active research direction in advanced batteries to make them flexible, which could lead to important applications in modern gadgets, such as roll up displays, wearable devices, radio-frequency identification tags, and integrated circuit smart cards.[28-30]

3.1.3.2   BIOMEDICAL APPLICATIONS

ECPs are biocompatible due to their organic nature. Moreover, the ability to entrap and controllably release biological molecules (i.e., reversible doping) as well as the ability to transfer charge from a biochemical reaction, have made them a popular choice for many biomedical applications, such as biosensors,[31-36] tissue engineering scaffolds,[37,38] neural probes,[39] drug delivery devices, and bioactuators.[40,41] Conducting polymers can also act as artificial nerves since they have the capacity to transport small electrical signals through the body. For example, PTh and PPy are capable of generating electrical signals by transferring electrons between different polymer chains. Work is also being done in modification of metal electrode materials used in active implantable devices, with conducting polymers[42] such as PPy and poly(3,4-ethylene dioxythiophene) (PEDOT) for improving the neural tissue–electrode interface and increasing the effective lifetime of these implants. Conductivity is also critical for tissue engineering and neural probe applications.[43-45]

3.1.3.3   ELECTRONIC DEVICES

Almost all levels of electronics are being produced using ECPs. These include the fabrication of individual electronic components such as resistors, capacitors, and transistors, so as to make better, faster, and smaller electronic devices for applications. PTh and its derivatives have been widely used to tune electronic properties via molecular structure.[46,47] The most common applications of electrochromic materials include a variety of displays, smart windows, spacecraft thermal control, optical shutters, optical switching devices, mirror devices, and camouflage materials.[48–52] Conjugated polymer-based electrochromic materials possess an ability to modify their structure to create multicolor electrochromes, and hence have gained popularity.[53–56] Recently, electrochromic devices with nanocomposite blends consisting of metal nanoparticles and conducting polymers poly(3,4-ethylene dioxythiophene):polystyrene sulfonate (PEDOT:PSS) were made.[57]

3.1.3.4　NANOELECTRONICS

Nanostructurization of conducting polymers has now combined the advantage of organic conductors with low-dimensional materials. Nanostructured conducting materials especially PANI, PP_y, and PEDOT have proved excellent candidates for application in nanoelectronic devices,[58–61] chemical and biological sensors[62–64] supercapacitors,[65,66] nanoconducting textiles,[67] and biomedical applications.[68,69]

3.2　STRATEGIES OF MOLECULAR DESIGNING

Highly conducting doped polymers have been synthesized far and wide. However, it is observed that the process of doping of ECPs is often the source of their chemical instability. Another problem often associated with doped polymers is their poor processability which is due to their insolubility and infusibility. Consequently, one of the fundamental challenges in the field of conducting polymers is to design low bandgap intrinsically conducting polymers so that there is no need to dope them. Therefore, molecular engineering of polymers is fast gaining popularity. Bandgap is simply reduced by either raising the HOMO or lowering the LUMO level of the polymer or by compressing the two levels closer together simultaneously. Recently, a series of thieno [3,4-b] thiophene derivatives were prepared by Park et al., wherein the bandgap of resulting polymers varied from 0.78 to 1.0 eV.[70] Several strategies are being used for designing polymers with tailor-made conduction properties these include:

3.2.1 SUBSTITUTION

This technique is based on the approach that those substituents should be attached to the conjugated polymer backbone which can contribute toward the delocalization in the chain. In this method, one starts with low bandgap polymers and then tries to modify their electronic properties by various substitution reactions, provided their chemical nature and experimental conditions allow such reactions. Steric effects of substituents also play a very important role during substitution process. The introduction of bulky groups in the polymer chain may introduce large nonbonding interactions between these groups and thus twist the polymer backbone chain. This may thereby lead to noncoplanarity and hence, a decrease in the extent of orbital overlap and effective conjugation length which further results in diminished carrier mobility and lower conductivities.[71,72] Several investigations have been made to study the effect of substituents on the band structure of PA[73] by analyzing the Bloch wave functions of the highest occupied and the lowest unoccupied bands of fluorinated PAs such as polymonofluoroacetylene (PMFA)[74,75] and polydifluoroacetylene (PDFA).[76] Both IP and EA values are found to be influenced, and hence the bandgap value is also affected. The application of dithieno[3,2-b:2′,3′-d] pyrroles (DTPs) in conjugated organic polymers has also resulted in a variety of materials with reduced and low bandgaps that exhibit high carrier mobilities.[77,78] These studies have shown how the electronic properties of polymers can be tuned by modifying the backbone chain or side groups.[79]

3.2.2 LADDER POLYMERIZATION

This strategy involves the building of ladder polymers, which are formed by joining simple polymers such as *cis* and/or *trans*-PA into symmetrical polymeric rings. The small energy gap in ladder polymers is a consequence of the direct interplay of electron–lattice and lattice–lattice interactions in them. This class of polymers, frequently referred to as *one-dimensional graphite family* (owing to their 1-D graphite-like configuration), includes polyacene (PAc), polyacenacene (PAcA), polyphenanthrene (PPh), poly-phenanthrophenanthrene (PPhP), and polyperinaphthalene (PPN). Recently, four ladder-type oligo-*p*-phenylenes have been synthesized by Katz et al.[80] Jacob et al. have developed a fresh route to ladder-type pentaphenylenes in which both good hole-accepting p-type as well as electron-accepting n-type materials can be prepared from a common intermediate.[81]

3.2.3 COPOLYMERIZATION

In this technique, quasi one-dimensional superlattices (or copolymers) of conducting polymers are generated in order to achieve the desired properties for specific applications. This strategy is a very efficient route for the synthesis as well as molecular designing of novel conducting polymers. The biggest advantage of this route is that the copolymers can be "tailor-made" by making an appropriate choice of homopolymers. Moreover, the electronic properties are largely affected by controlling the relative amounts of various components in the copolymer as well as their arrangement (periodic or aperiodic) in the conjugated chain. Depending upon the band alignment of the two constituent polymers, copolymers may be divided into four types:[82,83] type-I, where the energy gap of one component lies within the bandgap of the other component; type-II staggered, where the top of the valence band (VB) of one component lies within the bandgap of the other and the bottom of the conduction band (CB) of the second lies in the bandgap of the first; type-II misaligned, in which the CB minimum of one component is below the VB maximum of the second component; and lastly, type-III, in which one component is semimetallic while the other is a normal semiconductor. Based on this strategy, several interesting applications of ECPs have been reported in diverse fields. For example, the electrochemical copolymerization of 1-benzyl-2,5-di(thiophen-2-yl)-1H-pyrrole and EDOT (3,4-ethylenedioxythiophene) was carried out[84] in order to achieve a copolymer with multichromic properties, and the resultant copolymer displayed distinct color changes with high optical contrast.

3.2.4 DONOR–ACCEPTOR POLYMERIZATION

A donor–acceptor (D–A) polymer consists of a regular alternation of donor (electron rich) and acceptor (electron deficient) moieties along a conjugated backbone (Fig. 3.1). A regular arrangement of such repeat units in the π-conjugated chain significantly decreases the HOMO–LUMO separation. The bandgap of a D–A polymer is expected to be the lowest for that particular combination in which the electronegativity difference between the donor and acceptor moieties is the highest. The stability of such polymers is also believed to be considerably higher because of π-conjugation in the chain.

Moreover, substituents can be easily introduced in such systems via electrophilic substitution reactions. With the result, there is vast scope of chemical engineering at the molecular level (of the polymer) in order to fine tune the properties of the D–A polymer. The strategy of copolymerization may also be combined with that of D–A polymerization to investigate the conduction properties of copolymers of D–A polymers. Several copolymers have been successfully synthesized possessing bandgap around 1 eV.[85–87]

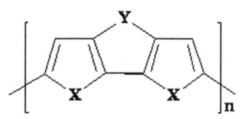

FIGURE 3.1 Structure of a donor–acceptor polymer with alternate donor (X) and acceptor (Y) moieties.

3.3 ARTIFICIAL INTELLIGENCE METHODS

Artificial intelligence (AI) has expanded its domain over the past few decades in terms of its versatile and comprehensive nature that has been capable of solving a variety of problems.[88,89] The underlying principle of AI research is knowledge-based computational intelligence combined with mathematical optimization, logic, methods based on probability and economics, and many others. Artificial optimization works to choose the best or a satisfactory solution from amongst the feasible solutions. The field of optimization has made a significant impact on many areas including engineering, management, physical sciences, social sciences, etc.

A salient aspect of the optimization algorithms used is that they are nature-inspired computing methodologies.[90,91] The idea from nature and biological processes has prompted development of experience-based algorithms for problem-solving. Various natural computing algorithms have been beneficially applied for solving complex problems.[92,93]

Genetic algorithm (GA) is one such technique[94–96] which is finding wide applicability due to its versatility and flexibility when it comes to problem-solving.

3.3.1 GENETIC ALGORITHM

GA belongs to the broader family of optimization algorithms called *evolutionary algorithms* that use techniques inspired by evolutionary biology such as inheritance, mutation, selection, and crossover (also known as recombination). It is one of the biology-inspired computing techniques that emulate the basic *Darwinian concept of natural selection*. Following these ideas, the algorithm allows us to use the computer to evolve automatic solutions over time. Because of its many positive attributes such as ease of use, flexibility, and versatility in application, GA has gained importance in materials research very rapidly ever since its inception by John Holland in 1975. GA, although in a major stage of theoretical development, has been successfully applied in the recent times in many fields such as production, designing, and processing of polymers[97–99] and other polymer as well as nonpolymer applications in science and technology.[100–103]

In this section, we first present the fundamental principles of GA to provide an idea of the modus operandi of this technique.

3.3.1.1 STEPS INVOLVED IN THE OPTIMIZATION PROCESS

The cycle of stages involved in a simple GA run is described herewith. The GA starts with a group of initial, randomly generated solutions. Each solution is represented as bit strings (sequences of zeros and ones) of specified length. Once the first population is generated, the fitness (how good is the proposed solution) of each individual (or *chromosome*) of the population is calculated through an evaluation function, also known as the *fitness function*.

The next populations (or *offsprings* arising from parents) are then generated by repeating the following steps until the convergence/optimization criterion is met.

1. Selection: includes selection of two parent chromosomes from a population based on their relative fitness values, analogous to "survival of the fittest" rule of nature. The better the fitness, more are the chances to be selected.
2. Crossover: involves a crossing-over between the two selected parents, so as to produce offsprings.
3. Accepting or placing the offsprings in a new population.

4. Use of the newly generated population for the next run of the algorithm.
5. If the termination criteria are fulfilled, the GA stops and returns the best solution in current population.

The purpose of our investigation is to find the optimal relative concentrations of the constituent homopolymers in the copolymer such that the copolymer so formed possesses a minimum bandgap value and maximum electronic delocalization, that is, the copolymer with maximum conducting ability. In our investigations, we have used a population consisting of *five* individuals for the sake of simplicity and hence the GA used is also called a micro-GA. Since, the number of chromosomes (size of population) is small; the population may become homogenous rapidly. If such a situation surfaces, we need to again randomly generate a few individuals of the existing population while keeping the best one from the last population. This is known as *elitism*. Therefore, best individuals from the previous generation are always included in the new population and hence the average fitness keeps on increasing by this procedure, from generation to generation. Moreover, the best solution found can survive to the end of the run. Elitism is a powerful strategy because crossover of the fittest chromosome is more valuable than crossover involving any other chromosome in the new population. To add to that, making a new population only by new offsprings may lead to loss of the best chromosome from the last population.

For example, considering a ternary copolymer $X_m Y_n Z_k$ constituted in the percentage ratio $m:n:k$ such that $m + n + k = 100$, a large number of random sequences are possible for a copolymer chain.

To achieve optimized results efficiently, we have combined meta-heuristic optimization algorithm with numerical methods, namely, negative factor counting (NFC) technique[104] and inverse iteration method (IIM).[105,106] The generic outline employed for designing low bandgap copolymers is as follows:

Step 1: Random generation[107] of a population of candidate solutions which represent the possible percentage combinations of homopolymers in a copolymer of specified length.

Step 2: Constructing the Hückel determinant for every copolymer sequence taking nearest neighbor interactions (tight binding approximation) into account,

$$
H(\lambda) =
\begin{vmatrix}
\alpha_1 - \lambda & \beta_2 & 0 & \cdots & 0 \\
\beta_2 & \alpha_2 - \lambda & \beta_2 & \cdots & 0 \\
0 & \beta_3 & \alpha_3 - \lambda & \beta_4 & 0 \\
0 & 0 & 0 & \ddots & \beta_N \\
0 & 0 & 0 & \cdots & \alpha_N - \lambda
\end{vmatrix} = 0,
$$

where, α's and β's are the diagonal (coulomb integral) and off-diagonal (resonance integral) elements and λ's are the eigenvalues. The α and β values are obtained from the corresponding band structures of the homopolymers. The electronic band structures of various homopolymers used in our calculations are obtained from ab initio Hartee–Fock crystal orbital (CO) method.[108–113]

Step 3: Solving the determinant using the NFC method to obtain the electronic properties namely IP, EA, and bandgap value corresponding to each copolymer.

Step 4: Using IIM to obtain the eigenvectors (coefficients C_{iv}) corresponding to the HOMO level of the VB for computation of inverse participation number (I_i, IPN) which is a measure of the extent of delocalization of a MO (molecular orbital) in the polymer chain[106] and is determined as,

$$
I_i = \frac{\sum_{v=1}^{n} |C_{iv}|^4}{\left(\sum_{v=1}^{n} |C_{iv}|^2 \right)^2}.
$$

IPN can assume values between 0 (maximum delocalization) and 1 (complete localization over one orbital).

A schematic description of the algorithm is presented in Figure 3.2. In our investigations, a population of five individuals (chromosomes) was used. Each chromosome was defined as a sequence of 7 bits which when converted to decimal form represented percentage concentrations of the respective homopolymers in the copolymer chain.[114–116] This is followed by construction of Hückel determinant and computation of eigenvalues and eigenvectors by NFC and IIM returned IP, EA, E_g, and IPN for each individual of the population. Using tournament selection criteria, two individuals were selected for crossover. Elitism technique was used, meaning, at least one best solution was copied without changes to a new population so that the best solution found survived to the end of the run. The population was allowed to evolve (iterate) until the convergence criterion was met with.

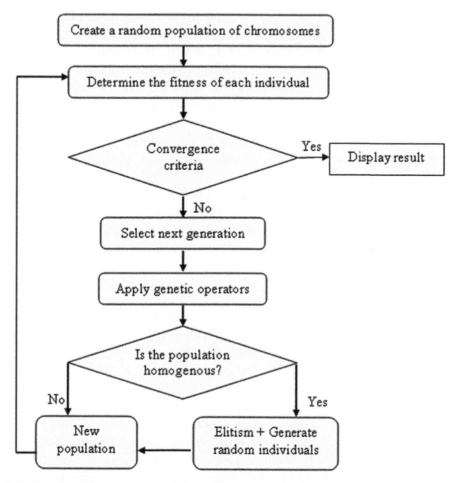

FIGURE 3.2 Schematic representation of the genetic algorithm.

3.4 THEORETICAL INVESTIGATIONS ON COPOLYMERS USING GA

3.4.1 NOVEL TERNARY COPOLYMERS OF DONOR–ACCEPTOR POLYMERS

In the present review, we have investigated three copolymer systems based on PPy (A), PTh (B), and polyfuran (PFu) (C) as repeat units using the GA. We have designed low bandgap copolymers based on PPy, PTh, and PFu skeletons, using alternate donor–acceptor moieties within the repeat units to maximize the extended π-conjugation. This arrangement may be capable of inducing charge transfer within the backbone chain. The presence of delocalization between the heterocyclic rings and substituents is expected to get enhanced by the electron-withdrawing character of the substituents, which in turn enhances the delocalization of the conjugated systems in the copolymer chain. The π-effects are significant for ECPs because the π-electron donating and accepting nature of the groups directly affects the electronic properties such as E_g, IP, and EA. The values of these parameters have been calculated using GA technique along with simple NFC and IIM.

System 1 (Fig. 3.3) consists of homopolymers PPy, PTh, and PFu as repeat units, while in systems 2 and 3 there is an alternation of donor and acceptor moieties because of bridging groups Y (>C=O, >C=CF$_2$). Introduction of >C=O groups in homopolymers PPy, PTh, and PFu leads to poly-4H-cyclopentadipyrrole-4-one (PCDP), poly-4H-cyclopentadithiophene-4-one (PCDT), and poly-4H-cyclopentadifuran-4-one (PCDF), respectively. While the introduction of >C=CF$_2$ groups gives us polydifluoromethylene cyclopentadipyrrole (PFPy), polydifluoromethylene cyclopentadithiophene (PFTh), and polydifluoromethylene cyclopentadifuran (PFFu), respectively. Their structures are shown in Figure 3.4. The electron-rich heterocyclic rings have been used as donor moieties while the carbonyl and difluoromethylene groups complete alternation by acting as acceptor units. The genetic operators which need to be essentially defined for a representation are: initialization, mutation, crossover, and comparison. The algorithm would rapidly converge to the optimum solution through intelligent searches, with selective sampling of values in the entire configuration space, that is, return the optimum relative concentration of the copolymer possessing minimum bandgap value and maximum electronic delocalization.

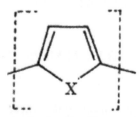

FIGURE 3.3 Schematic structure of unit cells of homopolymers PPy, PTh, and PFu, where X = NH, S, and O, respectively.

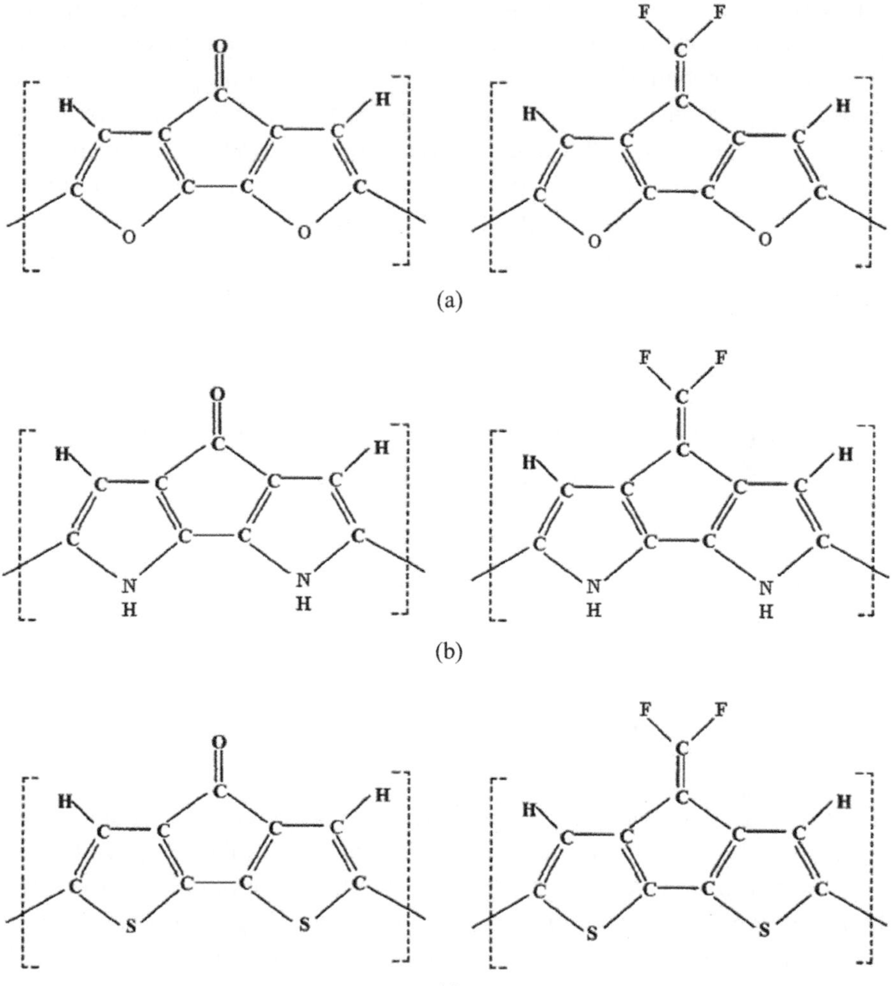

FIGURE 3.4 Schematic structures of unit cells of: (a) poly-4H-cyclopentadipyrrole-4-one (PCDP) and polydifluoromethylene cyclopentadipyrrole (PFPy), (b) poly-4H-cyclopentadithiophene-4-one (PCDT) and polydifluoromethylene cyclopentadithiophene (PFTh), and (c) poly-4H-cyclopentadifuran-4-one (PCDF) and polydifluoromethylene cyclopentadifuran (PFFu).

It is known that the negative of HOMO correlates very well with the vertical IP (Koopman's theorem). Table 3.1 shows the calculated electronic properties such as IP (IP corresponding to the top of the VB), EA (EA corresponding to the bottom of the CB), E_g (bandgap) of the random copolymer chains (of $A_xB_yC_z$ type), and the optimized percentage compositions (or optimized solutions) obtained from GA. These are the optimized values of the relative concentrations of the three constituent homopolymers in the resulting copolymer, which possesses minimum gap value and maximum level of delocalization in the chain. Molecular design and choice of specific substituents directly affect the properties of the polymer. Bandgap values are related to the conduction properties of the polymers in their pristine state, whereas IP and EA values ascertain the ability of the systems to form conducting polymers through oxidative and reductive doping, respectively.

TABLE 3.1 Calculated Electronic Properties—IP, EA, and E_g of the Random Ternary Copolymer Chains (of $A_xB_yC_z$ type) and the Optimized Percentage Compositions Obtained from GA. Y Denotes the Bridging Groups Introduced in the Carbon Backbone Chain of the Ternary Copolymer.

System	Y	IP (eV)	EA (eV)	E_g (eV)	Optimized composition
1	–	8.828	1.353	7.475	$A_{87}B_{12}C_1$
2	>C=O	7.412	3.425	3.987	$A_{86}B_1C_{13}$
3	>C=CF$_2$	7.084	3.010	4.074	$A_{84}B_1C_{15}$

3.4.1.1 RESULTS AND DISCUSSION

The system 1 under consideration is a ternary copolymer based on heterocyclic homopolymers PPy, PTh, and PFu. An analysis of the optimized solution obtained from GA ($A_{87}B_{12}C_1$) reveals that the homopolymer A (or PPy) should be present in maximum amount in the resulting copolymer so that the copolymer hence formed possesses maximum conducting ability and minimum bandgap. Here, it would be worth mentioning that out of the three homopolymers, A (or PPy) is the one with lowest value of IP (8.826 eV) as is evident from the band alignments (Fig. 3.5a). The component present in least amount, C (or PFu), has the highest IP value (9.740 eV). We therefore conclude that the resulting ternary copolymer with highest percentage of pyrrole skeleton is expectedly a suitable candidate for *p*-doping. The electronic density of states (DOS) for the above system corresponding to the optimized composition returned by GA is shown in Figure 3.6a. Since there is random placement of units in the copolymer sequence, their respective environments keep on changing and therefore, the DOS distribution consists of broad regions of allowed energy states.

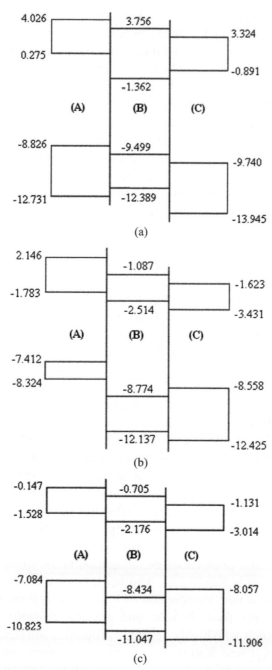

FIGURE 3.5 (a) Band alignments (system 1) of homopolymers PPy, PTh, and PFu, (b) band alignments (system 2) of homopolymers PCDP, PCDT, and PCDF, and (c) band alignments (system 3) of homopolymers PFPy, PFTh, and PFFu.

In the other two systems which we have studied, the electron-rich hetero-cycles act as donor units while the bridging groups act as acceptor moieties. System 2, a ternary copolymer of donor–acceptor-type polymers based on carbonyl substitution, comprises PCDP, PCDT, and PCDF as repeat units. The GA optimized solution is again found to contain the maximum percentage of pyrrole-based units PCDP, which have the lowest IP value 7.412 eV (Fig. 3.5b). On comparison of the electronic properties of this system with those of system 1, it is clear that the IP of the resulting ternary copolymer has decreased by ≈1.4 eV, while the EA has become more than double the previous value. As a result of this increase in EA value, the copolymer is expected to become a better candidate for *n*-doping (or reductive doping) because gain of electrons is now relatively easier. Moreover, an increase in the EA value decreases the bandgap drastically (by almost 3.5 eV). This implies that the introduction of electron-withdrawing carbonyl group in the backbone chain of the copolymer increases its intrinsic conductivity as well as dopant philicity because of extended π-conjugation in the chain, which in turn, is due to charge transfer within the backbone. The effect of substitution is also very clear from the DOS of the optimized solution (Fig. 3.6b). It can be seen that the separation between the cluster of VB and CB peaks has decreased significantly after substitution, thus enhancing the intrinsic conductivity of the copolymer.

When the difluoromethylene linkages are introduced into the polymer backbone chain, the ternary copolymer comprising PFPy, PFTh, and PFFu (system 3) is yet again found to contain maximum percentage of pyrrole-based units, PFPy, which have the lowest IP value (7.084 eV) amongst all three homopolymers (Fig. 3.5c). The trends in electronic properties are found to be more or less similar to those of the carbonyl-based ternary copolymers. The IP value has decreased by ≈1.7 eV, while the EA value has increased by ≈1.7 eV. As a result of this decrease of IP value, the copolymer is expected to become a better candidate for *p*-doping because the loss of electron is now relatively easier. The resulting copolymer has high intrinsic conductivity, again, because of extended π-conjugation in the chain. The trends in the DOS are shown in Figure 3.6c. Another important aspect related to this particular system is that, introduction of F atoms in organic molecules can profoundly influence their chemical and physical properties, leading to a range of compounds with highly desirable properties and remarkable physiological activity. These features are attributed to some unique properties of fluorine atom such as the largest electronegativity among all atoms and smallest van der Waal radius (apart from H), because of which fluorine is

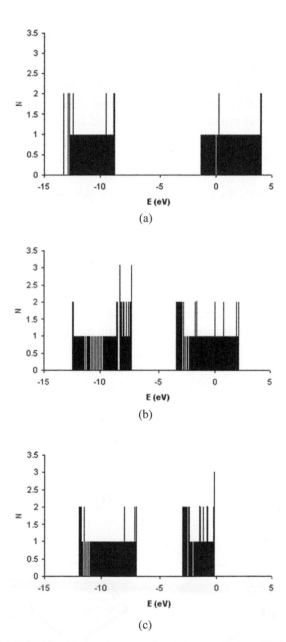

FIGURE 3.6 (a) DOS distribution (system 1) of the copolymer of PPy, PTh, and PFu corresponding to the optimized solution, (b) DOS distribution (system 2) of the copolymer of PCDP, PCDT, and PCDF corresponding to the optimized solution, and (c) DOS distribution (system 3) of the copolymer of PFPy, PFTh, and PFFu corresponding to the optimized solution.

not sterically demanding. Apart from these features, the C–F bond is also quite stable. These aspects can prove to be very useful during the synthesis of fluorine-containing organic polymers.

In general, we can say that, the electron acceptor substituents on heterocyclic rings in the copolymer backbone chain influence the LUMO energies (EA values) much more as compared to the HOMO energies (IP values) and thereby significantly lower the bandgap of the copolymer. The calculated gap values (from NFC) of such copolymers are in very good agreement with the expected values (if calculated from the band alignments directly). Thus, the trends in homopolymer units reflect the polymer properties.

3.4.2 NOVEL BINARY COPOLYMERS OF DONOR–ACCEPTOR POLYMERS

In the present review, we have investigated a novel binary copolymer of donor–acceptor-type polymers based on silane moiety and containing dicyanomethylene linkages. Silane containing units have been copolymerized with PFu, PPy, and PTh (discussed in section 3.4.1). The copolymer 1 consists of homopolymers PSICN ([A]$_x$) (Fig. 3.7) and PFUCN ([B]$_x$) (Fig. 3.8) as repeat units. The electron-rich heterocyclic rings have been used as donor moieties, while dicyanomethylene groups complete alternation by acting as acceptor units. To determine the electronic DOS, we have used an energy grid of 0.001 eV consistently in the calculations.

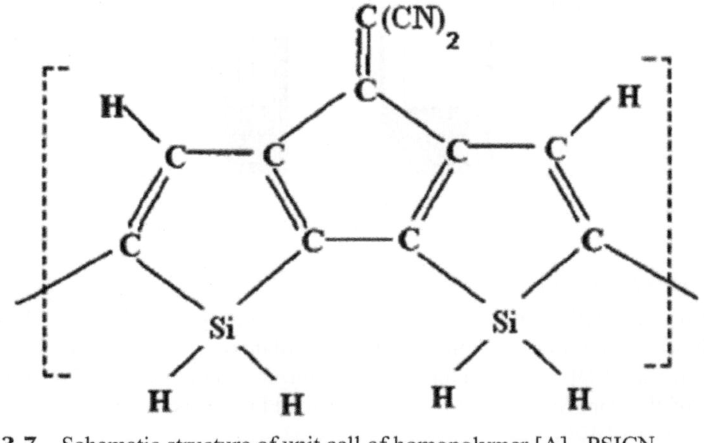

FIGURE 3.7   Schematic structure of unit cell of homopolymer [A]$_x$, PSICN.

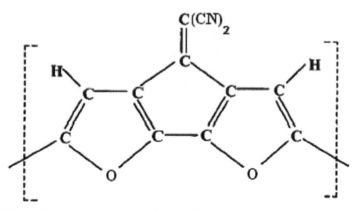

FIGURE 3.8 Schematic structure of unit cell of homopolymer [B]$_x$, PFUCN.

It is known that the negative of HOMO correlates very well with the vertical IP (Koopman's theorem). On basis of this, the various α and β values of the homopolymers, for both VB and CB, are listed in Table 3.2. These band structures used in our calculations have been obtained through ab initio Hartree–Fock CO method[117] using Clementi's minimal basis set. Consequently, the electronic properties might be slightly over estimated due to neglect of correlation effects. However, the results obtained are expected to provide able guidelines for synthesis of these copolymers with "tailor-made" electronic and conduction properties. All the parameters involved in the calculations remain exactly identical to what has been discussed in the previous calculations (section 3.4.1).

TABLE 3.2 The Ab Initio Electronic Properties (of Both Valence Band and Conduction Band) of the Donor–Acceptor Homopolymers.

Donor–acceptor polymer	Valence band		Conduction band	
	α (in eV)	β (in eV)	α (in eV)	β (in eV)
PSICN ([A]$_x$)	−10.655	0.829	−3.300	0.252
PFUCN ([B]$_x$)	−10.905	0.954	−3.749	0.212

3.4.2.1 RESULTS AND DISCUSSION

The results obtained from GA, for copolymer 1, suggest that the optimized solution would be the one which contains PSICN and PFUCN in the ratio 86:14 (A$_{86}$B$_{14}$) (Table 3.3). From this, we conclude that the homopolymer [A]$_x$ (or PSICN) should be present in maximum amount in the resulting copolymer so that the copolymer hence formed possesses maximum conducting ability and minimum bandgap. The electronic DOS for the abovementioned copolymer corresponding to the optimized composition returned by GA are shown in Figure 3.9. Since there is random placement of units in the copolymer sequence, their respective environments keep on changing and therefore, the DOS distribution consists of broad regions of allowed energy states.

TABLE 3.3 Calculated Electronic Properties—IP, EA, and E_g of the Random Binary Copolymer Along with the Optimum Percentage Composition as Obtained from GA. *PSICN denoted by $[A]_x$, PFUCN denoted by $[B]_x$.

Copolymer	IP (eV)	EA (eV)	E_g (eV)	Optimum composition
PSICN–PFUCN	8.996	4.171	4.825	$A_{86}B_{14}$

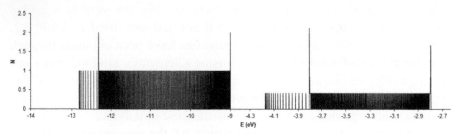

FIGURE 3.9   DOS distribution of copolymer 1.

3.5 CONCLUSIONS

Polymers play an indispensable role in our everyday life, ranging from the simplest organic conjugated polymer—polyethylene, to a variety of materials such as unbreakable utensils, LCD (liquid crystal display) TV sets, aerospace machinery, electrical sockets, or biopolymers such as DNA, proteins, and polypeptides that are essential for life.

Organic conjugated polymers based on the carbon backbone chain are basically insulators in their virgin state. Though the low electrical conductivity of polymers has found its immense use in the manufacture of insulators and dielectric substances, researchers have always been interested in producing polymers which exhibit conductivity similar to that of metals. The key characters of a potential ECP include low bandgap, high electronic delocalization, and ability to exhibit large electronic conductivity and reversibility with respect to doping and undoping cycles along with excellent stability in electrochemical environment and processes. Today, the study of these materials has become one of the foremost areas of research and their enormous applications have revolutionized the electronics industry, providing alternatives to silicon and germanium. The biggest application of ECPs lies in their future use as sensors and actuators, as well as in rechargeable batteries and microelectronic devices. Credit goes to the vast technological development in the electronics industry which has initiated extensive research in developing application-specific devices. The advantages of ECPs include—ease of fabrication, possibility of handling under ambient conditions, relatively large scale and inexpensive production, electronic tunability, possibility of making composites and blends with other polymers and inorganic materials, and tunable mechanical and chemical characteristics such as, solubility, strain, stress, and cross-linking properties.

During the last two decades, researchers have succeeded in preparing several conjugated polymers with high electrical conductivity. By bringing about subtle changes in the chemical structure of the molecules, the bulk electrical and optical properties of the material can be modified and regulated. Thus, for successful designing, it is necessary to have a complete understanding of the relationship between the chemical structure of polymer and its electronic and conduction properties such as IP, EA, and bandgap (E_g). The band structure can be tuned by altering either or both the electronic structure and stearics of the backbone.

To make ends meet, several strategies are being put to use, namely, substitution polymerization, ladder-growth polymerization, and topological growth and copolymerization including donor–acceptor moieties. Though several such routes are being followed for designing novel ECPs, a widely used approach is the donor–acceptor (D–A) polymerization technique. D–A polymers possess higher stability because of π-conjugation in the chain.

The basic aim of our study is the investigation of electronic structure and conduction properties of novel binary (two-component) and ternary (three-component) copolymers. When it comes to designing a polymer, there can be enormous ways in which the homopolymer units may be arranged to form a polymeric chain. As a result, the task of synthesizing a polymeric chain with desired properties becomes very cumbersome and expensive. Therefore, theoretical studies on polymers have a great worth and can serve as important guidelines in the designing and manufacturing process. The ambit of "theoretical designing of electrically conducting polymers" has been extended in the recent years with scientists venturing into the use of several optimization and search techniques available. Problem-specific algorithms and simulation techniques such as GA, ant-colony optimization, and traveling salesman problem are increasingly being used along with the conventional methods in order to evolve automatic solutions with the help of fast computing software available. In the present work, we have endeavored to the use of one such computational method of optimization for designing of low bandgap polymers, known as the GA. GA has proved to be a very versatile technique for theoretical designing of low bandgap polymers which possess high level of delocalization. The algorithm not only works toward finding the best solution but also improve the quality of the entire population at each successive iteration.

In our investigations on novel ternary copolymers of D–A polymers based on PPy, PTh, and PFu as repeat units, we have investigated the effect of substitution on the behavior and properties of the copolymers. For this purpose, we have introduced >C=O and then >C=CF$_2$ groups in PPy, PTh, and PFu. Again, GA helps us find out the optimum ternary copolymer with minimum bandgap. We arrive to the conclusion that in all the three copolymers, the optimized solution is found to contain the maximum percentage of pyrrole-based units. The introduction of electron-withdrawing carbonyl group in the backbone chain of the copolymer increases its intrinsic conductivity as well as dopant philicity because of extended π-conjugation in the chain, which in turn, is due to charge transfer within the backbone.

Hence, we conclude that copolymers of donor–acceptor polymers offer several advantages as compared to their carbon analogues. The extensive π-conjugation in the backbone adds to the stability of the copolymer. Substituents can be easily introduced in such systems via electrophilic substitution reactions. Therefore, chemical engineering at the molecular level of the polymer can be used along with theoretical designing in order to fine tune the properties of the polymer and serve as important aids in designing novel materials.

The results presented here contain useful theoretical predictions. We conclude that theoretical studies using AI methods can provide an insight into the relationship between the structure and electronic properties of conjugated polymers, thereby contributing to a better understanding of the various structural aspects of these materials which contribute to their favorable electronic properties and hence conductivity. GA technique drastically cuts down the time required for computation purpose and gives the optimized composition of the copolymer without compromising on accuracy. Competent algorithms such as GA adapted for molecular engineering of sustainable and efficient electronic materials can help streamline the meticulous experimental efforts on the choice and proportions of various constituents in a polymer, thereby providing an economical and potent passage for fruitful theoretical solutions. We are exploring the possibility of other heteroatom substitutions in order to make the polymers intrinsically more conducting. Investigations are also being made for suitable bridging substitutions that may be carried out on the backbone to make the polymers better candidates for *p*-doping and *n*-doping.

KEYWORDS

- polymers
- electronic structures
- conduction properties
- theoretical chemistry
- metaheuristic optimization algorithms
- artificial intelligence

REFERENCES

1. Mishra, S. P.; Palai, A. K.; Patri, M. *Synth. Met.* **2010**, *160*, 2422.
2. Gherras, H.; Hachemaoui, A.; Yahiaoui, A.; Benyoucef, A.; Belfedal, A.; Belbachir, M. *Synth. Met.* **2012**, *62*, 1759.
3. Lanzi, M.; Paganin, L.; Caretti, D.; Setti, L.; Errani, F. *React. Funct. Polym.* **2011**, *71*, 745.
4. Saini, G.; Jacob, J. *Macromol. Symp.* **2010**, *298*, 154.
5. Jo, Y.-R.; Lee, S.-H.; Lee, Y.-S.; Hwang, Y.-H.; Pyo, M.; Zong, K. *Synth. Met.* **2011**, *161*, 1444.
6. Hoyos, M.; Turner, M. L.; Navarro, O. *Curr. Org. Chem.* **2011**, *15*, 3263.
7. Yuan, C.; Hong, J.; Liu, Y.; Lai, H.; Fang, Q. *J. Polym. Sci. Part A: Polym. Chem.* **2011**, *49*, 4098.
8. Krüger, R. A.; Gordon, T. J.; Sutherland, T. C.; Baumgartner, T. *J. Polym. Sci. Part A: Polym. Chem.* **2011**, *49*, 1201.
9. Lu, B.; Liu, C.; Li, Y.; Xu, J.; Liu, G. *Synth. Met.* **2011**, *161*, 188.
10. Roncali, J. *Chem. Rev.* **1997**, *97*, 173.
11. Park, J. H.; Seo, Y. G.; Yoon, D. H.; Lee, Y.-S.; Lee, S.-H.; Pyo, M.; Zong, K. *Eur. Polym. J.* **2010**, *46*, 1790.
12. Aota, H.; Ishikawa, T.; Maki, Y.; Takaya, D.; Ejiri, H.; Amiuchi, Y.; Yano, H.; Kunimoto, T; Matsumoto, A. *Chem. Lett.* **2011**, *40*, 724.
13. Felix, J. F., Barros, R. A.; de Azevedo, W. M.; da Silva Jr., E. F. *Synth. Met.* **2011**, *161*, 173.
14. Saraswathi, R.; Gerard, M.; Malhotra, B. D. *J. Appl. Polym. Sci.* **1999**, *74*, 145.

15. Skotheim, T. A., Ed. *Handbook of Conducting Polymers;* Marcel Dekker: USA, 1986; Vol. 1.
16. Zinger, B.; Miller, L. L. *J. Am. Chem. Soc.* **1984,** *106*, 6861.
17. Min, G. *Synth. Met.* **2005,** *153*, 49.
18. Nigrey, P. J.; MacDiarmid, A. G.; Heeger, A. J. *J. Chem. Soc. Chem. Commun.* **1979,** *14*, 594.
19. MacInnes, D., Jr.; Druy, M. A.; Nigrey, P. J.; Nairns, D. P.; Mac Diarmid, A. G.; Heeger, A. J. *J. Chem. Soc. Chem. Commun.* **1981,** *7*, 317.
20. Song, H. K.; Palmore, G. T. R. *Adv. Mater.* **2006,** *18*, 1764.
21. Gofer, Y.; Sarker, H.; Killian, J. G.; Giaccai, J.; Poehler, T. O.; Searson, P. C. *Biomed. Instrum. Technol./Assoc. Adv. Med. Instrum.* **1998,** *32*, 33.
22. Colladet, K.; Nicolas, M.; Goris, L.; Lutsen, L.; Vanderzande, D. *Thin Solid Films* **2004,** *7, 451–452*.
23. Bundgaard, E.; Krebs, F. C. *Sol. Energ. Mat. Sol. C* **2007,** *91*, 954.
24. Jang, S.-Y.; Lim, B.; Yu, B.-K.; Kim, J.; Baeg, K.-J.; Khim, D. Y.; Kim, D.-Y. *J. Mater. Chem.* **2011,** *21*, 11822.
25. Haick, H.; Cahen, D. *Prog. Surf. Sci.* **2008,** *83*, 217.
26. Wagner, P.; Aubert, P.-H.; Lutsen, L.; Vanderzande, D. *Electrochem. Commun.* **2002,** *4*, 912.
27. Bakulin, A. A.; Elizarov, S. G.; Khodarev, A. N.; Martyanov, D. S.; Golovnin, I. V.; Paraschuk, D. Y.; Triebel, M. M.; Tolstov, I. V.; Frankevich, E. L.; Arnautov, S. A.; Nechvolodova, E. M. *Synth. Met.* **2004,** *147*, 221.
28. Nishide, H.; Oyaizu, K. *Science* **2008,** *319*, 737.
29. Nam, K. T.; Kim, D. W.; Yoo, P. J.; Chiang, C. Y.; Meethong, N. Hammond, P. T.; Chiang, Y. M. *Science* **2006,** *312*, 885.
30. Sugimoto, W.; Yokoshima, K.; Murakami, K.; Takasu, Y. *J. Electrochem. Soc.* **2006,** A*255*, 153.
31. Peng, H.; Soeller, C.; Vigar, N. A.; Caprio, V.; Travas-Sejdic, J. *Biosens. Bioelectron.* **2007,** *22*, 1868.
32. Cosnier, S. *Biosens. Bioelectron.* **1999,** *14*, 443.
33. Lewis, T. W.; Wallace, G. G.; Smyth, M. R. *Analyst* **1999,** *124*, 213.
34. Bidan, G.; Ehui, B.; Lapkowski, M. *J. Apys. D. Appl. Phys.* **1988,** *21*, 1043.
35. Gerard, M.; Chaubey, A.; Malhotra, B. D. *Biosens. Bioelect.* **2002,** *17*, 345.
36. Rahman, Md. A.; Kumar, P.; Park, D.-S.; Shim, Y.-B. *Sensors* **2008,** *8*, 118.
37. Schmidt, C. E.; Shastri, V. R.; Vacanti, J. P.; Langer, R. *Proc. Natl. Acad. Sci.* **1997,** *94*, 8948.
38. Kotwal, A.; Schmidt, C. E. *Biomaterials* **2001,** *22*, 1055.
39. Cui, X.; Wiler, J.; Dzaman, M.; Altschuler, R. A.; Martin, D.; *Biomaterials* **2003,** *24*, 777.
40. Guimard, N. K.; Gomez, N.; Schmidt, C. E. *Prog. Polym. Sci.* **2007,** *32*, 876.
41. Du, P.; Lin, X.; Zhang, X. *Sens. Actuators A* **2010,** *163*, 240.
42. Green, R. A.; Lovell, N. H.; Wallace, G. G.; Poole-Warren, L. A. *Biomaterials* **2008,** *29*, 3393.
43. Stauffer, W. R.; Cui, X. T. *Biomaterials* **2006,** *27*, 2405.
44. Cui, X.; Lee, V. A.; Raphael, Y.; Wiler, J. A.; Hetke, J. F.; Anderson, D. J. *J. Biomed. Mater. Res.* **2001,** *56*, 261.
45. Fonner, J. M.; Forciniti, L.; Nguyen, H.; Byrne, J.; Kou, Y.-F.; Syeda-Nawaz, J.; Schmidt, C. E. *Biomed. Mater.* **2008,** *3*, 034124.

46. Higashi, T.; Yamasaki, N.; Utsumi, H.; Yoshida, H.; Fujii, A.; Ozaki, M. *Appl. Phys. Exp.* **2011**, *4*, 091602.

47. Cheng, K.-F.; Chueh, C.-C.; Lin, C.-H.; Chen, W.-C. *J. Polym. Sci. A: Polym. Chem.* **2008**, *46*, 6305.

48. Sapp, S. A.; Sotzing, G. A.; Reynolds, J. R. *Chem. Mater.* **1998**, *10*, 2101.

49. Heuer, H. W.; Wehrmann, R.; Kirchmeyer, S. *Adv. Funct. Mater.* **2002**, *12*, 89.

50. Rosseinsky, D. R.; Mortimer, R. J. *Adv. Mater.* **2001**, *13*, 783.

51. Leventis, N.; Chen, M.; Liapis, A. I.; Johnson, J. W.; Jain, A. *J. Electrochem. Soc.* **1998**, *145*, 55.

52. Chandrasekhar, P.; Zay, B. J.; Birur, G. J.; Rawal, S.; Pierson, E. A.; Kauder, L.; Swanson, T. *Adv. Funct. Mater.* **2002**, *12*, 95.

53. Sonmez, G. *Chem. Commun.* **2005**, *42*, 5251.

54. Argun, A. A.; Aubert, P. H.; Thompson, B. C.; Schwendeman, I.; Gaupp, C. L.; Hwang, J.; Pinto, N. J.; Tanner, D. B.; MacDiarmid, A. G.; Reynolds, J. R. *Chem. Mater.* **2004**, *16*, 4401.

55. Sonmez, G.; Meng, H.; Wudl, F. *Chem. Mater.* **2004**, *16*, 574.

56. Argun, A. A.; Cirpan, A.; Reynolds, J. R. *Adv. Mater.* **2003**, *15*, 1338.

57. Namboothiry, M. A. G.; Zimmerman, T.; Coldren, F. M.; Liu, J.; Kim, K.; Carroll, D. L. *Synth. Met.* **2007**, *157*, 580.

58. Grimsdale, A. C.; Chan, K. L.; Martin, R. E.; Jokisz, P. G.; Holmes, A. B. *Chem. Rev.* **2009**, *109*, 897.

59. Boroumand, F. A.; Fry, P. W.; Lidzey, D. G. *Nano. Lett.* **2005**, *5*, 67.

60. Cho, S. I.; Kwon, W. J.; Choi, S. J.; Kim, P.; Park, S. A.; Kim, J.; Son, S. J.; Xiao, R.; Kim, S. H.; Lee, S. B. *Adv. Mater.* **2005**, *17,* 171.

61. Tseng, R. J.; Baker, C. O.; Shedd, B.; Huang, J.; Kaner, R. B.; Ouyang, J.; Yang, Y. *Appl. Phys. Lett.* **2007**, *90*, 053101–1.

62. Rajesh, Ahuja, T.; Kumar, D. *Sens. Actuators B* **2009**, *136*, 275.

63. Aussawasathien, D.; Dong, J. H.; Dai, L. *Synth. Met.* **2005**, *154*, 37.

64. Hangarter, M. C.; Chartuprayoon, N.; Hernández, S. C.; Choa, Y.; Myung, N. V. *Nano Today* **2013**, *8*, 39.

65. Gómez, H.; Ram, M. K.; Alvia, F.; Villalba, P.; Stefanakos, E. L.; Kumar, A. *J. Power. Sour.* **2011**, *196*, 4102.

66. Alvi, F.; Ram, M. K.; Basnayaka, P. A.; Stefanakos, E.; Goswami, Y.; Kumar, A. *Electrochim. Acta* **2011**, *56*, 9406.

67. Coyle, S.; Wu, Y.; Lau, K.-T.; De Rossi, D.; Wallace, G.; Diamond, D.; *Mrs. Bull.* **2007**, *32*, 434–442.

68. Abidian, M. R.; Kim, D. H.; Martin, D. C. *Adv. Mater.* **2006**, *18*, 405.

69. Kim, S.; Kim, J. H.; Jeon, O.; Kwon, I. C.; Park, K. *Eur. J. Pharm. Biopharm.* **2009**, *71*, 420.

70. Park, J. H.; Seo, Y. G.; Yoon, D. H.; Lee, Y.-S.; Lee, S.-H.; Pyo, M.; Zong, K. *Eur. Polym. J.* **2010**, *46*, 1790.

71. El-Nahas, A. M.; Mangood, A. H.; El-Shazly, T. S. *Comput. Theoret. Chem.* **2012**, *980*, 68.

72. Zgou, H.; Hamidi, M.; Bouachrine, M. *J. Mol. Str. (Theochem.)* **2007**, *814*, 25.

73. Bakhshi, A. K.; Ladik, J.; Seel, M. *Phys. Rev. B* **1987**, *35*, 704.

74. Bakhshi, A. K.; Ladik, J.; Liegener, C. M. *Synth. Met.* **1987**, *20*, 43.

75. Hayashi, S. I.; Aoki, Y.; Imamura, A. *Synth. Met.* **1990**, *36*, 1.

76. Salzner, U. *Synth. Met.* **2003**, *135*, 311.
77. Rasmussen, S. C.; Evenson, S. J. *Prog. Polym. Sci.* 2013 (http://dx.doi.org/10.1016/j.progpolymsci.2013.04.004).
78. Kroon, R.; Lundin, A.; Lindqvist, C.; Henriksson, P.; Steckler, T. T.; Andersson, M. R. *Polymer* **2013**, *54*, 1285.
79. Jaballah, N.; Chemli, M.; Hriz, K.; Fave, J.; Jouini, M.; Majdoub, M. *Eur. Poly. J.* **2011**, *47*, 78.
80. Zheng, Q.; Jung, B.; Sun, J.; Katz, H. E. *J. Am. Chem. Soc.* **2010**, *132*, 5394.
81. Jacob, J.; Sax, S.; Piok, T.; List, E. J. W.; Grimsdale, A. C.; Müllen, K. *J. Am. Chem. Soc.* **2004**, *126*, 6987.
82. Smith, D. L.; Maithio, C. *Rev. Mod. Phys.* **1990**, *62*, 173.
83. Esaki, L. *IEEE J. Quant. Elect.* **1986**, *22*, 1682.
84. Camurlu, P.; Tarkuç, S.; Şahmetlioğlu, E.; Akhmedov, İ. M.; Tanyeli, C.; Toppare, L. *Sol. Energ. Mat. Sol. C* **2008**, *92*, 154.
85. Radhakrishnan, S. Parthasarathi, R.; Subramanian, V.; Somanathan, N. *J. Chem. Phys.* **2005**, *123*, 164905.
86. Ferraris, Lambert, T. L. *J. Chem. Soc. Chem. Commun.* **1991**, *18*, 1268.
87. Lambert, T. L.; Ferraris, J. P. *J. Chem. Soc. Chem. Commun.* **1991**, *11*, 752.
88. Moore, J. H. *Comput. Meth. Prog. Bio.* **1995**, *147*, 73.
89. Arora, J. S.; Baenziger, G. *Comput. Method. Appl. Mech. Eng.* **1986**, *54*, 303.
90. Boussaïd, I.; Lepagnot, J.; Siarry, P. A Survey on Optimization Metaheuristics. *Inform. Sci.* 2013, doi:10.1016/j.ins.2013.02.041.
91. Decastro, L. *Phys. Life Rev.* **2007**, *4*(1), 1.
92. Franco, G.; Margenstern, M. *Theor. Comput. Sci.* **2008**, *404*(1–2), 88.
93. Masulli, F.; Mitra, S. *Inform. Fusion* **2009**, *10*(3), 211.
94. Goldberg, D. E.; *Genetic Algorithms in Search, Optimization and Machine Learning;* Addison-Wesley: New York, 1989.
95. Holland, J. *Adaptation in Natural and Artificial Systems*; MIT Press: Cambridge, 1975.
96. Cartwright, H. M. *Applications of Artificial Intelligence in Chemistry*; Oxford Chemistry Press: USA, 1994.
97. Sharma, R.; Nandy, S.; Chaudhury, P.; Bhattacharyya, S. P. *Mater. Manuf. Proc.* **2011**, *26*, 354.
98. Giro, R.; Cyrillo, M.; Galvão D. S. *Chem. Phys. Lett.* **2002**, *366*, 170.
99. Giro, R.; Cyrillo, M.; Galvão, D. S. *Mat. Res.* **2003**, *6*, 523.
100. Mohn, C. E.; Stølen, S.; Kob, W. *Mater. Manuf. Proc.* **2011**, *26*, 348.
101. Biswas, A.; Maitre, O.; Mondal, D. N.; Das, S. K.; Sen, P. K.; Collet, P. Chakraborti, N. *Mater. Manuf. Proc.* **2011**, *26*, 415.
102. Gramegna, N.; Corte, E. D.; Poles, S. *Mater. Manuf. Proc.* **2011**, *26*, 527
103. Chan, T. M.; Man, K. F.; Tang, K. S.; Kwong, S. *Comput. J.* **2005**, *48*, 749.
104. Ladik, J.; Seel, M.; Otto, P.; Bakhshi, A. K. *Chem. Phys.* **1986**, *108*, 203.
105. Wilkinson, J. H. *The Algebraic Eigenvalue Problem*; Clarendon Press: Oxford, 1965; p 633.
106. Bell, R. G.; Dean, P.; Butler, D. C. *J. Phys. C* **1970**, *3*, 2111.
107. Press, W. H.; Teukolsky, S. A.; Vetterling, W. T.; Flannery, B. P. *Numerical Recipes in Fortran: The Art of Scientific Computing*, 2nd ed.; Cambridge University Press: Cambridge, 1992; Vol I.
108. Pisani, C.; Dovesi, R.; Roetti, C. *Hartree-Fock Ab Initio Treatment of. Crystalline Systems;* Springer: New York, 1988.

109. Roothan, C. C. J. *Rev. Mod. Phys.* **1951,** *23*, 69.
110. Del, Re. G.; Ladik, J.; Biczo, G. *Phys. Rev. B* **1967,** *997,* 155.
111. Gouverneur, J. M. A.; Leroy, G. *Int. J. Quant. Chem.* **1967,** *1,* 427.
112. Duke, B. J.; Leary, B. O'. *J. Chem. Edu.* **1988,** *61*, 379.
113. Duke, B. J.; Leary, B. O'. *Int. J. Quant. Chem.* **1988,** *65*, 319.
114. Arora, V.; Bakhshi, A. K. *Chem. Phys.* **2010,** *373*, 307.
115. Arora, V.; Bakhshi, A. K. *Indian J. Chem.* **2011,** *50*A, 1555.
116. Kapoor, V.; Bakhshi, A. K.; *Superlattice Microst.* **2013,** *60*, 280.
117. Bakhshi, A. K.; Ladik, J.; Seel, M. *Phys. Rev. B* **1987,** *35*, 704.

CHAPTER 4

SPECTRAL TUNING OF P3HT/PTCDA BULK HETERO P–N JUNCTION BLEND FOR PLASTIC SOLAR CELL

ISHWAR NAIK[1,*], RAJASHEKHAR BHAJANTRI[2], and
JAGADISH NAIK[3]

[1]Department of Physics, Government Arts and Science College, Karwar, Karnataka, India

[2]Department of Physics, Karnatak University, Dharwad, Karnataka, India

[3]Department of Physics, Mangalore University, Mangalore, Karnataka, India

*Corresponding author. E-mail: iknaik@rediffmail.com

CONTENTS

ABSTRACT

The present work is focused to optimize the photoactive blend of poly[3-hexyl-thiophene-2,5-dily] (P3HT) and perylene-3,4,9,10-tetracarboxylic-dianhydride (PTCDA) for maximum absorption of the solar energy. P3HT:PTCDA blends of weight ratios 3:1, 1:1, and 1:3 are prepared in bromobenzene as the solvent and glass-coated samples are prepared by solution-cast method. Samples are analyzed by JASCO UV Vis NIR V 670 spectrometer. Spectral analysis revealed that the increased weight of PTCDA in the photoactive material will broaden the spectral region of absorption. The 1:1 blend of P3HT with PTCDA has a wide spectral sensitivity for absorption with a bandgap of 1.86 eV and is considered as the optimized photoactive material for construction of a plastic solar cell. The morphological study of the optimized blend has been carried out using scanning electron microscopy (SEM) and atomic force microscopy (AFM). SEM images reveal the formation of numerous hetero P–N junctions due to incorporation of N-type PTCDA in the P-type P3HT polymer matrix. Elemental analysis of the SEM images is also carried out using energy dispersion spectroscopy indicating the presence of only carbon and oxygen as expected. AFM studies remarked the presence of considerable surface roughness and the necessity of annealing for good contact with the electrodes and better conduction of the charges. Crystal structural investigations are made through X-ray diffraction studies where in the spectrum depicted the semicrystalline nature of the photoactive blend. Diffused bands indicate the amorphous state dominance in the blend. We end up with consideration of 1:1 P3HT:PTCDA blend to be the best active blend for practical construction of the solar cell and the work is in progress.

4.1   INTRODUCTION

Based on the inventional hierarchy, the solar cells are classified as first-generation, second-generation, third-generation, and fourth-generation solar cells. The first-generation solar cells are silicon-based P–N junctions with enough efficiency enabling them to be practically useful for the society. Broad spectral sensitivity and high charge mobilities have endowed them with good efficiency. The efficiency of these cells is also determined by the purity of the crystal used. The fabrication complications, installation cost, and material costs are the main drawbacks of these cells. The second-generation cells are the thin-film cells based on Si, CdTe, and CuInGaSe$_2$

(CIGS). The advantage of thin-film solar cells is that they can be fabricated on all substrates including flexible substrates and material cost is drastically reduced relative to the first-generation cells. Efficiency of these cells is lower than the first-generation solar cells and processing cost is more.

The third-generation solar cells make use of organic semiconductors such as conducting polymers or small molecules. The branch has been developed since the invention of conjugated polymers. The use of nanoparticles and dyes has improved the performance of the third-generation organic photovoltaic (OPV) devices. Lots of reports are available related to organic solar cells and the research on polymeric cells has taken a global importance in seeking a solution for energy crisis. Low cost, flexibility, tunability, and lightweight are the merits of these cells over the conventional solar cells but efficiency and stability must be highly improved to make it available for the public. Fourth-generation cells are the hybrid solar cells, fabricated from organic polymers along with inorganic semiconductors. The active material is prepared by incorporating inorganic semiconducting crystals into the conducting polymer matrix.

The present work is focused on the preparation and characterization of an organic solar cell active material so that a glance on organic conducting polymers is meaningful at this step. Generally, conventional polymers, such as plastic and rubber offer resistance to electrical conduction and are dielectrics. However, these beliefs have changed with the discovery of conducting polymers by Heeger, MacDiarmid, and Shikawa in 1976 who showed that the polyacetylene is conductive almost similar to a metal with conductivity increased by 10^9 times the original value, on doping with halogen and their work was awarded Nobel Prize for this discovery. Some of the examples of conducting polymers are polyacetylene, polyaniline, polythophene, PEDOT–PSS, etc. These conducting polymers can conduct electricity similar to semiconductors and metals. They are inexpensive and more advantageous due to lightweight, flexibility along with ease processing ability.[1,2,3] One early explanation of conducting polymers, used band theory as a method of conduction according to which half-filled valence band would be formed from a continuous delocalized pi system which is the ideal condition for conduction of electricity. Also, the conductivity can be increased by doping it with an electron acceptor or electron donor. This is similar to the doping of a silicon-based semiconductor with arsenic or boron. However, doping of silicon produces a donor energy level close to the conduction band or acceptor level close to the valence band, whereas this is not the case with conducting polymer. The high conductivity upon doping find wide application in organic electronic devices such as LED, solar cells, sensors, etc.

Range of conductivity values for good conductors, semiconductors, and insulators along with that of conducting polymers are depicted in the following schematic representation:

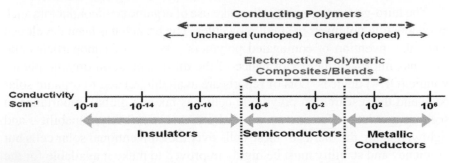

FIGURE 4.1 Classification of materials on the basis of conductivity values.

Table 4.1 indicates electrical conductivity values of some familiar polymers.[16]

TABLE 4.1 Electrical Conductivity Values of Polymers.[17]

Name of the polymer	Conductivity (S cm^{-1})
Polyacetylene	10^3–10^5
Polyphenylene	10^3
Polyphenylene vinylene	10^3
Polypyrrole	10^2
Polythiophenes	10^2
Polyaniline	10
PPS	10^2

4.1.1 ORGANIC SOLAR CELL AND TYPES

Organic solar cell (plastic solar cell or polymer solar cell) is a specialized P–N junction in which P and N-type semiconductors used are the organic conducting polymers, being able to convert light into direct current. These conducting polymers have energy gap [separation between highest occupied molecular orbit (HOMO) and lowest unoccupied molecular orbit (LUMO)] in the range 1–4 eV.[4] Donor (P-type) polymer absorbs visible light to form exciton which then splits into electron and a hole by the effective field across heterojunction (P–N junction) of active layer. Separated electrons and holes are collected by electrodes to generate electromotive force (emf).

(a) *Single-layer cell*—It is the simplest form of OPVs consisting of a layer of conducting polymer sandwiched between two metallic electrodes. When the P-type layer absorbs light, electrons will be excited to the LUMO and leave holes in the HOMO, thereby forming excitons. Exciton will dissociate into electron and hole, pulling electrons to the positive electrode and holes to the negative electrode. Single-layer organic solar cells do not work well. They have low quantum efficiencies (<1%) and low power conversion efficiencies (<0.1).[5,6]

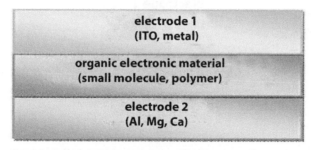

FIGURE 4.2   Graphical representation of single-layer solar cell[15].

(b) *Bilayer cell*—It contains heterojunction of donor and acceptor conductive polymer between two electrodes. Donor and acceptor layers are properly chosen to have internal field across the junction sufficient enough to create electron and hole separation. Efficiency is more than that of single-layer cells. In order to have exciton dissociation, the layer thickness must be in the range of about 10 nm or otherwise electron hole recombination will dominate the charge separation.

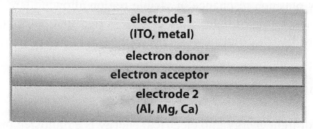

FIGURE 4.3   Graphical representation of bilayer solar cell.[15]

(c) *Bulk heterojunction cell*—Electron donor and acceptor are blended and interposed between electrodes. Numerous P–N junctions will be formed at the donor–acceptor interphases of the blend. The domain sizes of this blend are of the order of nanometers, allowing excitons to easily reach an interface and dissociate due to the large donor–acceptor interfacial area.[7] Bulk heterojunctions have an advantage over layered photoactive structures because they can be made thick enough for effective photon absorption. Most bulk heterojunction cells use two components, although three-component cells have been explored.

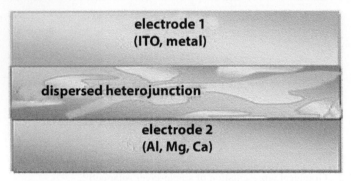

FIGURE 4.4   Graphical representation of bulk heterojunction cell.[15]

(d) *Multilayer or tandem solar cells*—These cells consist of multiple P–N junctions each one tuned to specific frequency of the spectrum. High bandgap solar cells are at the top followed by lower bandgap cells. Short wavelength is absorbed by top cell and longer wavelengths by successive lower cells. Efficiency can be crossed over Schockley–Queisser limit of single cell (two-layer cell).

(e) *Plasmonic enhanced tandem OPV*—This is based on plasmon resonance of metal nanoparticles such gold and silver. Metal nanospheres show plasmon resonance in visible region allowing large scattering and absorption of light. Yang et al. reported first plasmon enhanced cell with 20% efficiency.

4.1.2   WORKING PRINCIPLE

Conversion of light energy into electrical energy is a multistep process. The steps involved are:

- photon absorption,
- exciton formation,
- exciton migration,
- exciton dissociation,
- charge collection.

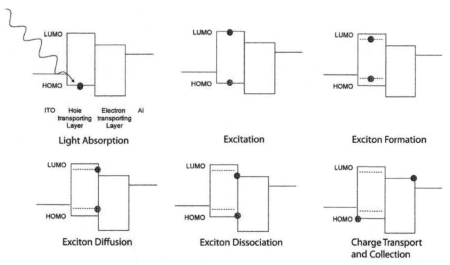

FIGURE 4.5 Steps involved in photovoltaic conversion.

(a) *Photon absorption*—It is the primary and most important aspect of the solar cell. The photoactive blend must have broad absorption spectrum for visible light. This can be achieved by proper selection of P- and N-type pair and also their weight ratio must be tuned for maximum absorption of the incident energy. Both P- and N-type materials should have low bandgap.

(b) *Exciton formation*—Incident photon will be absorbed by the P-type material due to which an electron in the HOMO will be excited to the LUMO. Unlike the case of inorganic system, here the hole and electrons are not free from each other but it is the bound state of electron and hole called exciton.

(c) *Exciton migration*—After the formation of exciton, it has to migrate toward the P–N junction for dissociation as free charges. The diffusion length of exciton is of the order 5–10 nm[20,21] and they have finite lifetime, so care must be taken to make them reach the junction before they decay or recombine.

(d) *Exciton dissociation*—As the exciton migrate and reach the P–N junction, it is influenced by the electric field of the depletion layer and gets dissociated as electron and hole. Electron will be in the LUMO of acceptor, holes in the HOMO of the donor.

(e) *Charge collection*—Free charges created will migrate toward the respective electrodes creating a potential difference to drive the current in the circuit. Charge collection at the electrodes will be efficient only when the mobilities of electron and holes are large enough. Therefore, choice of P- and N-type polymers having considerable mobilities of electron and holes is an important aspect in the construction of solar cell. As described above, dispersed heterojunctions of donor–acceptor organic materials have high quantum efficiencies compared to the planar heterojunction because in dispersed heterojunctions, it is more likely for an exciton to find an interface within its diffusion length.

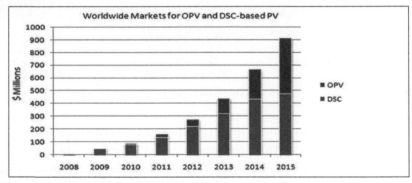

FIGURE 4.6   Worldwide markets for OPV- and DSC-based PV.[18]

4.2   MOTIVATION

The above discussions reveal that the first-generation silicon-based solar cells and the second-generation thin-film solar cells are not economical because of their material cost along with the manufacture and installation complications in spite of good efficiency. Organic electronics or the plastic electronics emerges as a promising field in looking for a cost-effective solar cell in this regard and the use of nanotechnology has further boosted this field of research. Organic solar cells seem to be advantageous and interesting because of their low-cost, simple processing, and flexibility. Low efficiency and low durability being the main drawback of these solar cells, the search for an efficient and durable cell has become a great challenge for researchers. The main difficulty with plastic solar cells is the electron–hole recombination before they migrate to the P–N junction interface to get separated as free electron and hole. The donor–acceptor system has proved remarkable progress in achieving the effective exciton diffusion mechanism and the charge collection rate but the blend must be properly tuned with respect to the choice of donor–acceptor pair as well as their weight percentage ratio to have a broad solar spectral response. Among the conducting polymers, P3HT has the absorption spectrum matching well with the strongest solar spectrum and also has good transport properties.

The work is focused to prepare an optimized active blend of the widely used P-type donor polymer poly[3-hexylthiophene-2,5-dily] (P3HT) and the N-type acceptor perylene-3,4,9,10-tetracarboxylic di anhydride (PTCDA). The strong absorption spectrum of P3HT in the visible region and the prominent absorption of PTCDA in UV–visible region along with the broad tail of absorption beyond the visible region are the key factors in selecting them as blend pair. The resulting active blend must show a broad spectral absorption for a proper composition between them. Blends of different donor–acceptor weight ratios are prepared in high boiling point solvent bromobenzene and their glass-coated samples prepared by the solution cast method are characterized by UV–visible spectra. The blend showing broad spectral sensitivity for absorption is selected as the best photoactive blend. Morphological study of the blend is also carried out.

4.3   EXPERIMENTAL

The P-type donor polymer poly(3-hexylthiophene-2,5-diyl) (P3HT) and the N-type material PTCDA are purchased from Sigma-Aldrich Corporation. The solvent bromobenzene is procured from Rankem Chemicals. All chemicals are used as received without further purification. The specifications of the chemicals are mentioned here.

P3HT: Molecular weight—15,000–45,000, HOMO—5.0 eV, LUMO—3.0 eV;

PTCDA: Functionalized Fullerene HOMO—6.8 eV, LUMO—4.7 eV;

Solvent used—bromobenzene, laboratory reagent, molecular weight—157.02.

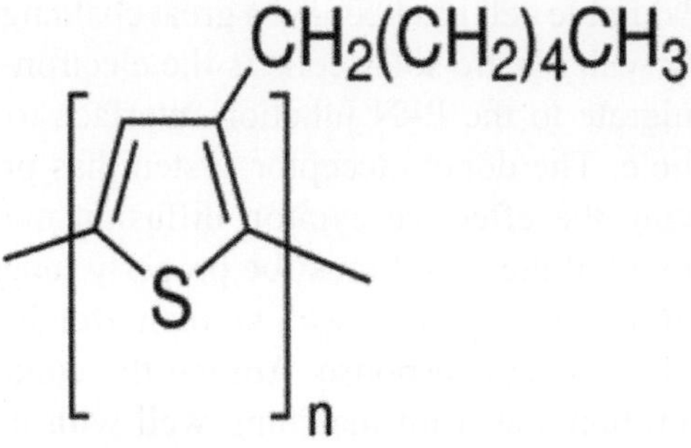

$$CH_2(CH_2)_4CH_3$$

FIGURE 4.7   Structure of P3HT.

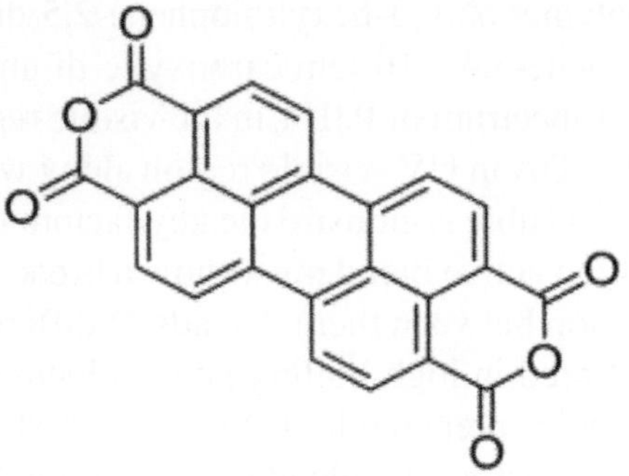

FIGURE 4.8   Structure of PTCDA.

P3HT and PTCDA each of 20 mg are dissolved in 100 and 50 cc of bromobenzene, respectively, in separate beakers, magnetically stirred for 48 h at room temperature until clear solutions are formed. The resulting solutions are of concentrations 0.2 mg/cc and 0.4 mg/cc, respectively. The solutions are blended with P3HT:PTCDA at weight ratios of 3:1, 1:1, and 1:3. The mixtures are magnetically stirred for 3 days at room temperature to ensure optimum blending, then transferred to 3 cm diameter petri plates, dried at room temperature and then at about 50°C in an oven.

TABLE 4.2   Sample Compositions.

P-type	N-type	P:N
2.0 mg (10 cc)		Pure P
1.5 mg (7.5 cc)	0.5 mg (1.25 cc)	3:1
1.0 mg (5 cc)	1.0 mg (2.5 cc)	1:1
0.5 mg (2.5 cc)	1.5 mg (3.75 cc)	1:3
	2.0 mg (5 cc)	Pure N

4.4 CHARACTERIZATION, RESULT, AND DISCUSSION

4.4.1 UV–VISIBLE SPECTROSCOPY

The UV–visible optical absorption spectroscopy is used for investigating optically induced transitions and helps to find out band structure, molecular constitution, and energy gap. It is based on the principle of electronic transitions in atoms or molecules on absorbing a proper radiation in the ultraviolet and visible region. The radiation that can be absorbed is specific for each element and varies with the chemical structure. A beam of UV–visible radiation is split into component wavelengths by grating or prism, then the monochromatic beam is split into two beams; one is passed through the reference sample and other through the sample. Some of the radiations will be absorbed resulting in the electron transition to a higher energy orbital. The spectrometer records the wavelengths at which absorption occurs, together with the degree of absorption at each wavelength. The resulting spectrum is presented as a graph of absorbance (A) versus wavelength λ. UV spectroscopy obeys the Beer–Lambert law A = log (I_0/I). A spectrophotometer can be either single beam or double beam. In a single-beam instrument, all of the light passes through the sample cell. I_0 must be measured by removing the sample. In a double-beam instrument, the light is split into two beams before it reaches the sample. One beam is used as the reference; the other beam passes through the sample. The reference beam intensity is taken as 100% transmission and the measurement displayed is the ratio of the two beam intensities. The absorption of UV or visible radiation corresponds to the excitation of outer electrons. The possible excitations are $s–s^*$ transitions, $n–s^*$ transitions, $n–p^*$, or $p–p^*$ transitions. Most absorption spectroscopy of organic compounds is based on transitions of n or p electrons to the p^* excited state. This is because, the absorption peaks for these transitions fall in an experimentally convenient region of the spectrum (200–700 nm).

In the present work, the samples are analyzed by JASCO UV Vis NIR V-670 spectrometer. The V-670 double-beam spectrophotometer utilizes a unique, single monochromator design covering a wavelength range from 190 to 2500 nm .A PMT detector is provided for the UV/VIS region and a Peltier-cooled PbS detector is employed for the NIR region. Both gratings and detector are automatically exchanged within the user selectable 750–900 nm range. Absorption spectra of the samples are presented below.

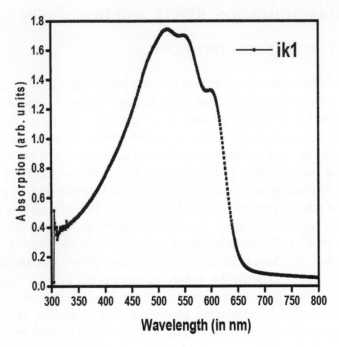

FIGURE 4.9　Pure P3HT film.

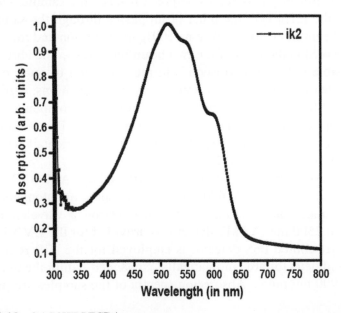

FIGURE 4.10　3:1 P3HT:PTCDA.

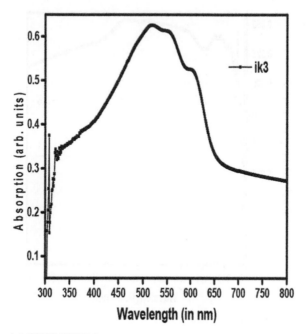

FIGURE 4.11   1:1 P3HT:PTCDA.

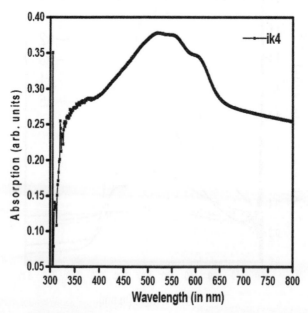

FIGURE 4.12   1:3 P3HT:PTCDA.

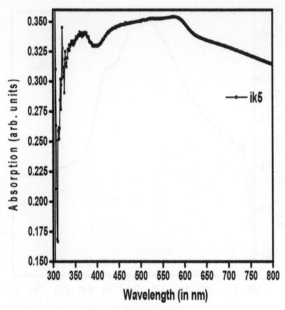

FIGURE 4.13   Pure PTCDA.

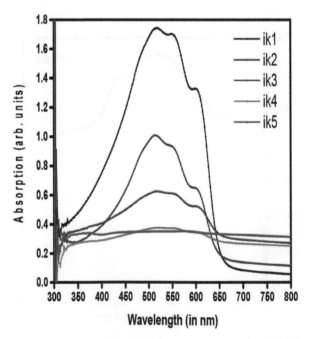

FIGURE 4.14   Overlay.

4.4.2   TAUC'S PLOT

A *Tauc plot* is used to determine the optical bandgap, or Tauc gap in semi-conductors. The Tauc gap is often used to characterize practical optical properties of amorphous materials. The absorption coefficient of amorphous material is given by Tauc and Davis–Mott model.

$$\alpha(v)hv = B \, (hv - E_{gap})^{m},$$

where E_{gap}, B, and hv are the optical gap, constant, and incident photon energy, respectively; $\alpha(v)$ is the absorption coefficient defined by the Beer–Lambert's law as $\alpha(v) = 2.303 \times A(\lambda)/d$, where d and A are the film thickness and film absorbance, respectively. In amorphous materials, four types of transitions occur, namely, indirect forbidden transition, indirect allowed, direct forbidden, and direct allowed transitions. These transitions are characterized by the m values 1/3, 1/2, 2/3, and 2, respectively.[12,13]

For indirect allowed transition, $m = 1/2$

$$\alpha(v)hv = B \, (hv - E_{gap})^{1/2}$$

on squaring,

$$[\alpha(v)hv]^2 = B \, (hv - E_{gap})$$

$$[\alpha(v) \, hC/\lambda]^2 = B \, (hC/\lambda - E_{gap})$$

$$[2.303 \times \text{Abs}(\lambda)/d \, hC/\lambda]^2 = B \, (hC/\lambda - E_{gap})$$

or

$$(A/\lambda)^2 = B \, (hC/\lambda) - B \, (hC/\lambda_{gap}).$$

A graph of $(A/\lambda)^2$ with $1/\lambda$ is plotted, If $(A/\lambda)^2 = 0$, then $B(hC/\lambda) - B(hC/\lambda_{gap}) = 0$ therefore, $1/\lambda = 1/\lambda_g$, that is, $\lambda = \lambda_{gap}$.

Thus, extrapolating the straight-line curve to zero absorption gives the onset wavelength λ_{gap}. Then energy gap is determined using the relation

$$E_{gap} = hC/\lambda_{gap}$$

where λ_{gap} is the onset wavelength

Tauc's plots are as shown below:

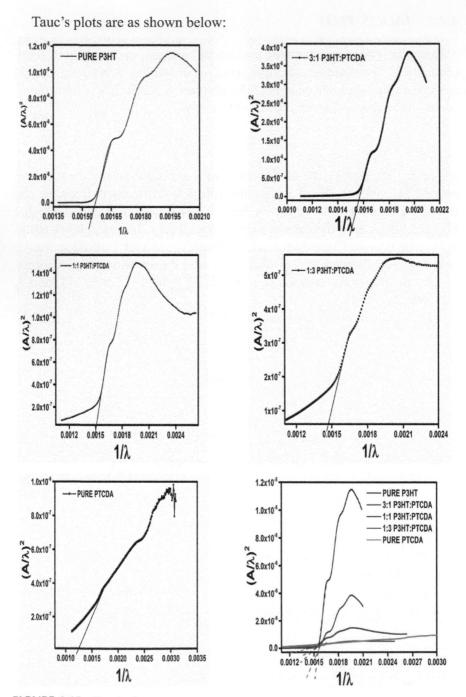

FIGURE 4.15　Tauc's plot.

The major problem in polymer solar cells is the insufficient absorption in the solar irradiance spectrum. Even the widely studied phenylene vinylene polymers have the band edges at 550 nm.[11] But pure P3HT film has absorption extending from 300 to 640 nm. The spectrum has two peaks at 520 and 560 nm arising from $\pi-\pi^*$ transition and the shoulder at 620 nm is due to interchain interactions.[11] The onset of absorption (absorption edge) is 640 nm matching well with the strongest solar spectrum (Fig. 4.1). The N-type material PTCDA is planar π-stacking organic molecule which is an excellent model compound for optoelectronic organic semiconductor thin films, particularly organic diodes. PTCDA is an acceptor and well known for high spectral selective absorption with good semiconducting property. It is a highly conjugated system with broadband of absorption with selective peaks between 400 and 600 nm having optical bandgap of about 2.2 eV. Molecule is almost as delocalized as the polymer and their $\pi-\pi^*$ transitions are similar.[12] Addition of PTCDA into P3HT polymer matrix generates multiple donor–acceptor P–N junctions assisting ease exciton migration and improved charge collection. Important features can be outlined from the overlay of spectrum. Relatively, 1:1 blend has wide spectral sensitivity and can be considered as the best photoactive blend among the samples prepared. The calculations from Tauc's plot show that the onset wavelength of the optimized blend is 666.67 nm which corresponds to a bandgap of 1.86 eV.

TABLE 4.3   Optical Parameters.

Sample	λ_{onset} (nm)	E_g^{opt} (eV)
Pure P	641	1.934
3:1	641.44	1.933
1:1	666.67	1.86

4.4.3   SCANNING ELECTRON MICROSCOPY

Scanning electron microscope (SEM) is a type of electron microscope that produces images of a sample by scanning it with a focused beam of electrons. The electrons interact with atoms in the sample, producing various signals that contain information about the sample's surface topography and composition. The electron beam is generally scanned in a raster scan pattern, and the beam's position is combined with the detected signal to produce an image. SEM can achieve resolution more than 1 nm. The most common SEM mode is the detection of secondary electrons emitted by atoms excited by the electron beam. By scanning the sample and collecting the secondary electrons that are emitted using a special detector, an image displaying the topography of the surface is created.

A normal SEM operates at a high vacuum. The electron beam is accelerated through a high voltage (e.g., 20 kV) and pass through a system of apertures and electromagnetic lenses to produce a thin beam of electrons, and then the beam scans the surface of the specimen by means of scan coils. Electrons are emitted from the specimen by the action of the scanning beam and collected by a suitably positioned detector.

SEM images are shown in Figure 4.16.

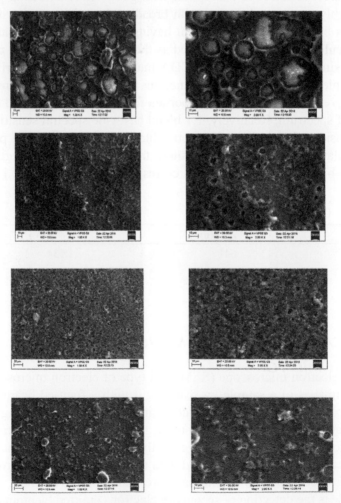

FIGURE 4.16   SEM images.

4.4.4   ENERGY DISPERSIVE SPECTROSCOPY

Energy dispersive spectroscopy (EDS) is a supplementary tool of SEM to determine the elemental analysis along with their composition in the SEM imaged sample. It is based on the principle that the electron beam hitting the sample will generate X-rays within the sample. The emitted X-rays have the varying energies determined by the elements from which it is emitted. A careful investigation of the energies of the emitted X-ray will reveal the elemental analysis along with the percentage composition.

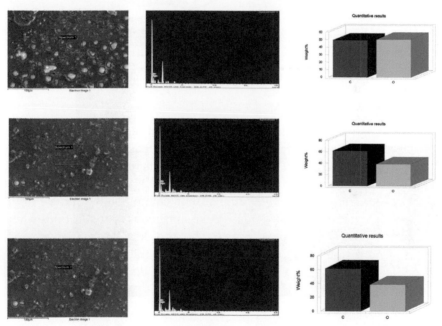

FIGURE 4.17   EDS results with SEM image.

4.4.5   ATOMIC FORCE MICROSCOPY

The atomic force microscopy (AFM) enables to get the three-dimensional surface profile, unlike SEM which provides two-dimensional projection. It has three major abilities: force measurement, imaging, and manipulation.

The reaction of the probe to the forces that the sample imposes on it is used to form an image of the three-dimensional shape (topography) of a sample. This is achieved by raster scanning the position of the sample. In addition to topographical images, other properties such as stiffness or adhesion strength, electrical properties such as conductivity or surface potential, can also be measured. An AFM can operate in different modes, such as contact mode, intermittent or tapping mode, and noncontact mode.

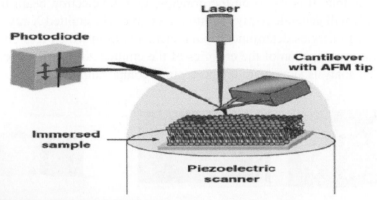

FIGURE 4.18 Working of an AFM.

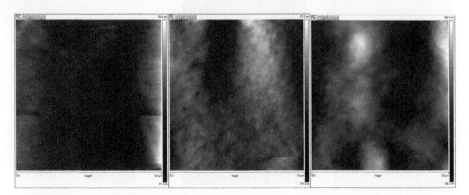

FIGURE 4.19 5 μm AFM images.

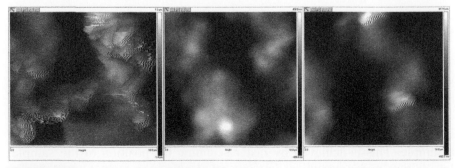

FIGURE 4.20 10 μm AFM images.

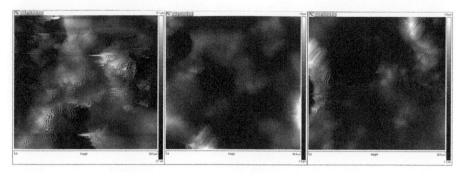

FIGURE 4.21 20 μm AFM images (two-dimensional).

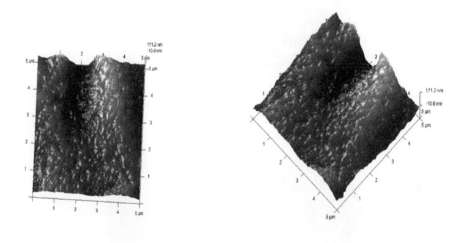

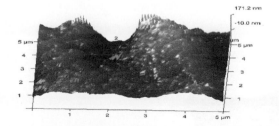

FIGURE 4.22 5 μm AFM images (three-dimensional).

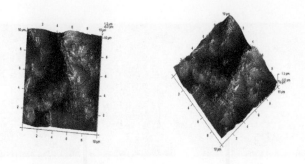

FIGURE 4.23 10 µm AFM images.

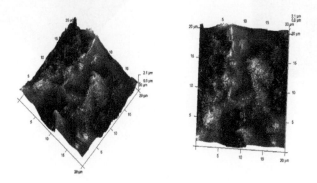

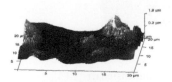

FIGURE 4.24 20 µm AFM images.

4.4.6 X-RAY DIFFRACTION

Polymers are not available in 100% crystalline state; they are available in amorphous state or as a mixture of both amorphous and crystalline state. They can be considered to be semicrystalline. X-ray diffraction (XRD) pattern of polymer contains both sharp as well as defused bands. Sharp bands correspond to crystalline orderly regions and defused bands correspond to amorphous regions.[14]

The fraction of the material that is crystalline, the crystallinity or crystalline index, is an important parameter in the two-phase model. Crystallinity can be determined from a wide-angle X-ray diffraction (WAXD) scan by comparing the area under the crystalline peaks to the total scattered intensity.[15]

Percentage of crystallinity $X_c\%$ is measured as the ratio of crystalline area to total area.

$$X_c\% = \{(A_c/A_a) + A_c\} \times 100 \ (\%).$$

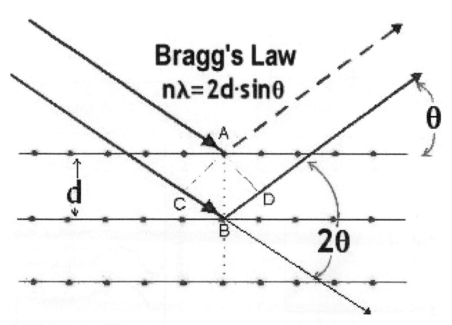

FIGURE 4.25 Bragg's law.

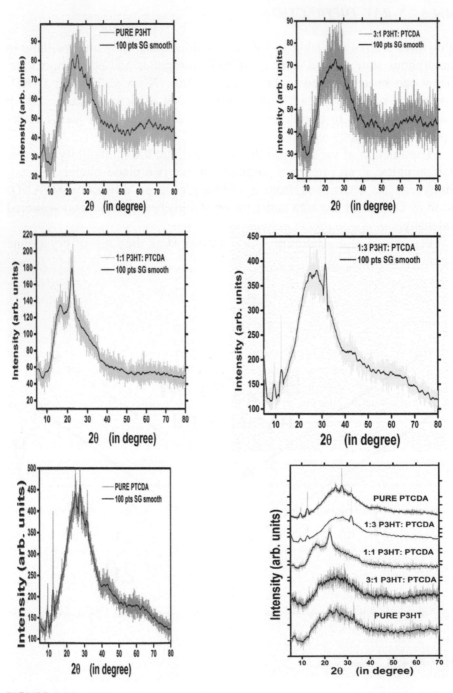

FIGURE 4.26　XRD spectrum.

TABLE 4.4 Parameters Calculated from X-ray Diffraction.

Sample	2θ	d-spacing	Size <P> (nm)	R (Å)
PURE P3HT	23.87	4.01	1.9099	4.6560
3:1 P3HT:PTCDA	16.3813	5.40681	0.838460163	6.758563148
–	22.1164	4.01601	1.903980402	5.020045888
1:1 P3HT:PTCDA	22.26	4.016	1.904465	4.98687
1:3 P3HT:PTCDA	12.3704	7.14942	59.27896782	8.936812449
–	31.6783	2.82224	56.36740201	3.52782019
PURE PTCDA	12.4074	7.12815	45.12187976	8.91026582
–	31.6845	2.82171	46.41906221	3.527147546

4.5 CONCLUSION

We have investigated UV–visible absorption spectra for 3:1, 1:1, and 1:3 blends of P3HT:PTCDA mixed P–N junction photoactive material along with their pristine glass-coated films (solution cast). Spectral analysis indicated that increased weight percentage of PTCDA in the blend has tuned the spectral absorption. 1:1 blend of P3HT:PTCDA shows a broad spectral absorption and selected as the best photoactive blend. The optimized blend has the least bandgap of 1.86 eV. The SEM images confirmed the formation of multi P–N junctions in the blend. AFM studies stress upon the need for annealing of the blend for optimal performance. XRD studies confirm the dominance of the amorphous nature of the photoactive blend. Finally, we conclude that 1:1 blend of P3HT:PTCDA can be used as the best photoactive material for constructing a plastic solar cell and the construction of the solar cell is under progress.

ACKNOWLEDGMENT

The author is grateful to UGC for financial assistance by sanctioning the Minor research project entitled "Construction and Characterization of an Organic Solar Cell (OPV) Devised from a Self-made, Low-cost Spin Coating Machine" No. 1419-MRP/14-15/KAKA088/UGC-SWRO, dated 4-2-2015.

KEYWORDS

- **P3HT**
- **PTCDA**
- **solar cell**
- **plastic electronics**
- **semiconductors**

REFERENCES

1. Alexander, P.; Nikolay, O.; Alexander, K.; Galina, S.; *Progr. Polym. Sci.* **2003**, *28*, 1701–1753.
2. Annamali, P. K.; Dilip, D.; Namrata, S. T.; Raj, P. S. *Prog. Polym. Sci.* **2009**, *34*, 479–515.
3. Khan, M. F. S.; Alexander, A. B. *Nano Lett.* **2012**, *12*, 861–867.
4. Rivers, P. N. *Leading Edge Research in Solar Energy*; Nova Science Publishers, 2007, ISBN 1600213367.
5. McGehee, D. G.; Topinka, M. A. *Nat. Mater.* **2006**, *5*(9), 675–676. Nelson, *J. Curr. Opin. Solid State Mater. Sci.* **2002**, *6*, 87–95.
6. Cao, W.; Xue, J. Recent Progress in Organic Photovoltaics: Device Architecture and Optical Design. *Energy Environ. Sci.* **2014**, *7*(7), 2123.
7. Kim, M.-S. Understanding Organic Photovoltaic Cells: Electrode, Nanostructure, Reliability, and Performance. Ph.D. Thesis, University of Michigan, 2009.
8. Nader, G. *Int. Nano Lett.* **2013**, *3*, 2.
9. Alias, A. N.; Kudin, T. I.; Zabidi, Z. M.; Harun, M. K.; Mali, A. M.; Yahya, M. Z. A. *Adv. Mater. Res.* **2013**, *652–654*, 527–513.
10. Mohamad, K. A.; Afishah, A.; Ismail, S.; Bablu, K. G.; Katsuhiro, U., et al. *J. Chem. Eng.* **2014**, *8*, 476–481.
11. GSoos, Z.; Hennessy, M. H. *Modelling PTCDA Spectra & Polymer Excitations;* NATO Science Series; Springer; Vol. 79, pp 11–323.
12. Morandat, S., et al. Atomic Force Microscopy of Model Lipid Membranes. *Anal. Bioanal. Chem.* **2013**, *405*(5), 1445–1461.
13. Gowariker, V. R.; Viswanathan, N. V.; Jayadev, S. *Polymer Science*; New Age International (P) Ltd: India, 1986; p 173.
14. Murthy, N. S. *Rigaku J.* **2004**, *21*(1), 15–24.
15. Bagher, A. M. Introduction to Organic Solar Cells. *Sustain. Energy* **2014**, *2*(3)85–90.
16. Kaur, G.; Adhikari, R.; Cass, P.; Bown, M.; Gunatilake, P. *RSC Adv.* **2015**, *5*, 37553–37567.
17. Sengupta, R.; Bhattacharya, M.; Bandyopadhyay, S.; Bhowmick, A. K. *Progr. Polym. Sci.* **2011**, *36*, 638–670.
18. https://juanbisquert.wordpress.com/2009/03/11/report-on-organic-solar-cells-opv-and-hybrid-organicinorganic-dye-sensitized-solar-cells-dsc/

CHAPTER 5

DESIGN STRATEGIES OF POLYMER FOR HIGH-PERFORMANCE ORGANIC PHOTOVOLTAICS

MEHA J. PRAJAPATI and KIRAN R. SURATI*

Department of Chemistry, Sardar Patel University, Vallabh Vidyanagar 388120, Anand, Gujarat, India

Corresponding author. E-mail: kiransurati@yahoo.co.in

CONTENTS

ABSTRACT

This chapter mainly deals with the background and brief history of photovoltaics and design strategies of organic photovoltaics. Our main focus is on desired band gap and photophysical properties of polymer via design strategies such as fused heterocycles, effect of substituent, effect of side chain, π-conjugation length.

5.1 INTRODUCTION

After the polyacetylene was found to exhibit profound increase in electrical conductivity when exposed to halogen vapor,[82] the researches on conducting polymers began in 1970s. In 2000, three scientists, Heeger, MacDiarmid, and Shirakawa, founders of the conjugated conducting polymer chemistry, won the Nobel Prize in chemistry for their discovery. After their pioneering work, several conjugated polymers were developed including polythiophene, polypyrrole, poly(paraphenylene), polyaniline, poly(phenylene vinylene), polyfluorene, etc.[83] These initial discoveries have led to a modern class of organic materials with the conductivity of classical inorganic systems. These discoveries accelerated development of nonclassical organic materials with the conductivity of classical inorganic system. Organic materials semiconducting in their natural state exhibit increased conductivity upon oxidation or reduction. Due to such properties, they have received considerable technological interest, leading to their current applications in sensors, organic field effect transistors (OFETs), organic photovoltaics (OPVs) device, electrochromic devices, and organic light-emitting diodes (OLEDs).[64,70,78]

Among these all these advanced devices, OPVs devices find attention due to potential alternative to inorganic solar cell because organic material offers several advantages such as high absorption, lower weight or flexibility, semitransparency, easily integration into other products, lower manufacture cost roll to roll process, short energy payback time, and lower environmental impact during the manufacture and operation.[28,52,88,93] Furthermore, due to the growth of human population on Earth and its increasing demand for energy, it is likely to utilize renewable energy resource such as solar energy with third-generation solar cell referred to as OPVs.

Unfortunately, OPVs have not reached widespread commercial viability due to their so far comparably low efficiencies, short device lifetimes, and stability. To solve these problems, it is important to engineer new organic materials starting on a fundamental, physical, chemical, and molecular/

structural level, keeping in mind the structure–property relationships at the heart of the molecular design process for light-harvesting photoconverting materials, as well as practical requirements for producing efficient devices. In view of these facts, here we summarized the design strategies approaches of organic materials for OPVs application and survey of all the work done toward this direction. Our objective is to provide the information to achieve high efficiency by molecular design strategies though the other point such as device engineering and state of art of device is also important. Figure 5.1 describes the molecular design approaches of polymer for OPVs. Such design approaches offer particular properties (thermoelectrical, morphology, charge transport, etc.). Here we also facilitate the structural property relationship to enhance the applicability and modification of scaffold with various approaches.

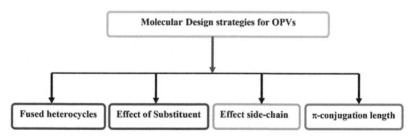

FIGURE 5.1   Molecular design approaches for OPVs.

Further before discussing about the molecular design approaches of organic materials for OPVs, it is really important to understand the basics of conduction mechanism in organic semiconductor.[8,19,63] Generally, conjugated polymers charges can be produced by chemical doping, light induce photo carrier generation, and injection of carrier by electrodes in an electric filed.[6] In general, charge transport depends on various mechanisms such as band transport, hopping transport, and tunneling transport mechanism.[6,50] Here the present chapter mainly deals with design strategies and survey of various polymer materials for OPVs.

5.2   HISTORICAL BACKGROUND OF SOLAR ENERGY AND SOLAR CELL

If we consider the year-wise development of solar energy and solar cell research summarized in Table 5.1, it could encourage the researchers and

TABLE 5.1 History of Solar Cell.

Sr. No.	Year	Details	References
1	1839	E. Becquerel discovered photovoltaic effect for the first time.	(Becquerel, 1839)
2	1860s	French mathematician August Mouchet planned an idea for solar-powered steam engines. In the following two decades; he and his assistant, Abel Pifre constructed the first solar powered engines which was used for variety of applications.	(https://en.wikipedia.org/wiki/Augustin_Mouchot)
3	1873	The photo conductivity of selenium was discovered by Willoughby Smith.	(https://en.wikipedia.org/wiki/Willoughby_Smith)
4	1876	W. G. Adams and D. E. Day observed photovoltaic effect in a sample of selenium placed between two metal electrode (This was the first solid PV device).	(Adams and Day, 1876)
5	1878	Samuel P. Langley, invents the bolometer. It was used to measure light from the faintest star and the sun's heat rays which consist of a fine wire connected to an electric circuit so when radiation falls on the wire, it becomes slightly warmer which increases the electrical resistance of the wire.	(https://en.wikipedia.org/wiki/Samuel_Pierpont_Langley)
6	1883	Charles Frittis developed the first large area photovoltaic device having < 1% (first solar cells made from selenium wafers).	(Van Nostrands Engineering Magazine, 1885) (http://en.wikipedia.org/wiki/Charles_Fritts)
7	1891	Baltimore inventor patented the first commercial solar water heater.	(http://www.historycommons.org/context.jsp?item=a1891kempwaterhtr)
8	1914	The existence of a barrier layer in PV devices was noted by Goldman and Brodsky.	(http://org.ntnu.no/solarcells/pages/history.php)
9	1916	Polish scientist Jan Czochralski developed a way to grow single crystal silicon.	(https://en.wikipedia.org/wiki/Czochralski_process)
10	1931	Bruno Lange built a solar cell panel out of selenium, but the panel generates tiny amount of electricity.	(RISE 2010)
11	1932	Audobert and Stora discover the PV effect in cadmium sulfide (CdS).	(https://en.wikipedia.org/wiki/Timeline_of_solar_cells)
12	1953	Dr. Dan Trivich, Wayne State University, makes the first theoretical calculations of the efficiencies of various materials of different band-gap widths based on the spectrum of sun.	(http://www.zoominfo.com/p/Dan-Trivich/11736085)

TABLE 5.1 *(Continued)*

Sr. No.	Year	Details	References
13	1954	Bell Labs first solar cell having 6% PCE capable of converting enough of the sun's energy into power to run everyday electrical equipment. (Chaplin, Fuller and Pearson)	(Chapin et al. 1954)
14	1956	Architect Fank Bridgers designed the world's first commercial office building using solar water heating and passive design.	(https://en.wikipedia.org/wiki/ Solar_Building)
15	1957	Hoffman Electronics achieved 8% efficient PV cells.	
16	1958	Hoffman Electronics achieved 9% efficient PV cells. In 1958, PV array- powered radios appeared on the US Vanguard I Satellite and this was the first time PV technology was practically utilized. Later that year, Explore III, Vanguard II and Sputnik-3 were launched with PV- powered systems on board.	(https://www1.eere.energy.gov/ solar/pdfs/solar_timeline.pdf)
17	1959	Hoffman Electronics achieves 10% efficient commercially available PV Cells.	
18	1959	A photovoltaic effect in a single crystal of anthracene was observed by Kallmann and Pope.	(Kallmann and Pope, 1959)
19	1960	Hoffman Electronics achieves 14% efficient PV cells. Silicon sensors, Inc., of Dodgeville, Wisconsin is founded. It starts producing selenium and silicon PV cells.	(https://www1.eere.energy.gov/ solar/pdfs/solar_timeline.pdf)
20	1980	At the University of Delaware, the 1st thin-film solar cell exceeds 10% efficiency using CuS/ CdS.	(https://www1.eere.energy.gov/ solar/pdfs/solar_timeline.pdf)
21	1984	The university of south wales breaks the 20% efficiency barrier for silicon solar cells under 1-sun conditions.	(https://www1.eere.energy.gov/ solar/pdfs/solar_timeline.pdf)
22	1984	The first OPV device based on poly(3-me-thylthiophene) was reported by Glenis et al. in 1984 are 0.007 % efficient.	(Glenis et al. 1984)

TABLE 5.1   *(Continued)*

Sr. No.	Year	Details	References
23	1986	The bilayer OPV cell structure was first reported by Tang using two-active layer structure of perylenedibenzimidazole and copper phthalocyanine.	(Tang, 1986)
		First organic thin-film solar cell having efficiency ~1% was reported by Tang.	
24	1991	Hiramoto reported the first dye-dispersed/bulk heterojunction in 1991, in which he found that the photocurrent was doubled as compared to the two-layered or bilayer cell	(Hiramoto et al. 1991) (Hiramoto et al. 1992)
25	1992	University of south Florida develops a 15.9% efficient thin-film PV Cells made of cadmium telluride, breaking the 15% barrier for the first time for this technology.	(https://www1.eere.energy.gov/solar/pdfs/solar_timeline.pdf)
26	1993	Karg et al. was the first to investigate PPV [poly(p-phenylene vinylene)] in ITO/PPV/Al LEDs and PV devices, the measured V_{OC} of 1V and power conversion efficiency of 0.1% under white light illumination.	(Karg et al. 1993)
27	1993	Sariciftci fabricated first bilayer OPV cell which uses a conjugated polymer, in which the hole transporting material was MEH-PPV and electron transporting material was C_{60}. This bilayer OPV cells had an PCE of 0.04%.	(Sacriciftci et al. 1993)
28	1993	Moriata et al. made a similar observation as that of Sacriciftci when working with poly (3-alkyl thiophene) (P3AT) and C_{60}.	(Morita et al. 1993)
29	1994	The National Renewable Energy laboratory develops a solar cell made from gallium indium phosphide and gallium arsenide that becomes the first one to exceed 30% conversation efficiency.	(Takamoto et al. 1997)
30	1994	Single layer OPV device structure was created by R. N. Marks et al. This low quantum efficiency (around 0.1%) resulted from intrinsically low mobility of charges through semiconducting organics.	(Marks, et al. 1994)
31	1994	Yu made the first dispered polymer heterojunction PV cell by spincoating on ITO covered glass from a solution of MEH-PPV and C_{60}.	(Yu et al. 1994)

TABLE 5.1   *(Continued)*

Sr. No.	Year	Details	References
32	1995	A phase separated polymer blend composite made of MEH-PPV as a donor and cyano-PPV as acceptor is investigated by Yu and Heeger for the first time having efficiency of 0.9 % which is 20 times larger than of diodes made with pure MEH-PPV.	(Yu and Heeger, 1995)
33	1996	Halls et al. made a very thorough study on PPV/C_{60} heterojunction and the cell had a fill factor of 0.48.	(Halls et al. 1996) (Halls and Pichler et al. 1996)
34	1996	The first reports of polymers bearing fullerene on the side chains came in 1996 by Benincori et al. and the polymers used was a poly thiophene.	(Benincori et al. 1996)
35	1997	Halls et al. investigated heterojunctions where perylene was the electron acceptor, bis(phenethylimido)perylene was vacuum sublimed onto PPV films spincoated on ITO covered glass and Al was used as a counter electrode material, PCE of 1% and high FF value of 0.6 was reported.	(Halls and Friend, 1997)
36	1999	Tada et al. fabricated a heterojunction cell consisting of the donor polymer (P3HT) and acceptor polymer (PpyV). The photocurrent increased some three orders of magnitude when the donor layer was present indicating efficient charge transfer at the interface between the layers.	(Tada et al. 1999) (Tada et al. 1996)
37	2001	Schmidt-Mende et al. fabricated the PV cells of self-organized discotic liquid crystals of hexaphenyl-substituted hexabenzocoronene (HBC) and perylene. The cell had a power conversion efficiency of 2%.	(Schmidt-Mende et al. 2001)
38	2001	Ramos et al. reported the first use of double-cable polymers in PV devices in 2001.	(Ramos et al. 2001) (Ramos and Rispens et al. 2001)
39	2003	Krebs et al. synthesized a block copolymer consisting of an electron acceptor block and an electron donor block.	(Krebs and Jorgensen, 2003)
40	2003	Peumans and Forrest replaced perylene tetracarboxylic derivative with C_{60} as an acceptor in the device structure and the device produced 3.5% power conversion efficiency.	(Peumans and Forrest, 2003)

TABLE 5.1 *(Continued)*

Sr. No.	Year	Details	References
41	2004	Cao et al. demonstrated a polymer containing alternating carbazole and thiophene-BT-thiophene units that achieved a V_{OC} of 0.95 V, single-cell device showed only 2.2% efficiency.	(Cao et al. 2004)
42	2005	Forrest et al. replaced perylenedibenzimidazole in the bilayer device of tang with C_{60} in bulk heterojunction PV devices and reported efficiency of 5%.	(Forrest et al. 2005)
43	2005	Li et al. fabricated a blend of OPV cell using P3HT [poly(3-hexyl thiophene)] and PCBM (phenyl-C61-butyric acid methyl ester) and the best performance from the optimized blend OPV cell was 4.4%.	(Li et al. 2005)
44	2006	Boer and Janseen et al. demonstrated first polymer tandem solar cell consisting of two sub-cells with different materials in 2006 which are 0.57% efficient.	(Boer and Janseen et al. 2006)
45	2007	Kim et al. successfully demonstrated improved performance of an OPV cell with 6.5% power conversion efficiency using tandem cell structure.	(Kim et al. 2007)
46	2009	Similarly Heeger et al., Liang et al., Chen et al. in 2009 reported bulk heterojunction PV devices having efficiencies 6–7%.	(Heeger, 2009); (Liang, 2009); (Chen, 2009)
47	2010	Solarmer Energy Inc. reported an efficiency of 8.13% at the end of 2010, in which PEDOT: PSS layer provides an improved interface between the active layer and the electrode which improves the performance of devices.	(http://www.Solarmer.com/newsevents.php)
48	2013	A german company "Heliatek" in January 2013; reported 12% power conversion efficiency in organic BHJ devices for small molecules organic solar cells.	(National Centre for Photovoltaics Home Page http:/www.nrel.gov/ncpv) (accessed March 13, 2014).
			(Heliatek newscenter http://www.heliatek.com/newscenter/latest_news/neuer-weltrekord-fur-organische-solarzellen-heliatek-behauptet-sich-mit-12-zelleffizienz-als-technologiefuhrer/?lang=en) (accessed March 13, 2014).

TABLE 5.1 *(Continued)*

Sr. No.	Year	Details	References
49	2015	Zhou et al. has reported a high power conversion efficiency of 11.3% in which two similar BHJ layer were stacked in a series connection using novel interconnecting layer in tandem cells.	(Zhou et al. 2015)

further provide information of stepwise development in this area. The discovery of photovoltaic (PV) effect[4] by Becquerel was the first move in this direction. After that in 1860s, Augustin Mouchet (French mathematician) proposed an idea for solar-powered steam engines. He and his assistant, Abel Pifre, constructed the first solar-powered engines and used them for a variety of applications.[32] During the era of 1873–1891s, photoconductivity of selenium (Se), Se-based solid PV device, invention of bolometer, large scale Se-based solar cell, and solar water heater were discovered (Table 5.1).[1,33,35,38,40,97,] Subsequently, in 1914–1953s, many attempts on silicon (Si), Se and cadmium sulfide (CdS) based PV devices were discovered.[34,37,39,43,77] In 1954, Bell labs develop first solar cell capable of converting enough of the sun's energy into power to run everyday electrical equipment.[11] Afterward, inorganic-based solar cell researches were established which achieved efficiency in the range of 6–20%.[36,41,48] In such type of solar cell, various inorganic compounds from silicon wafer to gallium arsenide (GaAs), copper sulfide (CuS), and cadmium sulfide (CdS were used as active materials. After the three decades of inorganic solar cell, the first OPV device-based on poly(3-methylthiophene) was reported by Glenis et al. in 1984 with less efficiency.[18] Then after, OPV research grew and large numbers of review articles came after the research and development of OPV was accelerated by Tang et al. and he reported 1% power conversion efficiency (PCE) by pn heterojunction stacking-type cell.[93] In 1991, Hiramoto et al. first reported the dye-dispersed/bulk heterojunction (BHJ) concept and they found the efficiency double as compared with bilayer.[29,30] A large number of work have been done in the field of OPVs (see serial no. 26–50 of Table 5.1) which are summarized in Table 5.1.

Looking to these significant progresses is the result of multidisciplinary results ranging from chemistry, physics, materials science, and engineering. Moreover, that is also due to the high-performance measurements system and advancement in instruments techniques such as AFM and SEM that provide the information about the materials in atomic level to subatomic level and computational stimulation tools such as density functional theory (DFT), and time dependent-density functional theory (TD-DFT).

5.3　DESIGN STRATEGY

In the development of the polymers for OPVs, several factors that control the bandgap of polymer materials need to be discussed. These factors are:

1. fused heterocycles,
2. effect of substituent,
3. effect of side chain,
4. π-conjugation length.

5.3.1　FUSED HETEROCYCLES

Resulting from the union of an aromatic ring with a quinoid ring, fused heterocycles are formed which represent a significant class of building blocks to build low-bandgap polymers for PV applications. Normally, conjugated polymers have two resonances structure, that is, aromatic form and quinoid form (Fig. 5.2). As discussed in the literature, the energy of the quinoid form is lower than that of the aromatic form[109] which helps in stabilizing the quinoid form, results in reducing the bandgap of conjugated polymers.

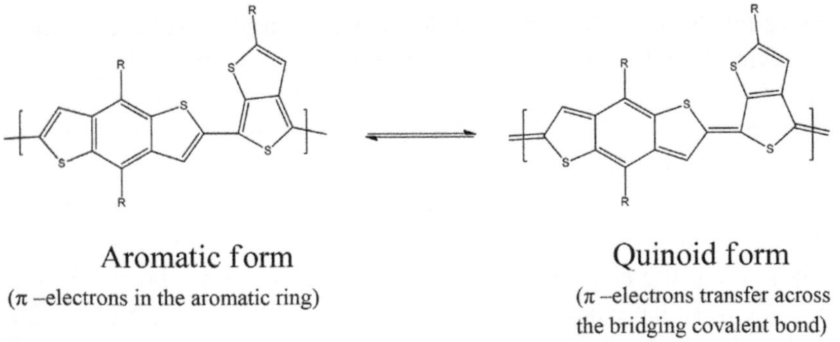

Aromatic form
(π –electrons in the aromatic ring)

Quinoid form
(π –electrons transfer across the bridging covalent bond)

FIGURE 5.2　Aromatic and quinoid form—two resonance structure.

The quinoid form of polyisothianaphthelene (PITN) (Fig. 5.3) has different energy levels when compared to the aromatic form.[99] The bandgap of PITN was found to be 1 eV and is the first narrow bandgap polymer.[54]

In Figure 5.3b, we can see that benzene part gains aromaticity but the thiophene part loses the aromaticity which stabilizes the quinoid form and hence results in lowering the bandgap.[96]

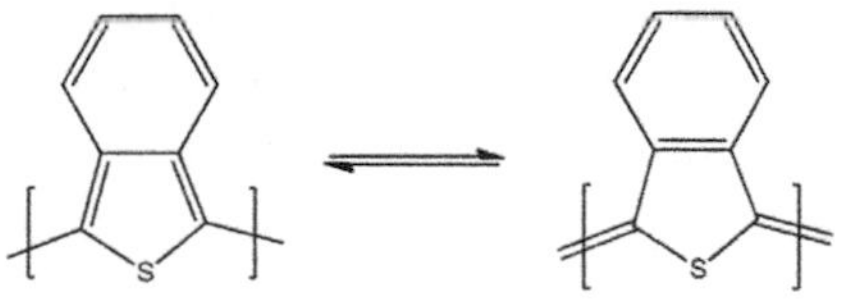

(a)   (b)

FIGURE 5.3   Resonance structure of poly(isothianapthalene)—(a) aromatic form and (b) quinoid form.

Polymers-based on thienothiophene (TT) also forms stable quinoid structure (Fig. 5.4) and forms a series of low-bandgap polymers when attached with benzodithiophene (BDT) moieties well reported in literature.[31,60,61,87]

(a)   (b)

FIGURE 5.4   (a) Aromatic and (b) quinoid form of poly(thieno[3,4-b]thiophene).

In addition to these, there are several others moieties (benzobisthiadiazole, dithienosilole, thienopyrazine, carbazole, etc.) which also possess the stable quinoid structure which is considered as important building blocks for low-bandgap polymers.

Another method reported in literature is push–pull system or the donor–acceptor approach.[96] Therefore, polymers for solar cells were mainly designed having conjugated structure with electron-rich and electron-poor moieties along their main backbone chain which is called as low-bandgap polymers.[79] This donor–acceptor strategy was planned in 1993[2,24] and Zhang and Tour best illustrated this donor–acceptor approach using a copolymer of 3,4-nitrothiophene and 3,4-aminothiophene which has a bandgap of 1 eV.[106]

Many low-bandgap polymers were designed using this method, yet it is difficult to predict the exact theoretical calculations of a newly designed conjugated structure.

Due to enhanced intramolecular charge transfer (ICT) from donor to acceptor results in lower bandgap by increasing highest occupied molecular orbit (HOMO) of a donor and low energy level of lowest unoccupied molecular orbit (LUMO) of the acceptor.[46,53] The strength of ICT ensuing from the donor–acceptor alteration can be simply modulated by the electron-rich and electron-poor units constituting the donor–acceptor building blocks, thereby alternating strong donor with a strong acceptor results in low bandgap.

When alternating cyclopentadithiophene—a strong donor with 2, 1, 3-benzothiadiazole (BT)—a strong acceptor results in polymer PCPDTBT having a bandgap of 1.4 eV (Fig. 5.5).[66]

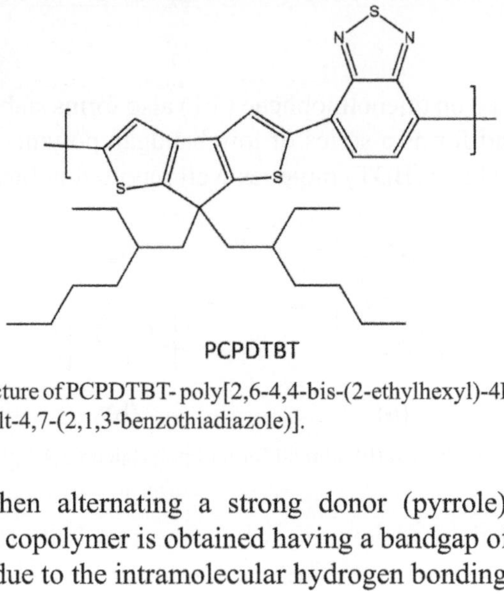

PCPDTBT

FIGURE 5.5   Structure of PCPDTBT- poly[2,6-4,4-bis-(2-ethylhexyl)-4H-cyclopenta[2,1-b; 3,4-b']dihiophene-alt-4,7-(2,1,3-benzothiadiazole)].

Similarly, when alternating a strong donor (pyrrole) with a strong acceptor (BT), a copolymer is obtained having a bandgap of 1.1 eV and this low bandgap is due to the intramolecular hydrogen bonding (Fig. 5.6).[94]

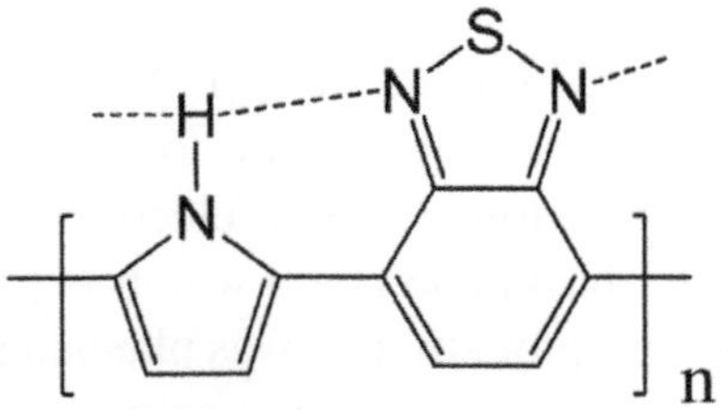

FIGURE 5.6   Copolymer of pyrrole and 2,1,3-benzothiadiazole.

One of the distinctive properties of donor–acceptor polymer is that their energy levels (HOMO and LUMO) are largely localized on the donor and acceptor moiety which offers an important advantage of tuning the bandgap. HOMO level and LUMO levels of a conjugated polymer can be adjusted by varying the electron-donating ability of the donor moiety and the electron-withdrawing ability of the acceptor moiety.[7,107]

To develop ideal conjugated polymers, one must have low-lying HOMO energy levels to have a high V_{OC}, J_{SC}, and narrow bandgap. These parameters can be tuned by using donor–acceptor approach which offers the unique feature of tuning the energy levels and the bandgap. As reported in the literature, a "weak donor–strong acceptor" strategy has been proposed to maintain a low-lying HOMO energy level and hence results in narrow bandgap.[107]

In view of these, here we have given the energy level diagram (Fig. 5.7) in which HOMO and LUMO energy levels of ideal polymer and $PC_{61}BM$ have shown. Also, the HOMO energy levels for various donors are shown which can help in choosing the donor to construct novel low-bandgap polymers.

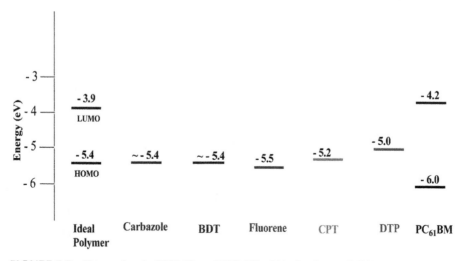

FIGURE 5.7 Energy levels (HOMO and LUMO) of ideal polymer, PC61BM and HOMO levels of various donors are shown.

Most of the conjugated polymers reported so far are based on this donor–acceptor concept showing up to 11% efficiency.[9,15,16,44,45,59,85,103,110]

Recently, Zhou and co-workers has reported very high PCE of 11.3% using a novel inter connecting layer (ICL) in which two similar BHJ layers were stacked in a series connection in tandem cells.[110]

Different efficient donors and acceptors used to construct low-bandgap polymers are listed below and their structure are shown in Figures 5.8 and 5.9.

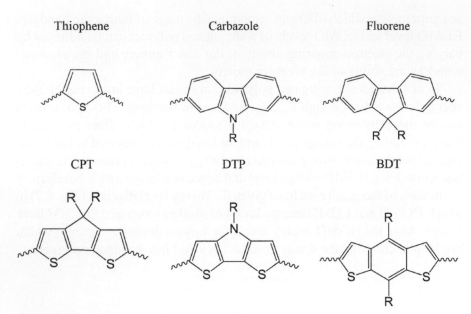

FIGURE 5.8   Structure of efficient donors.

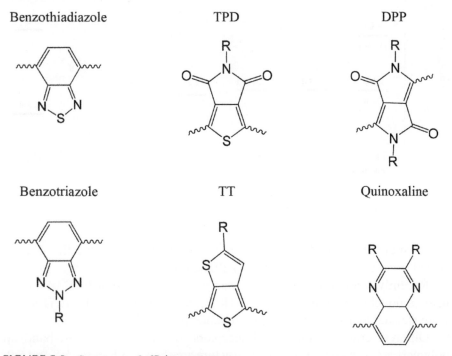

FIGURE 5.9   Structures of efficient acceptors.

Donors	**Acceptors**
Thiophene	Benzothiadiazole (BT)
Carbazole	Diketopyrrolopyrrole (DPP)
Fluorene	Benzotriazole
Cyclopentadithiophene (CPT)	Quinoxaline
Benzodithiophene (BDT)	Thienothiophene (TT)
Dithienopyrrole (DTP)	Thienopyrrolodione (TPD)

5.3.2 EFFECT OF SUBSTITUENT

Another important factor to achieve a low-bandgap polymer is to increase the strength of donor and the acceptor units which are efficiently achieved by attaching substituents on the polymer backbone. Substituents are used to tune some electronic properties such as bandgap, energy levels, charge mobility, molecular interactions, and sometimes even morphology.

Substituents can influence the position of HOMO and the LUMO level (via mesomeric effect or inductive effect) and is achieved by attaching electron donating group (EDG) and electron withdrawing group (EWG) on the polymer backbone. Therefore, by attaching EDG on the donor, the energy level of the HOMO of a donor can be increased. In a similar manner, when by attaching EWG on the acceptor the energy level of the LUMO of the acceptor is lowered.[2]

Attaching methoxy group (EDG) to the benzene unit of poly-phenylene vinylene (PPV) raises the HOMO energy level, while attaching cyano group (EWG) to the vinylene unit of PPV lowers the LUMO energy level and was experimentally observed too.[27]

Fluorine is another substituent most widely used in the polymer backbone for tuning the electronic properties without introducing the steric hindrance. Fluorine is the most electronegative element in the periodic table and is a smallest EWG, therefore introduces less steric hindrance; also the fluorine substituted molecules are explored because of its thermal and oxidative stability.[100] In addition to the above unique features, fluorine atoms provide intra- as well as intermolecular interactions, that is, C–F–H, F–S, and C–F–πF.[76,98]

In order to increase the electron withdrawing capacity of BT unit, Zhou et al. introduced two fluorine atoms on the BT unit to form difluorobenzothiadiazole (DFBT) (Fig. 5.10).[108] Fluorine atoms in DFBT lower the HOMO level of the polymers from −5.18 eV to −5.34 eV with increase in V_{oc} (0.85) and PCE (5.6%).

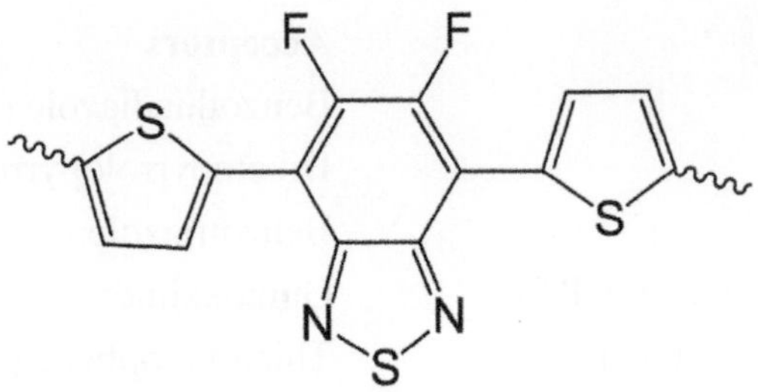

FIGURE 5.10 Structure of DFBT with conjugated polymer.

Similarly, incorporating two F-atoms on the benzotriazole (TAZ) units (Fig. 5.11) reduces the HOMO level to −5.36 eV from −5.29 eV (TAZ without F-atoms) having a higher V_{OC} (0.79 V) and PCE (7.10%). Due to higher hole mobility, its fill factor (FF) is higher (72.9%).[73]

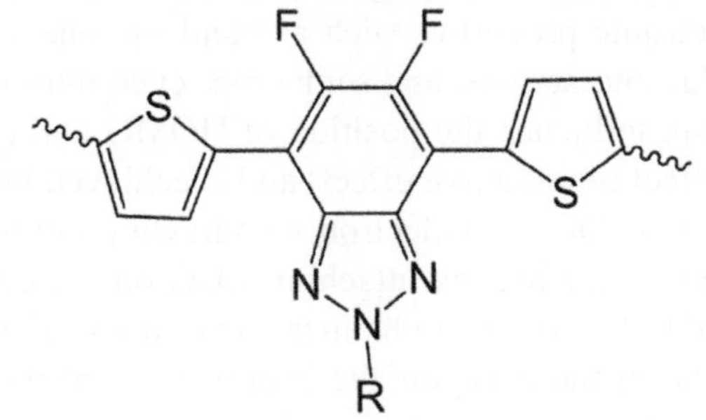

FIGURE 5.11 Structure of fluorinated-TAZ with conjugated polymer.

Also, the position of the fluorine atom on the polymer backbone has a great influence on the performance of related solar cell and is discussed by Son et al.[86] Fluorination effect was also observed on the TT moiety with BDT in the conjugated backbone lowering the HOMO energy level.[12,60,86]

The number of fluorine atom has an effect on the energy levels. As the number of fluorine atoms increases both HOMO and LUMO energy levels are lowered.[3,60,86]

5.3.3 EFFECT OF SIDE CHAIN

Effect of side chain on the polymeric backbone plays an important role in improving the molecular weight, solubility, processability, and morphology (miscibility with fullerene derivatives).[102] The very first important parameter is the side chain which is to be attached on the polymeric backbone which must make the polymer soluble in a suitable solvent which allows solution processability. Once the processability is achieved, it becomes easy to develop flexible solar cells in future with low-cost mass production. In addition to the position of the side chain,[46,47] length and shape of the side chain are equally important parameters to be taken into account.

Increase in the bandgap is observed when the side chain is attached on the nitrogen atom of the pyrrole ring.[17,94,95] When octyl side chain was attached on the thiophene units to increase the solubility, resulted in an increase in the bandgap due to steric hindrance.[46]

When the two side groups (i.e., alkyl and alkoxy) were compared, it was found that the alkoxy side groups are more electron-rich, which help to raise the HOMO level of the conjugated polymer. For example, poly(3,4-ethylene dioxy-thiophene) has a bandgap of 1.5 eV which is less than that of parent polythiophene because of the attachment of electron-rich alkoxy group.[72]

Recently, Peng et al. worked on the different positions of the side chains (i.e., alkyl and alkoxy) on dibenzo[a,c]phenazine (BPz) derivatives.[69] Alkyl groups were substituted on 2 and 7 position while the alkoxy groups were substituted on 11 and 12 position of the BPz moiety. As the alkoxy groups were more electron-rich, the polymer PBDT-OBPz (Fig. 5.12) has higher lying HOMO energy levels as well as LUMO energy levels than the corresponding polymer PBDT-BPzC (Fig. 5.13) with alkyl side chain. Therefore, PBDT-BPzC shows higher V_{OC} of 0.90 V, FF of 56.6%, and J_{SC} of 8.74 mAcm^{-2} with PCE of 4.44% which is 0.49% higher than PBDT-OBPz.

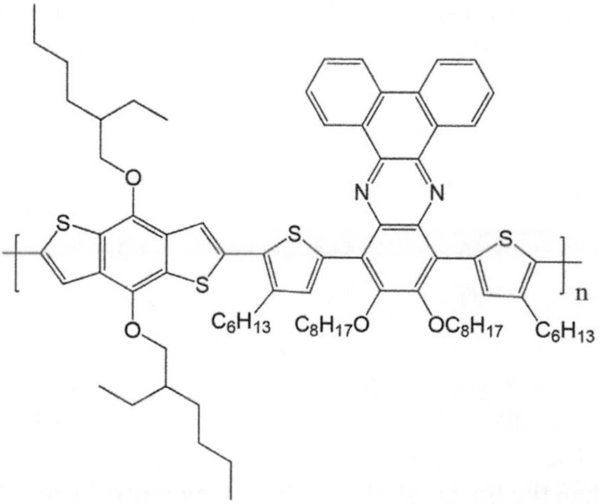

FIGURE 5.12 Chemical structure of PBDT-OBPz.

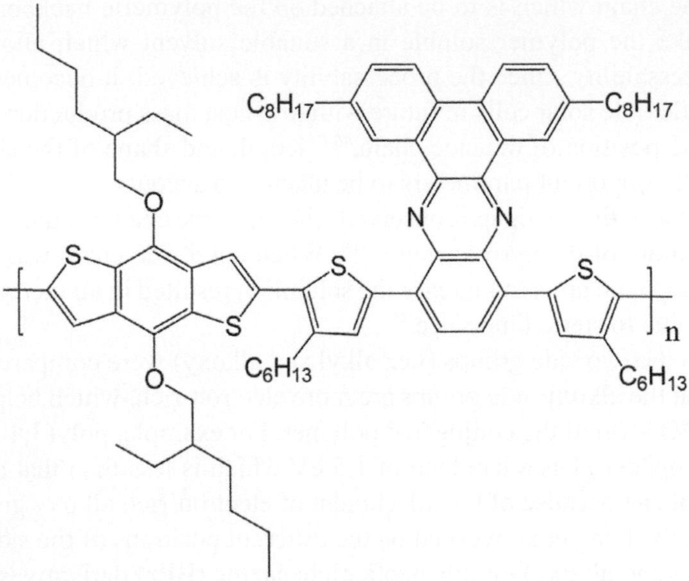

FIGURE 5.13 Chemical structure of PBDT-BPzC.

The same group has also synthesized a series of polymer with 2, 7-dioctyl as side chain on BPz derivatives as electron-poor units with an indacenodithiophene (IDT) as electron-rich units. When 2, 7-dioctyl as side chain is introduced on BPz derivatives, solubility is enhanced and shows high hole mobilities. One of the polymers PIDT-OFBPz shows high PCE of 5.13%. For all the copolymers devices based on polymer:$PC_{61}BM$ blend ratio of 1:2 exhibited superior performances to the other ratios (1:1, 1:3) having device structure ITO/ZnO/PFN/Polymer:$PC_{61}BM$/MoO_3/Al (Table 5.2).[26]

TABLE 5.2 Performance of Polymer Solar Cell.

Polymer	V_{OC} (V)	J_{SC} (mAcm^{-2})	FF (%)	PCE (%)	Blend ratio D/A (W/W)	Reference
PIDT-OHBPz	0.83	8.00	56.5	3.75	1:2	(He et al. 2014)
PIDT-OFBPz	0.97	8.96	58.99	5.13	1:2	
PIDT-OBPQ	0.86	8.97	50.90	3.93	1:2	

When the linear alkyl side chain is compared with the branched alkyl side chains, it was observed that the branched side chains cause an increase in the π–π stacking than the linear alkyl side chain which results in a decrease in the efficiency. This effect was observed by Szarko and co-workers on a series of polymers having TT and BDT units in their conjugated backbone.[89] Caban-etos et al. also modified the structure by incorporating side chain in the BDT and thienopyrrolodione (TPD) unit.[9] They found that when branched side chain is incorporated on the BDT unit, the polymers adopts face on orientation and when linear side chain is attached, no specific orientation was observed, also when n-hexyl side chain is replaced by n-heptyl side chain in the TPD unit resulted in polymer solubility. Therefore, when a polymer structure is made by incorporating 2-ethylhexyl on BDT unit, n-heptyl side chain on TPD unit (Fig. 5.14) resulted in high PCE of 8.5% and FF of 70%.

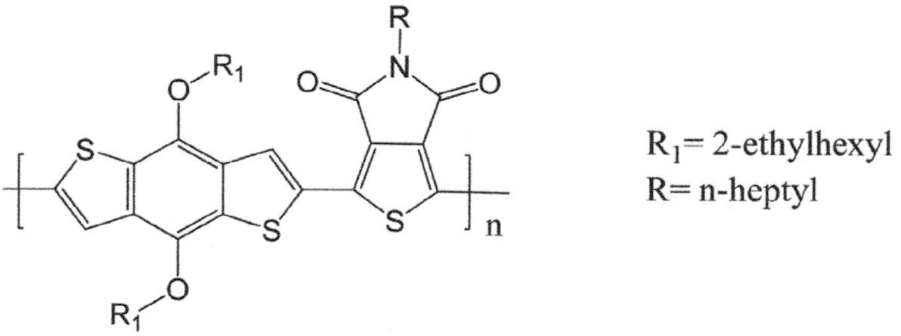

R_1= 2-ethylhexyl
R= n-heptyl

FIGURE 5.14 Chemical structure of polymer having BDT and TPD units with side chain.

Recently, Constantinou et al. have shown the effect of side chain on the polymer having BDT and TPD units in their backbone (Fig. 5.15).[13] They reported that while going from linear alkyl side chain to the branched alkyl side chain have remarkably shown decrease in their J_{SC} and FF values and as a result shows negative impact on the overall device efficiency (Table 5.3). Linear alkyl side chains have a higher degree of charge transfer, closer π–π stacking distance, and the ordered structure. On the other hand, the branched alkyl side chain has shown an increase in the π–π stacking which results in lower hole mobility and highly energetic disordered structure.

Deng et al. also synthesized four donor–acceptor conjugated polymers based on dithienocarbazole (DTC) and diketopyrrolopyrolle (DPP) units having four different alkyl (straight/branched) side chains on the N-atom of the DTC unit.[14] All the polymers, that is, P-C8C8, P-C5C5, P-C12, and P-C10 (Fig. 5.16) are soluble in various organic solvents. They concluded that the polymers with the straight alkyl side chains show higher hole mobility and much better miscibility with $PC_{71}BM$ when compared to branched alkyl side chains. Out of the four polymers, P-C10 which is having straight side chain exhibits best performance having V_{OC} of 0.72 V, J_{SC} of 13.4 mAcm^{-2}, FF of 62% and PCE of 5.9%.

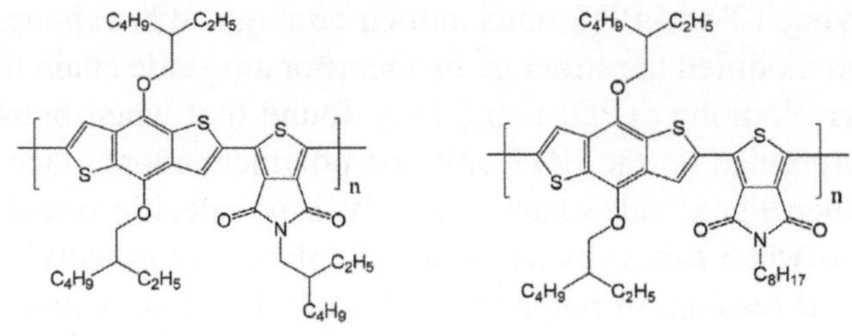

PBDT (EtHex)-TPD (EtHex)

(a)

PBDT (EtHex)-TPD (Oct)

(b)

FIGURE 5.15 Chemical structure of polymers showing difference in the side-chain (a) PBDT (EtHex)-TPD (EtHex) and (b) PBDT (EtHex)-TPD (Oct).

TABLE 5.3 Average Performance of the Polymers.

Polymer/$PC_{70}BM$	J_{SC} (mA/cm^2)	V_{OC} (V)	FF (%)	PCE (%)	Reference
PBDT (EtHex)-TPD (Oct)	11.10 ± 0.5	0.89 ± 0.01	60 ± 2.0	5.9 ± 0.6	(Constantinou et al. 2015)
PBDT (EtHex)-TPD (EtHex)	± 0.6	0.98 ± 0.01	51 ± 3.0	1.95 ± 0.4	

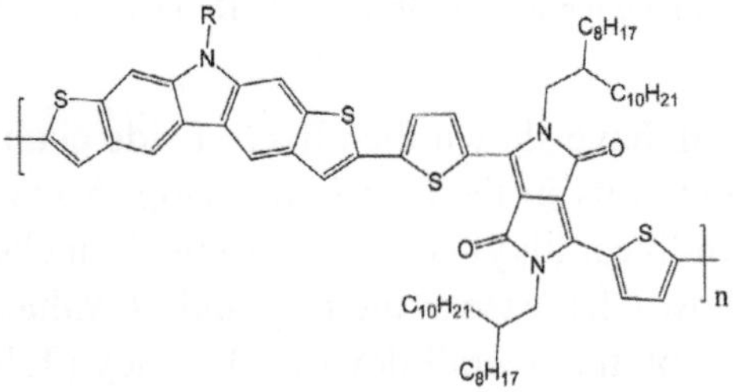

P-C8C8; R= 1-Octylnonyl
P-C5C5; R= 1-Pentylhexyl
P-C12; R= n-dodecyl
P-C10; R= n-decyl

FIGURE 5.16 Chemical structure of DTC and DPP polymers having different R-groups.

Li et al. also introduced long alkyl side chain (2-decyltetradecyl) on the DPP unit (Fig. 5.17) to improve the solubility and molecular weight. The result of the device performance shows J_{SC} of 14.8 mAcm^{-2}, FF of 70% and PCE of 6.9%.[58]

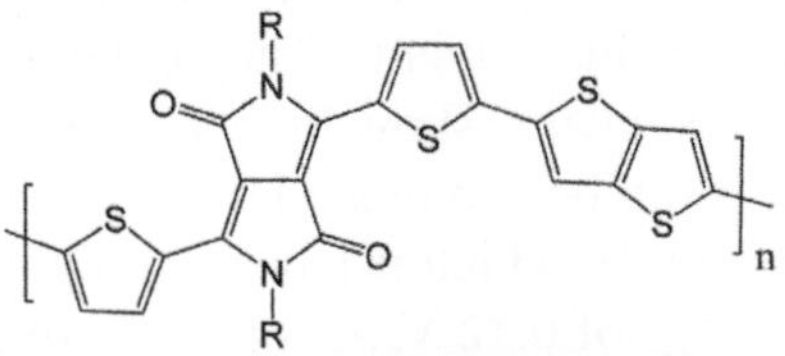

R=2-decyltetradecyl

FIGURE 5.17 Chemical structure of DPP derivatives.

Campos et al. also synthesized the polymer PB3OTP (Fig. 5.18) having octyl group as the side chain at the 3 position of the thiophene moiety to increase the solubility of the polymer for its processability in the BHJ solar cells.[10]

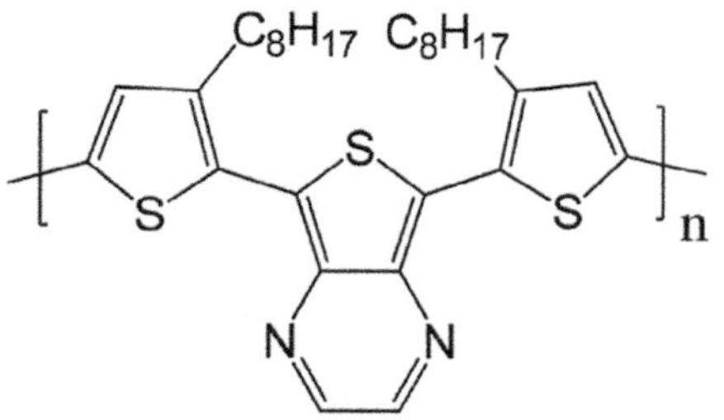

FIGURE 5.18 Chemical structure of PB3OTP.

Liang et al. also reported that by incorporating long alkyl side chain on the TT moiety enhances the solubility and the miscibility of the polymer with fullerene derivatives. They introduced n-octyl carboxylate group as side chain into TT unit (Fig. 5.19) to enhance the polymer properties.[60]

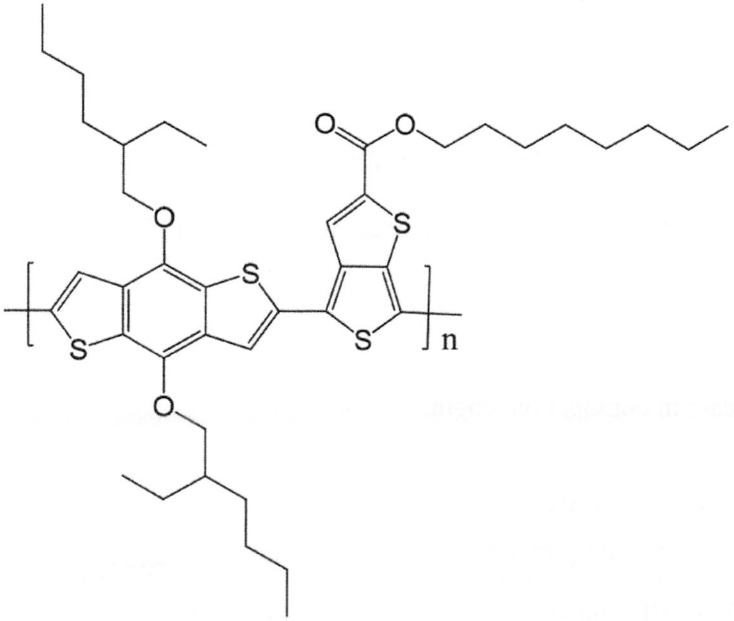

FIGURE 5.19 Chemical structure of BDT and TT units having n-octyl carboxylate as the side chain.

By replacing carboxylate group by a ketone side chain on the TT unit (Fig. 5.20), it was found that HOMO level moved down to −5.2 eV and high V_{OC} of 0.76 V is obtained with high PCE of 7.73%.[12]

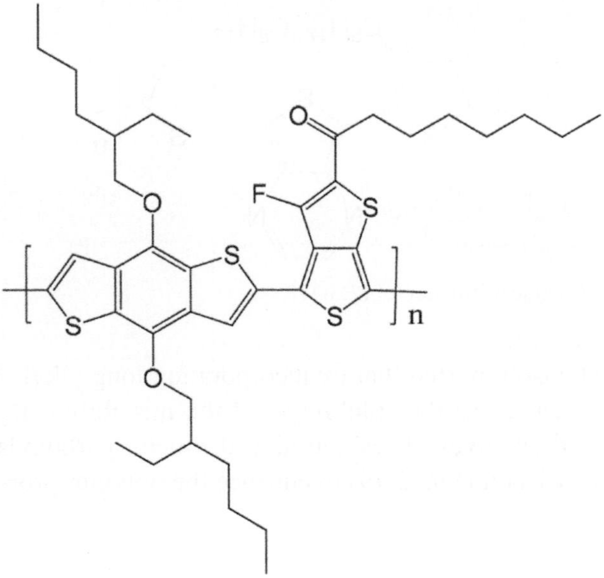

FIGURE 5.20   Chemical structure of BDT and TT units having ketone as the side chain.

5.3.4   π-CONJUGATION LENGTH

As discussed in the literature, longer conjugation length lowers the bandgap and when the torsion in the polymer backbone disrupts, the conjugation results in an increase in the bandgap.[84,99]

It was also found that the bandgap is also affected by the molecular weight of the polymer because increase in molecular weight lowers the bandgap due to increase in conjugation length.[56]

5.4   CONCLUSION

In this article, we describe the brief introduction of OPVs and historical background of solar cells and solar energy. Our main emphasis is on design strategies and its effect on final properties of the materials. We come across to the following conclusions:

- We reviewed the design strategy to develop polymers along with the existing literature and also different donors and acceptors units are examined and discussed.
- Energy level diagram for various efficient donors is shown with their HOMO levels which can help in choosing the donor to construct new polymers for OPVs.
- The effect of substituents and side chain is also discussed. It was concluded that the substituents and side chain may not have the same effect on the conjugated backbone rather the effect changes with the change in the donor–acceptor unit. So having an understanding of structure–property relationship is a key challenge in this growing field.

ACKNOWLEDGMENT

Authors are thankful to Science and Engineering Research Board (SERB), New Delhi for financial support under "Empowerment and Equity Opportunities for Excellence in Science" program reference no. SB/EMEQ-204/2014. One of the authors (M. J. P.) gratefully acknowledges financial support as INSPIRE fellowship (DST/INSPIRE Fellowship/2015/IF 150296) from the Department of Science and Technology (DST), New Delhi.

KEYWORDS

- **electrical conductivity**
- **electrochromic devices**
- **photovoltaics**
- **polymers**
- **OPVs**

REFERENCES

1. Adams, W. G.; Day, R. E. *Proc. R. Soc. Lond.* **1876,** *25*, 113–117.
2. Ajayaghosh, A. *Chem. Soc. Rev.* **2003,** *32*(4), 181–191.
3. Albrecht, S.; Janietz, S.; Schindler, W.; Frisch, J.; Kurpiers, J.; Kniepert, J.; Inal, S.; Pingel, P.; Fostiropoulos, K.; Koch, N.; Neher, D. *J. Am. Chem. Soc.* **2012,** *134*, 14932–14944.

4. Becquerel, A. E. *C. R. Acad. Sci.* (Paris) **1839**, *9*, 561–567.
5. Benincori, T.; Brenna, E.; Sannicolo, F.; Trimarco, L.; Zotti, G. *Angew. Chem.* **1996**, *108*, 718.
6. Brabec, C. J.; Dyakonov, V.; Parisi, J.; Sariciftic, N. S. *Organic Photovoltaics: Concepts and Realization;* Springer-Verlag: Berlin, Heidelberg, Germany, 2003.
7. Bredas, J. L.; Norton, J. E.; Cornil, J.; Coropceanu, V. *Acc. Chem. Res.* **2009**, *42*(11), 1691–1699.
8. Bruting, W. *Physics of Organic Semiconductor;* Wiley-VCH: Germany, 2005.
9. Cabanetos, C.; Labban, A. E.; Bartelt, J. A.; Douglas, J. D; Mateker, W. R.; Frechet, J. M. J.; McGehee, M. D.; Beaujuge, P. M. *J. Am. Chem. Soc.* **2013**, *135*, 4656.
10. Campos, L. M.; Tontcheva, A.; Gunes, S.; Sonmez, G.; Neugebauer, H.; Sariciftci, N. S.; Wudl, F. *Chem. Mater.* **2005**, *17*(16), 4031.
11. Chapin, D. M.; Fuller, C. S.; Pearson, G. L. *J. Appl. Phys.* **1954**, *25*, 676.
12. Chen, H. Y.; Hou, J.; Zhang, S.; Liang, Y.; Yang, G.; Yang, Y; Yu, L.; Wu, Y.; Li, G. *Nature Photonics* **2009**, *3*, 649–653.
13. Constantinou, I.; Lai, T. H.; Klump, E. D.; Goswami, S.; Schanze, K. S.; Franky, S. *ACS Appl. Mater. Interfaces* **2015**, *7*, 26999–27005.
14. Deng, Y.; Chen, Y.; Liu, J.; Liu, L.; Tian, H.; Xie, Z.; Geng, Y.; Wang, F. *ACS Appl. Mater. Interfaces* **2013**, *5*, 5741–5747.
15. Dou, L.; Chen, C. C.; Yoshimura, K.; Ohya, K.; Chang, W. H.; Gao, J.; Liu, Y.; Richard, E.; Yang Y. *Macromolecules* **2013**, *46*, 3384.
16. Dou, L.; Cheng, W. H.; Gao, J.; Chen, C. C.; You, J.; Yang, Y. *Adv. Mater.* **2012**, *25*, 825.
17. Edder, C.; Armstrong, P. B.; Prado, K. B.; Frechet, J. M. *J. Chem. Commun.* **2006**, *18*, 1965.
18. Glenis, S.; Horowitz, G.; Tourillon, G.; Garnier F. *Thin Solid Films* **1984**, *111*(2), 93–103.
19. Gutman, F.; Lyons, L. E. *Organic Semiconductors, Part A;* Robert, E., Ed.; Krieger Publishing Company: Malabar, Florida, 1980; p 251.
20. Hadipour, A.; De Boer, B.; Wildeman, J.; Kooistra, F.; Hummelen, J. C.; Turbiez, M. G. R.; Wienk, M. M.; Janssen, R. A. J.; Blom, P.W. M. *Adv. Funct. Mater.* **2006**, *16*, 1897–903.
21. Halls, J.; Pichler, K.; Friend, R.; Moratti, S.; Holmes, A. *Synth. Met.* **1996**, *77*, 277.
22. Halls, J. J. M.; Friend, R. H. *Synth. Met.* **1997**, *85*, 1307.
23. Halls, J. J. M.; Pichler, K.; Friend, R. H.; Moratti, S. C.; Holmes, A. B. *Appl. Phys. Lett.* **1996**, *68*, 3120–3122.
24. Havinga, E. E.; Ten Hoeve, W.; Wynberg, H. *Synth. Met.* **1993**, *55*(1), 299–306.
25. Heliatek Newscenter. http://www.heliatek.com/newscenter/latest_news/neuer-weltrek ordfur-organische-solarzellen-heliatek-behauptet-sich-mit-12-zelleffizienz-als-technologiefuhrer/?lang=en (accessed March 13, 2014).
26. He, R. F.; Yu, L.; Cai, P.; Peng, F.; Xu, J.; Ying, L.; Chen, J. W.; Yang, W.; Cao, Y. *Macromolecules* **2014**, *47*, 2921–2928.
27. Helbig, M.; Horhold, H. H. *Makromol. Chem.* **1993**, *194*(6), 1607–1618.
28. Helgesen, M.; Søndergaard, R.; Krebs, F. C. *J. Mater. Chem.* **2010**, *20*, 36.
29. Hiramoto, M.; Fujivara, H.; Yokoyama, M. *Appl. Phys. Lett.* **1991**, *58*, 1062.
30. Hiramoto, M.; Fukusumi, H.; Yokoyama, M. *Appl. Phys. Lett.* **1992**, *61*, 2580.
31. Hong, S. Y.; Marynick, D. S. *Macromolecules* **1992**, *25*(18), 4652–4657.
32. https://en.wikipedia.org/wiki/Augustin_Mouchot.

33. http://en.wikipedia.org/wiki/Charles_Fritts.
34. https://en.wikipedia.org/wiki/Czochralski_process.
35. https://en.wikipedia.org/wiki/Samuel_Pierpont_Langley.
36. https://en.wikipedia.org/wiki/Solar_Building.
37. https://en.wikipedia.org/wiki/Timeline_of_solar_cells.
38. https://en.wikipedia.org/wiki/Willoughby_Smith.
39. http://org.ntnu.no/solarcells/pages/history.php.
40. http://www.historycommons.org/context.jsp?item=a1891kempwaterhtr.
41. https://www1.eere.energy.gov/solar/pdfs/solar_timeline.pdf.
42. http://www.Solarmer.com/newsevents.php.
43. http://www.zoominfo.com/p/Dan-Trivich/11736085.
44. Huang, Y.; Guo, X.; Liu, F.; Huo, L.; Chen, Y.; Russell, T. P.; Han, C. C.; Li, Y.; Hou, J. *Adv. Mater.* **2012**, *24*, 3383.
45. Huo, L.; Zhang, S.; Guo, X.; Xu, F.; Li, Y.; Hou, J. *Angew. Chem. Int. Ed.* **2011**, *50*, 1.
46. Jayakannan, M.; Van Hal, P. A.; Janssen, R. A. J. *J. Pol. Sci. A Pol. Chem.* **2002**, *40*, 251.
47. Jayakannan, M.; Van Hal, P. A.; Janssen, R. A. J. *J. Pol. Sci. A Pol. Chem.* **2002**, *40*, 2360.
48. Kallmann H.; Pope M. *J. Phys. Chem.* **1959**, *30*, 585–586.
49. Karg, S.; Riess, W.; Dyakonov, W.; Schwoerer, M. *Synth. Met.* **1993**, *54*, 427.
50. Karl, N. *Synth. Met.* **2002**, *133*, 649–657.
51. Kim, J. Y.; Lee, K.; Coates, N. E.; Moses, D.; Nguyen, T. Q.; Dante, M.; Heeger, A. J. *Science* **2007**, *317*, 222.
52. Kippelen, B.; Bredas, J. L. *Energy Environ. Sci.* **2009**, *2*, 251–261.
53. Kitamaru, C.; Tanaka, S.; Yamashita, Y. *Chem. Mater.* **1996**, *8*, 570.
54. Kobayashi, M.; Colaneri, N.; Boysel, M.; Wudl, F.; Heeger, A. J. *J. Chem. Phys.* **1985**, *82*, 5717.
55. Krebs, F. C.; Jorgensen, M. *Polym. Bull.* **2003**, *50*, 359.
56. Kroon, R.; Lenes, M.; Hummelen, J. C.; Blom, P. W. M.; De Boer, B. *Polym. Rev.* **2008**, *48*, 531–582.
57. Li, G.; Shrotriya, V.; Huang, J.; Yao, Y.; Moriarty, T.; Emery, K.; Yang, Y. *Nature Mater.* **2005**, *4*, 864–868.
58. Li, W.; Hendriks, K. H.; Christian Roelofs, W. S.; Kim, Y.; Wienk, M. M.; Janssen, R. A. J. *Adv. Mater.* **2013**, *25*, 3182.
59. Liao, S. H; Jhou, H. J.; Cheng, Y. S.; Chen, S. A. *Adv. Mater.* **2013**, *25*(34), 4766–4771.
60. Liang, Y.; Feng, D.; Wu, Y.; Tsai, S. T.; Li, G.; Ray, C.; Yu, L. *J. Am. Chem. Soc.* **2009**, *131*, 7792.
61. Liang, Y.; Xu, Z.; Xia, J.; Tsai, S. T.; Wu, Y.; Li, G.; Ray, C.; Yu, L. *Adv. Mater.* **2010**, *22*, 1.
62. Marks, R. N; Halls, J. J. M.; Bradley, D. D. C.; Friend, R. H.; Holmes, A. B. *J. Phys. Condens. Matter* **1994**, *6*, 1379.
63. Meller, G.; Grasser, T. *Organic Elelctronics;* Springer-Verlag: Berlin, Heidelberg, Germany, 2010.
64. Monk, P.; Mortimer, R.; Rosseinsky, D. *Electrochromism and Electrochromic Devices;* Cambridge University Press: University of Exeter, 2007.
65. Morita, S.; Zakhidov, A. A.; Yoshino, K. *Jpn. J. Appl. Phys.* **1993**, *32*, 873.
66. Muhlbacher, D.; Scharber, M.; Morana, M.; Zhu, Z.; Waller, D.; Gaudiana, R.; Brabec, C. *Adv. Mater.* **2006**, *18*, 2884–2889.
67. National Centre for Photovoltaics Home Page. http:/www.nrel.gov/ncpv (accessed March 13, 2014).

68. Park, S. H.; Roy, A.; Beaupré, S.; Cho, S.; Coates, N.; Moon, J. S.; Moses, D.; Leclerc, M.; Lee, K.; Heeger, A. J. *J. Nature Photon* **2009**, *3*, 297.
69. Peng, F.; Zhao, B.; Xu, J.; Zhang, Y.; Fang, Y.; He, R.; Wu, H.; Yang, W.; Cao, Z. *Org. Electron.* **2016**, *29*, 151–159.
70. Perepichka, I. F.; Perepichka, D. F., Eds.; *Handbook of Thiophene Based Materials;* John Wiley & Sons: Hoboken, NJ, 2009.
71. Peumans, A. Y. P.; Forrest, S. R. *J. Appl. Phys.* **2003**, *93*, 3693.
72. Pie, Q.; Zuccarello, G.; Ahlskog, M.; Inganas, O. *Polymer* **1994**, *35*, 1347.
73. Price, S. C.; Stuart, A. C.; Yang, L.; Zhou, H.; You, W. *J. Am. Chem. Soc.* **2011**, *133*, 4625.
74. Ramos, A. M.; Rispens, M. T.; Hummelen, J. C.; Janssen, R. A. J. *Synth. Met.* **2001**, *119*, 171.
75. Ramos, A. M.; Rispens, M. T.; van Duren, J. K. J.; Hummelen, J. C.; Janssen, R. A. J. *J. Am. Chem. Soc.* **2001**, *123*, 6714.
76. Reichenbacher, K.; Suss, H. I.; Hulliger, J. *Chem. Soc. Rev.* **2005**, *34*, 22.
77. RISE. Small Photovoltaic Arrays. Research Institute for Sustainable Energy, Murdoch University (retrieved Feb, 5 2010).
78. Roncali, J. *Chem. Rev.* **1992**, *92*, 711.
79. Roncali, J. *Chem. Rev.* (Washington, DC, U.S.) **1997**, *97*(1), 173–205.
80. Sacriciftci, N. S.; Braun, D. ; Zhang, C.; Srdanov, V. I.; Heeger, A. J. *Appl. Phys. Lett.* **1993**, *62*, 585–587.
81. Schmidt-Mende, L; Fechtenkötter, A.; Müllen, K.; Moons, E.; Friend, R. H.; MacKenzie, J. D. *Science* **2001**, *293*, 1119.
82. Shirakawa, H.; Louis, E. J.; MacDiarmid, A. G.; Chiang, C. K.; Heeger, A. J. *J. Chem. Soc. Chem. Commun.* **1977**, *474*, 578.
83. Skotheim, T. A.; Reynolds, J. R., Eds.; *Handbook of Conducting Polymers,* 3rd ed.; CRC Press: Boca Raton, FL, 2007.
84. Sivula, K.; Luscombe, C. K.; Tompson, B. C.; Frechet, J. M. *J. Am. Chem. Soc.* **2006**, *128*, 13988.
85. Son, H. J.; Lu, L.; Chen, W.; Xu, T.; Zheng, T.; Carsten, B.; Strzalka, J.; Darling, S. B.; Chen, L. X.; Yu, L. *Adv. Mater.* **2013**, *25*, 838.
86. Son, H. J.; Wang, W.; Xu, T.; Liang, Y.; Wu, Y.; Li, G.; Yu, L. *J. Am. Chem. Soc.* **2011**, *133*(6), 1885–1894.
87. Sotzing, G. A.; Lee, K. H. *Macromolecules* **2002**, *35*(19), 7281–7286.
88. Spanggaard, H.; Krebs, F. C. *Sol. Energy Mater. Sol. Cells.* **2004**, *83*, 125–146.
89. Szarko, J. M.; Guo, J.; Liang, Y.; Lee, B.; Rolczynski, B. S.; Strzalka, J.; Xu, T.; Loser, S.; Marks T. J.; Yu, L.; Chen, L. X. *Adv. Mater.* **2010**, *22*(48), 5468–5472.
90. Tada, K.; Onoda, M; Nakayama, H.; Yoshino, K. *Synth. Met.* **1999**, *102*, 982.
91. Tada, K.; Onoda, M.; Zakhidov, A. A.; Yoshino, K. *Jpn. J. Appl. Phys.* (Part 2-Lett.) **1996**, *36*, L306–L309.
92. Takamoto, T.; Ikeda, E.; Kurita, H.; Ohmori M. *Appl. Phys. Lett.* **1997**, *70*, 381–383.
93. Tang, C. W. *Appl. Phys. Lett.* **1986**, *48*, 183.
94. Van Mullekom, H. A. M.; Vekemans, J. A. J. M.; Meijer, E. W. *Chem. Commun.* **1996**, *18* 2163.
95. Van Mullekom, H. A. M.; Vekemans, J. A. J. M.; Meijer, E. W. *Chem. Eur. J.* **1998**, *4*, 1235
96. Van Mullekom, H. A. M; Vekemanas, J. A. J. M.; Havinga, E. E.; Meijer, E. W. *Mater. Sci. Eng. R Rep.* **2001**, *32*, 1–40.
97. *Van Nostrands Eng. Mag.* **1885**, *32*, 388–395.

98. Wang, Y.; Parkin, S. R.; Gierschner, J.; Watson, M. D. *Org. Lett.* **2008,** *10*(15), 3307–3310.
99. Winder, C.; Sariciftci, N. S. *J. Mater. Chem.* **2004,** *14*, 1077.
100. Wong, S.; Ma, H.; Jen, A. K. Y.; Barto, R.; Frank, C. W. *Macromolecules* **2003,** *36*(21), 8001–8007.
101. Xue, J.; Rand, B. P.; Uchida, S.; Forrest, S. R. *J. Appl. Phys.* **2005,** *98*, 124903.
102. Yang, L.; Zhou, H.; You, W. *J. Phys. Chem. C* **2010,** *114*(39), 16793–16800.
103. You, J.; Dou, L.; Yoshimura, K.; Kato, T.; Ohya, K.; Moriarty, T.; Emergy, K.; Chen, C. C.; Gao, J.; Li, G.; Yang, Y. *Nature Commun.* **2013,** *4*, 1446.
104. Yu, G.; Heeger, A. J. *J. Appl. Phys.* **1995,** *78*, 4510.
105. Yu, G.; Pakbaz, K.; Heeger, A. J. *Appl. Phys. Lett.* **1994,** *64*, 3422.
106. Zhang, Q. T.; Tour, J. M. *J. Am. Chem. Soc.* **1998,** *120*(22), 5355–5362.
107. Zhou, H.; Yang, L.; Stoneking, S.; You, W. *ACS Appl. Mater. Interfaces* **2010,** *2*(5), 1377–1383.
108. Zhou, H.; Yang, L.; Stuart, A. C.; Price, S. C.; Liu, S; You, W. *Angew. Chem. Int. Ed.* **2011,** *50*(13), 2995–2998.
109. Zhou, H.; Yang, L.; You, W. *Macromolecules* **2012,** *45*, 607.
110. Zhou, H.; Zhang, Y.; Mai, C. K.; Collins, S. D.; Bazan, G. C.; Nguyen, T. Q.; Heeger, A. J. *Adv. Mater.* **2015,** *27*, 1767–1773.
111. Zhou, Q.; Hou, Q.; Zheng, L.; Deng, X.; Yu, G.; Cao, Y. *Appl. Phys. Lett.* **2004,** *84*, 1653–1655.

CHAPTER 6

AN OVERVIEW ON LI–S BATTERY AND ITS CHALLENGES

RAGHVENDRA KUMAR MISHRA*, ASWATHY VASUDEVAN, and SABU THOMAS

International and Inter University Centre for Nanoscience and Nanotechnology, Mahatma Gandhi University, Priyadarshini Hills P.O., Kottayam, Kerala 686560, India

Corresponding author. E-mail: raghvendramishra4489@gmail.com

CONTENTS

ABSTRACT

Lithium/sulfur batteries are the new-generation efficient rechargeable battery systems, based on their high cell energy density in comparison to state of the art lithium-ion battery that could make most promising long-range candidates for the new-generation energy storage, with high energy density and low-cost application. This study is based on lithium, which works as a negative electrode and solid sulfur (sulfur with a theoretical specific capacity of 1675 mAh g^{-1}), which works as one of the most promising positive electrode materials. Lithium sulfide (Li$_2$S) is the final discharge product and the only thermodynamically stable binary Li–S phase. However, the Li–S battery technology has addressed several challenges such as the inherent high electronic resistivity of sulfur and the higher order polysulfide's (PS) shuttle mechanism in organic electrolytes, in between the anode and cathode, leads to active material loss due to solubility of lithium PS formed during the charging process in which long-chain lithium PS produced at the cathode, dissolve into the electrolyte, migrate and react with the lithium anode electrode, by that way, generate lower order PS in a parasitic manner, which leads to rapid reduction the capacity degradation and short cycle life performance. This review aims to cover recent advances in the Li–S battery, providing an overview of the Li-ion battery applications in energy storage by eliminating PS shuttle and maximizing the utilization of the PS, contributing to enhance energy density and columbic efficiency.

6.1 GENERAL INTRODUCTION

Recently, the massive demand of energy has increased the attention of researchers toward battery power systems in order to reduce the use of these fossils and overcome the problem such as global warming, air pollution, and environmental pollution that are becoming a serious threat. To prevent these issues and reduce the use of fossil fuel, alternative of these energies based on renewable sources need to be developed and acquired. The rechargeable batteries are electrochemical energy storage devices which are most promising candidates for electrical energy storage. Able to achieve higher capacity of cost-efficient and high energy storage characteristics, the secondary batteries have been developed to provide as energy storage system for a wide range of applications including hybrid electric vehicles (HEVs) and plug-in HEVs, electric utility industry, consumer electronics, medical electronics, and military/defense devices. In terms of advancement of electronic devices and energy triggered devices, lead–acid, nickel–cadmium, nickel–metal hydride, and lithium-ion batteries are being developed as power assists for automobile, electronics system, electrical engineering, etc., where the energy, power, cycle life, safety, and environmental compatibility are the most important parameters. This is extremely useful in achieving energy density and development of advanced electrode materials.

A recent report suggested that the high energy/power density-based energy storage system will play a dominating role in the future.[1] The world market for batteries has been growing rapidly. Nevertheless, in order to improve the performance of the state-of-the-art rechargeable battery to satisfy the rapidly increasing performance demands for the various applications. Among all the batteries, lithium batteries have showed more effective alternatives for energy storage over the past two decades. Lithium batteries have much specific energy (energy per unit weight), energy density (energy per unit volume), and more environmental friendliness than any other rechargeable systems shown in Figure 6.1.[2]

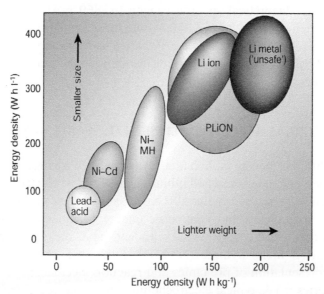

FIGURE 6.1 Typical volumetric and gravimetric energy density behavior of different battery technologies.

Lithium (Li) metal has superior theoretical specific capacity over those currently used in rechargeable batteries. This chapter summarizes the recent research and technology in developing Li–S batteries for energy storage application and we summarize the recent advances in the structural and functional evolution of in Li–S batteries. In this chapter, a variety of methods that have been used with the specific aim of improving the battery performance will be discussed. Also, the developments in Li–S batteries design will be considered to describe the behavior of materials, as well as drawback in the Li–S batteries for practical applications.

6.2 LITHIUM (LI) METAL-BASED BATTERIES

The rechargeable lithium-ion battery was commercialized by Sony in the early 1990s; they utilized a negative electrode (anode), positive electrode (cathode) material, and an electrolyte to fabricate the lithium ion. They inserted a negative electrode (anode) and a positive electrode (cathode) in an electrolyte. The electrodes are an essential component in terms of theoretical specific capacity and play an important role in determining energy and power densities of batteries. Lithium (Li) metal is an important material because of its ultrahigh theoretical specific capacity (3860 mAh g^{-1}), the lowest negative reduction potential (−3.040 V) versus a standard hydrogen electrode, and its low density (0.59 g cm^{-3}). Nowadays life application, lithium metal-based batteries have gained significant attention at both research and industry levels.[3–5]

6.3 HISTORICAL ASPECTS OF LI RECHARGEABLE BATTERY

Typically, lithium rechargeable batteries consist of three essential components, as per abovementioned and lithium is the lightest electrical conductor and low-density electrode material. Due to above facts, lithium batteries have higher gravimetric and volumetric energy densities compared to the other secondary battery systems such as lead–acid, nickel–cadmium (Ni–Cd), and nickel–metal hydride (Ni–MH). Figure 6.2 shows a schematic pictorial representation of principles and reaction mechanism of secondary Li-ion batteries.[20] The typical electrochemical reactions are given below:[6]

$$Li_x CoO_2 \underset{\text{Discharge}}{\overset{\text{Charge}}{\underset{\longleftarrow}{\longrightarrow}}} Li_x CoO_2 + Li^+ + xe^- \text{(cathode)} \tag{6.1}$$

$$\text{C} + x\text{Li}^+ + x\text{e}^- \quad \overset{\text{Charge}}{\underset{\text{Discharge}}{\rightleftarrows}} \quad \text{Li}_x\text{C} \quad (\text{anode}). \tag{6.2}$$

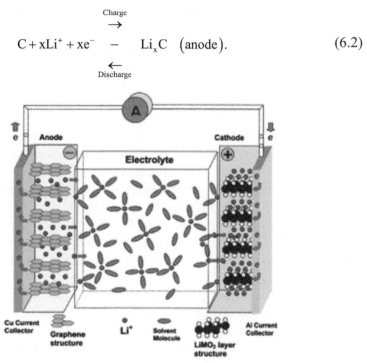

FIGURE 6.2 Schematic representation and operating principles of rechargeable Li-ion batteries.

Overall performance of the batteries is a critical parameter. To maximize the cell voltage, the cathode and anode must have highest and lowest voltage, respectively, and electron transfer through the external circuit of batteries is usually evaluated by the flow of ions through the electrolyte, which is associated with the reduction/oxidation reaction which varies with the type of ionic conductive/good chemical stability of electrolyte and types of electrode materials. In lithium batteries, the lithium insertion/extraction process helps the flow of lithium ions through the electrolyte, which refers to the electron transfer through the external circuit via oxidation and reduction process. Specifically, owning to Li battery performance, typical electrode reactions of a Li-ion cell with $LiCoO_2$ as the cathode and graphite as the anode are shown by eqs 6.1 and 6.2.[6] Like a conventional rechargeable battery, Li–metal oxide acts as cathode materials and metallic lithium acts as anode materials; it has one of the highest capacity among anode materials. However, the limited electrochemical stability window of the liquid electrolytes caused to increase the cathode operating voltage beyond ~4.3 V,[3–5] and the formation of dendrites upon continuous cycling can cause short circuit and other safety concerns, whereas prevents the use of lithium as anode in the lithium battery.[8] Furthermore, the internal chemical reaction is accompanied by gas generation, results in hazardous pressure buildup and escalating failure within the cells. In recent years, lot of research has been performed to develop different types of Li batteries to improve the performance in terms of capacity, fast diffusion of Li ions into the anode, and carbon nanotubes (CNTs), carbon nanofibers, mesoporous carbons, graphene-based anode materials for capacity as well as great retard of surface degradation during charge/discharge cycles.[9–13] Comparing with different types of Li batteries, Li batteries-based storage system possess abovementioned problems. Therefore, new lithium battery systems are needed to be developed to overcome this problem. With this point of view, $Li-O_2$ and Li–S batteries with high energy-based research are being intensively widely pursued.[2–3] Recently, Li–S battery has received significant interest due to synergetic properties of sulfur, which has almost the higher theoretical capacity 1675 mAh g^{-1} and energy density 3500 Wh kg^{-1} than the transition metal oxide cathodes with low cost.[14,15] It has been widely accepted that Li–S batteries can achieve the theoretical energy density (2600 Wh kg^{-1}) that can be obtained by the two-electron redox process in it, which is at many times greater than that any rechargeable other commercial lithium batteries. This conclusion coincides with the importance of sulfur that has been pointed out previously confirming that the solid sulfur with a theoretical specific capacity of 1675 mAh g^{-1} is a major reason responsible for high-performance Li battery. These characteristics make sulfur-based cathodes to the frontier of a battery research. These rechargeable Li–S battery has been utilized to overcome the limitation of Li ion and other batteries. The rechargeable Li–S battery is participating reversible redox process between sulfur and lithium to form Li_2S which offers unique theoretical capacity (1675 mAh g^{-1}sulfur), using elemental sulfur as a cathode material to directly react with lithium metal.[16,17] In addition, S is also low in cost, abundant, and nontoxic. Sulfur is blended with conductive additives (e.g., carbon) to produce cathode, and conductivity of the electrodes (cathode) is achieved by the conductive additives. However, It is well known that the Li–S batteries particularly suffer from long-chain reduced lithium (poly) sulfide products (shuttle mechanism) that gradually decrease the cycle stability of lithium–sulfur cells. The lithium (poly) sulfide product (shuttle mechanism) does trigger the formation of insoluble lithium PS precipitates on both cathode and anode surface leading to parasitic self-discharge. The formation of intermediate lithium PS during the conversion reaction creates challenges. In the recent years, reversible, stable, novel nanocomposites, efficient electrolytes, configurations, efficient sulfur cathodes, and novel cell have also been develop to reduce the drawback of Li–S batteries. The main reason for this development is to decrease the limitation of Li–S batteries which include huge volumetric expansion of sulfur during lithiation, degradation of sulfur cathode under mechanical strain, and the formation of soluble long-order PS backward and forward between the electrodes lead to corrosion of the lithium anode and result in a relatively short battery life, conversely lower efficiency, and minimal active material utilization.[18–21] It has been indicated that the shuttling process is driven by the concentration gradient of PS and on advantageous point of view, it protects the Li–S batteries from overcharge.[21–24] Therefore, the majority of recent research activities in the field of Li–S batteries have been committed to facilitate the efficient electrochemical performances with improving the Li–S system and host matrix, which is useful to full conversion of sulfur into Li_2S during the reduction process.[25–27] Table 6.1 displays the development in the field of Li–S battery.

6.4   PRINCIPLES OF LI–S BATTERIES

The overview of Li–S battery has been discussed in the various previous studies. According to that sulfur-based positive electrode material was introduced to fabricate electrochemical process in 1960s by Herbert and Ulam.[28] Up to this time, quite many attempts have been undertaken to develop the metal–sulfur batteries, some of them mainly focusing on primary Li–S cells.[29] But 2008 onward, very rapid increase in the development of emerging applications [such as electric vehicles (EV), military power supplies, and stationary storage systems for renewable energy] provoked even higher demand for high-performing batteries.[30,31] In particular, the increasing market of EV seems to be the strongest motivation for making Li/S batteries as a system of choice for the future. Indeed, the ever-increasing attention of the Li/S cells can be also seen in terms of published papers number, with practically exponential growth since 2010. Because sulfur offers one of the highest theoretical specific capacity (1675 mAh g^{-1}) among all existing positive electrode materials, which is an order of magnitude higher than what is currently proposed by the transition metal oxide cathodes.[32] The system could theoretically deliver energy density of ~2600 Wh kg^{-1}. Moreover, sulfur is an abundant element, nontoxic, and extremely cheap, which may drastically decrease the final cost of the battery. A realistic target of practical energy density is in the range of 400–600 Wh kg^{-1}, which is higher as compared

TABLE 6.1 Important Advances in the Development of Li–S Batteries.[84]

Anode	Cathode	Date and source of research	Specific capacity (after cycling) and specification
Lithium	Sulfur–graphene oxide nanocomposite with styrene-butadiene-carboxymethyl cellulose copolymer binder	Lawrence Berkeley National Laboratory, United States 2013	700 mAh g^{-1} at 1500 cycles (0.05 C discharge) 400 mAh g^{-1} at 1500 cycles (0.5 C charge/1 C discharge) Lithium bis(trifluoromethanesulfonyl)imide dissolved in a mixture of nmethyl-(n-butyl) pyrrolidinium bis(trifluoromethanesulfonyl)-imide (PYR14TFSI), 1,3-dioxolane (DOL), dimethoxyethane (DME) with 1 M lithium bis-(trifluoromethylsulfonyl)imide (LiTFSI), and lithium nitrate (LiNO3). High-porosity polypropylene separator. Specific energy is 500 Wh/kg (initial) and 250 W·h/kg at 1500 cycles (C = 1.0)
Lithiated graphite	Sulfur	Pacific Northwest National Laboratory, United States 2014	400 cycles. Coating prevents polysulfides from destroying the anode.
Lithiated graphene	Sulfur/lithium–sulfide passivation layer	OXIS Energy 2014	240 mAh g^{-1} (1000 cycles) 25 Ah/cell Passivation layer prevents sulfur loss
Lithium sulfur batteries	Carbon nanotube/sulfur	Tsinghua University, China 2014	15.1 mAh·cm^{-2} at a sulfur loading of 17.3 mgs·cm^{-2} Short MWCNTs served as the short-range electrical conductive network and super-long CNTs acted as both the long-range conductive network and intercrossed binders.
	Glass-coated sulfur with mildly reduced graphene oxide for structural support	University of California, Riverside, US 2015	700 mAh·g^{-1} (50 cycles) Glass coating prevents lithium polysulfides from permanently migrating to an electrode
Lithium	Sulfur	LEITAT 2016	500 W·h/kg Li–S battery for cars with new components and optimized regarding anode, cathode, electrolyte and separator

with the classical Li-ion cells, and which would allow to extend the range of EVs automotive industry to 500 km.[33] There is absolutely no doubt that Li/S batteries are considered as a very realistic candidate for the next-generation energy storage systems. Significant progresses were made during past few years, which are shortly presented in further section. However, there is still a long way to go and many challenges to overcome, since the system suffers from several unresolved drawbacks, which simply induce a large gap between reality and expectations. Typically, Li/S batteries operate unlike conventional classical Li-ion cells. Typically, Li/S cell consist of sulfur as positive electrode, a metallic lithium such as negative electrode, both separated with a porous polymeric separator(s) soaked with an organic electrolyte. A schematic representation of Li–S batteries is shown in Figure 6.3.

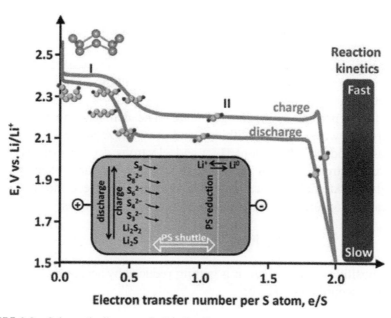

FIGURE 6.3 Schematic diagram of a Li–S cell.

Sulfur is a better active material for batteries electrode. It is desirable to combine a conductive carbon additive with whole electrode for achieving the electronic percolation of the electrodes. Combining the highly polymeric binder with electrodes is employed to provide the bonding all the electrode components and also current collector. Considering the properties, as mentioned in various studies, sulfur is a yellow nonmetallic solid, a cyclic molecule composed of eight atoms, so-called S_8. Sulfur forms more than 30 different allotropes.[35] Comparing to allotropes, orthorhombic alpha-sulfur (α-S_8) have more thermodynamically stability than other one at room temperature (RT) but sulfur has relatively lower melting point around 115°C, and it can get easily converted from solid to vapor (sublimation). Generally, orthorhombic α-sulfur form is used as sulfur electrodes materials. Another allotropic form, monoclinic beta-sulfur (β-S_8) is rather known from being stable at the temperatures higher than 95.5°C, and it is likely formed by slowly cooling down the melted sulfur solution.[36-38] Some recent studies have been demonstrated that the unusual formation of this allotropic sulfur form in the Li/S system is found at the end of charge,[39-41] or as a starting material in the positive electrode (obtained through infiltration of elemental sulfur into CNT structure.[42]

6.5   ELECTROCHEMICAL REACTION IN LI–S BATTERY

According to the electrochemical reaction mechanism of Li–S system, when sulfur reacts with lithium, it tends to form lithium sulfide as a reaction product.

Sulfur is present in its charged state. The cell reaction begins from discharge; however, metallic lithium tends to oxidize at the negative electrode and produce lithium ions (Li^+) and electrons. Li^+ ions diffuse to the positive electrode through the separator/electrolyte, and electrons pass through the external electrical circuit toward positive electrode. The oxidation reaction occurring at the negative electrode is as follows[20]:

$$16LiO \rightarrow 16Li^+ + 16e^- \tag{6.3}$$

On the positive electrode, sulfur is getting reduced by accepting Li^+ ions and electrons, according to the eq:

$$S_8 + 16e^- \rightarrow 8Li_2S \tag{6.4}$$

The overall reaction occurring inside the Li/S cell, at the voltage of ~2.15 V, is as described by eq:

$$16Li + S_8 \rightarrow 8Li_2S \tag{6.5}$$

The element sulfur has been considered as a suitable material for electrode because of sulfur having light weight than transition metals and at least electron presence in each sulfur atom and as abovementioned theoretical specific capacity of sulfur, working mechanism this battery is neither an intercalation reaction nor a direct reaction. The reduction of solid sulfur is based on the formation of a group of lithium polysulfides (PS) species (Li_2S_x, $3 < x \leq 8$), these species are soluble in the organic electrolyte. The characteristics and behavior of polysulfide depend on its chain length.[43] A formed polysulfide have different properties such as "high-order" or "long-chain" PS (Li_2S_8, Li_2S_6), "medium-order" PS (Li_2S_4 easily solubilized), and "lower order" or "short-chain" PS (Li_2S_3, Li_2S_2, and Li_2S, less soluble when present in the form of solid product). During the discharge process, from beginning cyclocta molecules of sulfur (cyclic structure with eight atoms) are started to reduce and it causes to open the rings of cyclocta, and these molecule of sulfur start react with reaction between Li^+, resultant different types of lithium sulfides and linear chain of lithium polysulfide are produced. The length and order of lithium PS is measured by the duration of discharging process, as the discharge proceeds, the length of PS are steadily decreased at end of end discharging process through the reduction process, and polysulfide are mostly formed in lower order or shorten length (Li_2S), which is insoluble and settled in between two electrodes. After lot of studies on reactions mechanism for Li–S battery, it was noticed that the "global" model applicable to all Li/S cells does not exist practically, and reaction mechanism refers to the several parameters such as types of solvents in electrolytes, cathode microstructure, cycling characteristics, etc. In addition to, various functional groups in Li–S batteries offer distinct aspect in enhancing the electrochemical performance of devices, as mentioned in Table 6.2.

6.6 TECHNICAL CHALLENGES

In order to Li–S have great potential to fulfill the future energy storage problem. However, it has some limitation such as the active material can easily diffuse to the electrolyte, some other problem are mentioned in above section. The main problem with Li–S battery is very poor electronic conductivity of the sulfur and was synthesized by a scalable solution-based oxidation process; to rise the conductivity, additional conducting fillers are acquired.[44,45] It has been also found that a higher loading amount of carbon in sulfur matrix[1] may reduce the final volumetric or gravimetric energy densities and also, carbon/binder agglomerates can reduce the collector performance.[46] Sulfur is easily soluble in most of the organic solvents which are used to make the electrolyte, it can explore self-discharge process.[47,48] In

TABLE 6.2 The Performance of Different Functional Groups.

Atom as dopant	Type of materials	Approach	Types of bonds	Doping amount (at%)	Initial capacity/ rate (mAh g^{-1}/C)	Retain capacity/ rate (mAh g^{-1}/ cycles)	References
Nitrogen	Hollow porous carbon bowls doped by Nitrogen	SiO$_2$ template approach	LiS$_n$Li$^+$.......N	5.1	1192/0.2	1020/50	85–86
Nitrogen, sulfur	Sulfur codoped graphene sponge and 3D nitrogen	Hydrothermal reaction and freeze-drying approach	S–S, S–C, and S–O, C–N	S: 3.9 and N: 5.4	1200/0.2	822/100	87–88
Oxygen	Graphene oxide	Chemical reaction deposition method	C=O, O–C=O, – S–C, S–O		1320/0.02	954/more than 50	89–90
Boron	Graphene aerogel based on 3D boron doped	Single-step hydrothermal approach	BCO$_2$, BC$_2$O, BC$_3$	1.7	1290/0.2	994/100	91–92
Boron and nitrogen	Boron codoped carbon layer and graphene-supported nitrogen	Double-step annealing process	N=B/N–B, C–N	B: 2.9 and N: 5.3	829/0.5C	627/300	93
Nitrogen, phosphorous	Nitrogen and phosphorus codoped porous graphene	Thermal annealing and hydrothermal reaction route	C–P, P–O	N: 4.38 P: 1.93	1158.3/1	638/500	94–95

general, significant positive electrode morphology changes are occurred during cycling which tend to limitation of Li–S batteries.

It has been reported that when battery starts to operate, shuttle mechanism occurs by removal of active materials from the carbon/binder matrix and dissolution in the electrolyte in the form PS.[49] During discharge process, Li_2S solid products are formed, which are in insoluble and has insulating in nature; it may be deposited on electrode and forms a passive layer, which covers electrode surface. Once a high-order PS is formed at the positive electrode, a concentration gradient is created in between positive and negative electrode. This concentration gradient causes the diffusion process and polysulfide diffuse toward the Li metal interface through electrolyte, where they can again reduce and form lower order PS. This process is continuously repeated by generation of concentration gradient in between electrodes.[50–57] To consider the another problem, Li batteries offer unwanted dendrites growth, however, some authors have reported that the polysulfide formation can slow down the dendrites growth in the Li electrode, it is due to reactivity of polysulfide with lithium.[58,59] To mention another drawback, Li is very reactive with air, water, and organic species, this reactivity tend to formation of solid electrolyte interphase (SEI) during the addition of electrolyte aging, etc. The main disadvantage of this SEI is that it affects the irreversible capacity loss and lithium consumption of lithium batteries.[58–63]

6.7 CURRENT SCENARIO OF LI–S BATTERY COMPONENTS

6.7.1 CATHODE MATERIALS FOR LI–S BATTERY

In order to reduce the insulating property of active material and the polysulfide shuttle effect in Li–S batteries, a variety of methods have been carried out such as using a proper electrolyte or additives and development of excellent conducting active materials based on polymer composites and nanocomposites to improve the electron transport between the electrical conductor and the active material. To improve the conductivity of active materials, various types of conducting fillers are being utilized, including CNTs, graphene, carbon black, carbon nanofibers, etc., to make sulfur-based conducting polymer composites and nanocomposites, and these materials are utilized as cathode materials in Li–S batteries. In particular, these carbon-based micro and nanomaterials have high surface area, porosity, high conductivity, large pore volume, and electrical conductivity, which can absorb the active material and help to reduce the polysulfide dissolution. It has been reported that sulfur–porous carbon (PC) composites were synthesized for improving the cathode conductivity and also for enhancing the storage of active material in porosity of PC.[64,65] The PC is effective for enhancing the cyclability compared to the absence of PC in sulfur cathode, although sulfur–mesoporous carbon composite could improve the discharge capacity and stable cyclability by utilizing the efficient sulfur.[66] In addition to recent reports, ordered mesoporous space showed confinement of active material and suppresses the free migration of polysulfide species within the porous structure of composites, which improved relatively good cycling stability as the cathode. Several excellent cycling performances have been reported on the reasonable sulfur content to carbon ratio in cathode; however, these studies also suggest the drawback of the S–C composite cathode.[67,68] In this system, the morphology of sulfur may be altered by the surrounding conductive carbon fillers. A new study is proposed to prepare polymers as a conductive matrix to block the polysulfide (PS). Unique sulfur and metal oxide core–shell composites is synthesized as the TiO_2–sulfur yolk–shell composite.[83] This system existed long cyclability around 1000 cycles. Yolk–shell structure stopped the breakage of TiO_2 spheres framework, and providing a support to TiO_2 spheres in the case of active materials volume expansion, and these outer shell and inner core are highly porous in nature, these porous structures are providing a barrier for dissolution of lithium PS.[69] An overview of various nanostructured materials for sulfur-cathode in Li–S batteries is mentioned in Table 6.3.

TABLE 6.3 Overview of Different Nanostructured Materials for Sulfur-based Cathodes in Li–S Batteries.[95]

Nanostructure	Remarks
Sulfur infused into porous carbon matrices	Carbonaceous materials with high porosity can act as conductive agents and as cathode hosts/matrices to trap the active polysulfide species in the cathode during cell operation and hence improved life time. Enhanced cell performance, including the plateau voltage, active material utilization, rate capability, discharge capacity, and capacity retention.
	Example: microporous, mesoporous and macroporous carbon–sulfur composites.
Sulfur infused into anisotropic carbon nanostructures	Excellent electrochemical performance of the battery is attributed to their high-quality fiber morphology, controlled porous structure, large surface area, and good electrical conductivity, enhanced and long stability of the sulfur cathode.
	Example: mutiwalled carbon nanotubes (MWCNTs)

TABLE 6.3 *(Continued)*

Nanostructure	Remarks
Sulfur infused into hollow carbon nanostructures	Prevent polysulfide dissolution, the double-shell structure facilitates ion and electron transport as well.
	Example: core/shell carbon framework
Interlayer and interfacial coatings	Improves cycle life and performance of Li–S cells. It can suppress polysulfide dissolution resulting in an initial discharge capacity of 1367 mAh g^{-1} of sulfur at a rate of 1 C.
	Example: a bifunctional microporous carbon paper.

6.7.2 BINDERS FOR LI–S BATTERY

Polymeric materials such as poly-(tetrafluoroethylene) (PTFE) or PVdF, polypyrrole, polyaniline, poly(3,4-(ethylenedioxy)thiophene) (PEDOT), poly(ethylene oxide) (PEO)-, or PEG-modified cells have attracted tremendous attention as binder in both conventional Li ion and other batteries applications. These materials are of high interest to hold the active material and carbon additives together without sacrificing any capacity; these materials play a crucial role to influence the electrochemical properties and cell performance in Li–S cell.

In brief, during the structure and morphology changes occurred, conventional binders are not capable to hold all the active materials and PS. Therefore, advance binders are current requirements to preventing the formation of "dead" sites during the electrochemical reactions between lithium and sulfur. As a results, a variety of sulfur–polymer hybrid materials have been developed with the target of reducing the dissolution and shuttle effect of PS and facilitate the electronic conductivity, and controlling the microstructure of sulfur–polymer composite electrodes. PEO as a binder for sulfur cathodes has facilitated the longer cycle life, electrochemical reversibility, and suppressed passivation of the cathode at the end of discharge process.[70–72] In some of the study, the binder was used which is based on the natural polymer such as gelatin. It should be noted that the gelatin is able to improve the adhesion and redox reversibility of sulfur cathodes. Moreover, freeze-drying-treated sulfur–gelatin cathodes can show better performance than PEO because freeze-drying facilitate the porosity in electrode.[72–75]

6.7.3 ELECTROLYTES AND SEPARATORS FOR LI–S BATTERY

The electrolytes commonly used as ion-transport medium between the anode and cathode. Liquid electrolytes are commonly used in batteries, which have high ionic conductivity. Moreover, owing to the importance of electrolytes in Li–S batteries, Li–S batteries electrolytes are medium, which are helpful for dissolution of PS and shuttle between the both electrodes. The solubility parameter of PS in a liquid electrolyte affect the rate/cycling performance.

A solid-state electrolyte work as better candidates for electrolytes for Li–S battery, it is an ideal candidate for the reducing the dissolution and shuttle of PS. However, low ionic conductivity and interfacial instability is major problem in the case of solid electrolytes-based Li–S battery system. Porous polymers membranes are the better candidate as a separator for Li battery system but in the case of Li–S battery, ion selective functionalized separators are the type of porous membrane, and that has been widely used in Li–S battery due to its merits of reducing the shuttle of PS.

6.7.4 ANODES FOR LI–S BATTERY

Recent research has showed that the long-term cycle stability of Li–S batteries based on the stability of the anode. This is because the metallic lithium is unstable, when it comes in contact with organic electrolytes, which create hazardous problems. Up to date, many strategies have been used to control the dendrite formation and low-lithium cycling efficiency, dendrites formation causes the cycling stability, short circuit, explosion, etc.[76–79.]Moreover, dendrite formation are accompanied by the instability of the passivation layer (SEI layer) on the metallic lithium anode during cycling. In contrast, dendrites formation is a dissolution and nonuniform lithium deposition on lithium electrode surface.[76] In addition to Li–S batteries participate the dissolution of intermediate PS in an organic electrolytes, these process tends to more reactive on the lithium metal surface and formation of stable passivation layer metallic lithium anode. This mechanism reduces the decreases of the practical energy density of Li–S batteries. Recently, studies suggested that the carbon-fiber paper current collector with SiO_2,[78] graphene sheets-based 3D current collectors usually act as the barrier for dendrites formation by decreasing the effective current density.[79,80] Some of studies suggested that carbon-based anodes in Li–S batteries can solve this issue because of good stability of carbon anode in ether-based electrolytes.[81,82]

6.8   CONCLUSION

Lithium/sulfur batteries is new-generation efficient rechargeable battery systems, based on their high cell energy density in comparison to state of the art lithium-ion battery that could make most promising long-range candidates for the new-generation energy storage, with high energy density and low-cost application. However, the Li–S battery technology has been facing several challenges such as the inherent high electronic resistivity of sulfur and the higher order PS shuttle mechanism in organic electrolytes in between the anode and cathode leads to active material loss due to solubility of lithium PS formed during the charging process in which long-chain lithium PS produced at the cathode, dissolve into the electrolyte, migrate and react with the lithium anode electrode, by that way generate lower order PS in a parasitic manner, which leads to rapid reduction the capacity degradation and short cycle life performance.

KEYWORDS

- Li–S
- sulfur
- lithium
- polysulfides
- battery

REFERENCES

1. Page, K. A.; Soles, C. L.; Runt, J., Eds. *Polymers for Energy Storage and Delivery: Polyelectrolytes for Batteries and Fuel Cells*; American Chemical Society, 2012.
2. Armand, M.; Tarascon, J.-M. Building Better Batteries. *Nature* **2008**, *451*(7179), 652–657.
3. Whittingham, M. S. Preparation of Stoichiometric Titanium Disulfide. U.S. Patent 4,007,055, February 8, 1977.
4. Whittingham, M. S. Chemistry of Intercalation Compounds: Metal Guests in Chalcogenaide Hosts. *Prog. Solid State Chem.* **1978**, *12*, 41–99.
5. Whittingham, M. S. Intercalation Chemistry and Energy Storage. *J. Solid State Chem.* **1979**, *29*(3), 303–310.

6. Xu, K. Nonaqueous Liquid Electrolytes for Lithium-based Rechargeable Batteries. *Chem. Rev.* **2004,** *104*(10), 4303–4418.

7. Orsini, F.; Du Pasquier, A.; Beaudouin, B.; Tarascon, J. M.; Trentin, M.; Langenhuizen, N.; De Beer, E.; Notten, P. In Situ SEM Study of the Interfaces in Plastic Lithium Cells. *J. Power Sources* **1999,** *81*, 918–921.

8. Hassoun, J.; Scrosati, B. Moving to a Solid State Configuration: A Valid Approach to Making Lithium–Sulfur Batteries Viable for Practical Applications. *Adv. Mater.* **2010,** *22*(45), 5198–5201.

9. Goriparti, S.; Miele, E.; De Angelis, F.; Di Fabrizio, E.; Zaccaria, R. P.; Capiglia, C. Review on Recent Progress of Nanostructured Anode Materials for Li-ion Batteries. *J. Power Sources* **2014,** *257*, 421–443.

10. Landi, B. J.; Ganter, M. J.; Cress, C. D.; DiLeo, R. A.; Raffaelle, R. P. Carbon Nanotubes for Lithium Ion Batteries. *Energy Environ. Sci.* **2009,** *2*(6), 638–654.

11. Kim, C.; Yang, K. S.; Kojima, M.; Yoshida, K.; Kim, Y. J.; Ahm Kim, Y.; Endo, M. Fabrication of Electrospinning Derived Carbon Nanofiber Webs for the Anode Material of Lithium Ion Secondary Batteries. *Adv. Funct. Mater.* **2006,** *16*(18), 2393–2397.

12. Hou, J.; Shao, Y.; Ellis, M. W.; Moore, R. B.; Yi, B. Graphene-based Electrochemical Energy Conversion and Storage: Fuel Cells, Supercapacitors and Lithium Ion Batteries. *Phys. Chem. Chem. Phys.* **2011,** *13*(34), 15384–15402.

13. Zhou, H.; Zhu, S.; Hibino, M.; Honma, I.; Ichihara, M. Lithium Storage in Ordered Mesoporous Carbon (CMK-3) with High Reversible Specific Energy Capacity and Good Cycling Performance. *Adv. Mater.* **2003,** *15*(24), 2107–2111.

14. Manthiram, A. Materials Challenges and Opportunities of Lithium Ion Batteries. *J. Phys. Chem. Lett.* **2011,** *2*(3), 176–184.

15. Hassoun, J.; Agostini, M.; Latini, A.; Panero, S.; Sun, Y.-K.; Scrosati, B. Nickel-layer Protected, Carbon-coated Sulfur Electrode for Lithium Battery. *J. Electrochem. Soc.* **2012,** *159*(4), A390–A395.

16. Bruce, P. G.; Freunberger, S. A.; Hardwick, L. J.; Tarascon, J.-M. Li–O2 and Li–S Batteries with High Energy Storage. *Nat. Mater.* **2012,** *11*(1), 19.

17. Ji, X.; Nazar, L. F. Advances in Li–S Batteries. *J. Mater. Chem.* **2010,** *20*(44), 9821–9826.

18. Mikhaylik, Y. V.; Akridge, J. R. Polysulfide Shuttle Study in the Li/S Battery System. *J. Electrochem. Soc.* **2004,** *151*(11), A1969–A1976.

19. Marmorstein, D.; Yu, T. H.; Striebel, K. A.; McLarnon, F. R.; Hou, J.; Cairns, E. J. Electrochemical Performance of Lithium/Sulfur Cells with Three Different Polymer Electrolytes. *J. Power Sources* **2000,** *89*(2), 219–226.

20. Akridge, J. R.; Mikhaylik, Y. V.; White, N. Li/S Fundamental Chemistryand Application to High-performance Rechargeable Batteries. *Solid State Ion.* **2004,** *175*(1), 243–245.

21. Meyer, W. H. Polymer Electrolytes for Lithium-ion Batteries. *Adv. Mater.* **1998,** *10*(6), 439–448.

22. Balach, J.; Jaumann, T.; Klose, M.; Oswald, S.; Eckert, J.; Giebeler, L. Improved Cycling Stability of Lithium–Sulfur Batteries Using a Polypropylene-supported Nitrogen-doped Mesoporous Carbon Hybrid Separator as Polysulfide Adsorbent. *J. Power Sources* **2016,** *303*, 317–324.

23. Narayanan, S. R.; Surampudi, S.; Attia, A. I.; Bankston, C. P. Analysis of Redox Additive-Based Overcharge Protection for Rechargeable Lithium Batteries. *J. Electrochem. Soc.* **1991,** *138*(8), 2224–2229.

24. Golovin, M. N.; Wilkinson, D. P.; Dudley, J. T.; Holonko, D.; Woo, S. Applications of Metallocenes in Rechargeable Lithium Batteries for Overcharge Protection. *J. Electrochem. Soc.* **1992,** *139*(1), 5–10.

25. Richardson, T. J., Ross, P. N. Overcharge Protection for Rechargeable Lithium Polymer Electrolyte Batteries. *J. Electrochem. Soc.* **1996,** *143*(12), 3992–3996.

26. Evers, S.; Nazar, L. F. New Approaches for High Energy Density Lithium–Sulfur Battery Cathodes. *Acc. Chem. Res.* **2012,** *46*(5), 1135–1143.

27. He, G.; Evers, S.; Liang, X.; Cuisinier, M.; Garsuch, A.; Nazar, L. F. Tailoring Porosity in Carbon Nanospheres for Lithium–Sulfur Battery Cathodes. *ACS Nano* **2013,** *7*(12), 10920–10930.

28. Tao, X.; Wang, J.; Ying, Z.; Cai, Q.; Zheng, G.; Gan, Y.; Huang, H., et al. Strong Sulfur Binding with Conducting Magnéli-Phase Ti n O2 n − 1 Nanomaterials for Improving Lithium–Sulfur Batteries. *Nano Lett.* **2014,** *14*(9), 5288–5294.

29. Herbert, D. U. J. Electric Dry Cells and Storage Batteries. U.S. Patent 3,043,896, 1962.

30. Rauh, R. D.; Abraham, K. M.; Pearson, G. F.; Surprenant, J. K.; Brummer, S. B. A Lithium/Dissolved Sulfur Battery with an Organic Electrolyte. *J. Electrochem. Soc.* **1979,** *126*(4) 523–527.

31. Bruce, P. G. Energy Storage Beyond the Horizon: Rechargeable Lithium Batteries. *Solid State Ion.* **2008,** *179*(21), 752–760.

32. Choi, N. S.; Chen, Z.; Freunberger, S. A.; Ji, X.; Sun, Y. K.; Amine, K.; Yushin, G.; Nazar, L. F.; Cho, J.; Bruce, P. G. Challenges Facing Lithium Batteries and Electrical Double Layer Capacitors. *Angew. Chem. Int. Ed.* **2012,** *51*(40), 9994–10024.

33. Manthiram, A.; Fu, Y.; Su, Y.-S. Challenges and Prospects of Lithium–Sulfur Batteries. *Acc. Chem. Res.* **2012,** *46*(5), 1125–1134.

34. Bruce, P. G.; Freunberger, S. A.; Hardwick, L. J.; Tarascon, J.-M. Li–O2 and Li–S Batteries with High Energy Storage. *Nat. Mater.* **2012,** *11*(1), 19.

35. Wang, D.-W.; Zeng, Q.; Zhou, G.; Yin, L.; Li, F.; Cheng, H.-M.; Gentle, I. R.; Lu, G. Q. M. Carbon–Sulfur Composites for Li–S Batteries: Status and Prospects. *J. Mater. Chem. A* **2013,** *1*(33), 9382–9394.

36. Greenwood, N. N.; Earnshaw, A. *Chemistry of the Elements;* Elsevier, 2012.

37. Templeton, L. K.; Templeton, D. H.; Zalkin, A. Crystal Structure of Monoclinic Sulfur. *Inorg. Chem.* **1976,** *15*(8), 1999–2001.

38. Pastorino, C.; Gamba, Z. Toward an Anisotropic Atom–Atom Model for the Crystalline Phases of the Molecular S 8 Compound. *J. Chem. Phys.* **2001,** *115*(20), 9421–9426.

39. Pastorino, C.; Gamba, Z. Test of a Simple and Flexible Molecule Model for α-, β-, and γ-S 8 Crystals. *J. Chem. Phys.* **2000,** *112*(1), 282–286.

40. Waluś, S.; Barchasz, C.; Colin, J.-F.; Martin, J.-F.; Elkaïm, E.; Leprêtre, J.-C.; Alloin, F. New Insight into the Working Mechanism of Lithium–Sulfur Batteries: In Situ and Operando X-ray Diffraction Characterization. *Chem. Commun.* **2013,** *49*(72), 7899–7901.

41. Kulisch, J.; Sommer, H.; Brezesinski, T.; Janek, J. Simple Cathode Design for Li–S Batteries: Cell Performance and Mechanistic Insights by in Operando X-ray Diffraction. *Phys. Chem. Chem. Phys.* **2014,** *16*(35), 18765–18771.

42. Villevieille, C.; Novák, P. A Metastable β-sulfur Phase Stabilized at Room Temperature During Cycling of High Efficiency Carbon Fibre–Sulfur Composites for Li–S Batteries. *J. Mater. Chem. A* **2013,** *1*(42), 13089–13092.

43. Moon, S.; Jung, Y. H.; Jung, W. K.; Jung, D. S.; Choi, J. W.; Kim, D. K. Encapsulated Monoclinic Sulfur for Stable Cycling of Li–S Rechargeable Batteries. *Adv. Mater.* **2013,** *25*(45), 6547–6553.

44. Patel, M. U. M; Demir-Cakan, R.; Morcrette, M.; Tarascon, J. M.; Gaberscek, M.; Dominko, R. Li–S Battery Analyzed by UV/V is in Operando Mode. *ChemSusChem* **2013,** *6*(7), 1177–1181.

45. Gao, J.; Lowe, M. A.; Kiya, Y.; Abruña, H. D. Effects of Liquid Electrolytes on the Charge–Discharge Performance of Rechargeable Lithium/Sulfur Batteries: Electrochemical and In-situ X-ray Absorption Spectroscopic Studies. *J. Phys. Chem. C* **2011,** *115*(50), 25132–25137.

46. Li, Z.; Yuan, L.; Yi, Z.; Sun, Y.; Liu, Y.; Jiang, Y.; Shen, Y.; Xin, Y.; Zhang, Z.; Huang, Y. Insight into the Electrode Mechanism in Lithium–Sulfur Batteries with Ordered Microporous Carbon Confined Sulfur as the Cathode. *Adv. Energy Mater.* **2014,** *4*(7).

47. Busche, M. R.; Adelhelm, P.; Sommer, H.; Schneider, H.; Leitner, K.; Janek, J. Systematical Electrochemical Study on the Parasitic Shuttle-effect in Lithium–Sulfur cells at Different Temperatures and Different Rates. *J. Power Sources* **2014,** *259*, 289–299.

48. Ryu, H. S.; Ahn, H. J.; Kim, K. W.; Ahn, J. H.; Cho, K. K.; Nam, T. H. Self-discharge Characteristics of Lithium/Sulfur Batteries Using TEGDME Liquid Electrolyte. *Electrochim. Acta* **2006,** *52*(4), 1563–1566.

49. Ryu, H. S.; Ahn, H. J.; Kim, K. W.; Ahn, J. H.; Lee, J. Y.; Cairns, E. J. Self-discharge of Lithium–Sulfur Cells Using Stainless-steel Current-collectors. *J. Power Sources* **2005,** *140*(2), 365–369.

50. Barchasz, C.; Leprêtre, J.-C.; Alloin, F.; Patoux, S. New Insights into the Limiting Parameters of the Li/S Rechargeable Cell. *J. Power Sources* **2012,** *199*, 322–330.

51. Cheon, S.-E.; Choi, S.-S.; Han, J.-S.; Choi, Y.-S.; Jung, B.-H.; Lim, H. S. Capacity Fading Mechanisms on Cycling a High-capacity Secondary Sulfur Cathode. *J. Electrochem. Soc.* **2004,** *151*(12), A2067–A2073.

52. Cheon, S.-E.; Ko, K.-S.; Cho, J.-H.; Kim, S.-W.; Chin, E.-Y.; Kim, H.-T. Rechargeable Lithium Sulfur Battery I. Structural Change of Sulfur Cathode During Discharge and Charge. *J. Electrochem. Soc.* **2003,** *150*(6), A796–A799.

53. Zhang, S. S. Liquid Electrolyte Lithium/Sulfur Battery: Fundamental Chemistry, Problems, and Solutions. *J. Power Sources* **2013,** *231*, 153–162.

54. He, X.; Ren, J.; Wang, L.; Pu, W.; Jiang, C.; Wan, C. Expansion and Shrinkage of the Sulfur Composite Electrode in Rechargeable Lithium Batteries. *J. Power Sources* **2009,** *190*(1), 154–156.

55. Yin, Y.-X.; Xin, S.; Guo, Y. G.; Wan, L. J. Lithium–Sulfur Batteries: Electrochemistry, Materials, and Prospects. *Angew. Chem. Int. Ed.* **2013,** *52*(50), 13186–13200.

56. Kolosnitsyn, V. S.; Karaseva, E. V. Lithium–Sulfur Batteries: Problems and Solutions. *Russ. J. Electrochem.* **2008,** *44*(5), 506–509.

57. Diao, Y.; Xie, K.; Xiong, S.; Hong, X. Shuttle Phenomenon—The Irreversible Oxidation Mechanism of Sulfur Active Material in Li–S Battery. *J. Power Sources* **2013,** *235*, 181–186.

58. Aurbach, D.; Pollak, E.; Elazari, R.; Salitra, G.; Kelley, C. S.; Affinito, J. On the Surface Chemical Aspects of Very High Energy Density, Rechargeable Li–Sulfur Batteries. *J. Electrochem. Soc.* **2009,** *156*(8), A694–A702.

59. Mikhaylik, Y. V.; Kovalev, I.; Schock, R.; Kumaresan, K.; Xu, J.; Affinito, J. High Energy Rechargeable Li–S Cells for EV Application: Status, Remaining Problems and Solutions. *Ecs Trans.* **2010,** *25*(35), 23–34.

60. Hofmann, A. F.; Fronczek, D. N.; Bessler, W. G. Mechanistic Modeling of Polysulfide Shuttle and Capacity Loss in Lithium–Sulfur Batteries. *J. Power Sources* **2014,** *259*, 300–310.

61. Bauer, I.; Thieme, S.; Brückner, J.; Althues, H.; Kaskel, S. Reduced Polysulfide Shuttle in Lithium–Sulfur Batteries Using Nafion-based Separators. *J. Power Sources* **2014**, *251*, 417–422.

62. Barchasz, C. Développement d'accumulateurs lithium/soufre. Ph.D. Diss., Thèse de doctorat de l'Université de Grenoble, 2011.

63. Li, Z.; Huang, J.; Liaw, B. Y.; Metzler, V.; Zhang, J. A Review of Lithium Deposition in Lithium-ion and Lithium Metal Secondary Batteries. *J. Power Sources* **2014**, *254*, 168–182.

64. Wang, J. L.; Yang, J.; Xie, J. Y.; Xu, N. X.; Li, Y. Sulfur–Carbon Nano-composite as Cathode for Rechargeable Lithium Battery Based on Gel Electrolyte. *Electrochem. Commun.* **2002**, *4*(6), 499–502.

65. Wang, Y.-X.; Huang, L.; Sun, L.-C.; Xie, S.-Y.; Xu, G.-L.; Chen, S.-R.; Xu, Y.-F., et al. Facile Synthesis of a Interleaved Expanded Graphite-embedded Sulphur Nanocomposite as Cathode of Li–S Batteries with Excellent Lithium Storage Performance. *J. Mater. Chem.* **2012**, *22*(11), 4744–4750.

66. Ji, X.; Lee, K. T.; Nazar, L. F. A Highly Ordered Nanostructured Carbon–Sulphur Cathode for Lithium–Sulphur Batteries. *Nat. Mater.* **2009**, *8*(6), 500.

67. Zhang, S. S.; Tran, D. T. A Proof-of-concept Lithium/Sulfur Liquid Battery with Exceptionally High Capacity Density. *J. Power Sources* **2012**, *211*, 169–172.

68. Zhang, S. S. Liquid Electrolyte Lithium/Sulfur Battery: Fundamental Chemistry, Problems, and Solutions. *J. Power Sources* **2013**, *231*, 153–162.

69. Yang, Y.; Zheng, G.; Misra, S.; Nelson, J.; Toney, M. F.; Cui, Y. High-capacity Micrometer-sized Li_2S Particles as Cathode Materials for Advanced Rechargeable Lithium-ion Batteries. *J. Am. Chem. Soc.* **2012**, *134*(37), 15387–15394.

70. Zhang, S. S. Binder Based on Polyelectrolyte for High Capacity Density Lithium/Sulfur Battery. *J. Electrochem. Soc.* **2012**, *159*(8), A1226–A1229.

71. Lacey, M. J.; Jeschull, F.; Edström, K.; Brandell, D. Why PEO as a Binder or Polymer Coating Increases Capacity in the Li–S System. *Chem. Commun.* **2013**, *49*(76), 8531-8533.

72. Sun, J.; Huang, Y.; Wang, W.; Yu, Z.; Wang, A.; Yuan, K. Application of Gelatin as a Binder for the Sulfur Cathode in Lithium–Sulfur Batteries. *Electrochim. Acta* **2008**, *53*(24), 7084–7088.

73. Sun, J.; Huang, Y.; Wang, W.; Yu, Z.; Wang, A.; Yuan, K. Preparation and Electrochemical Characterization of the Porous Sulfur Cathode Using a Gelatin Binder. *Electrochem. Commun.* **2008**, *10*(6), 930–933.

74. Huang, Y.; Sun, J.; Wang, W.; Wang, Y.; Yu, Z.; Zhang, H.; Wang, A.; Yuan, K. Discharge Process of the Sulfur Cathode with a Gelatin Binder. *J. Electrochem. Soc.* **2008**, *155*(10), A764–A767.

75. Wang, Q.; Wang, W.; Huang, Y.; Wang, F.; Zhang, H.; Yu, Z.; Wang, A.; Yuan, K. Improve Rate Capability of the Sulfur Cathode Using a Gelatin Binder. *J. Electrochem. Soc.* **2011**, *158*(6), A775–A779.

76. Wang, Q.; Wang, W.; Huang, Y.; Wang, F.; Zhang, H.; Yu, Z.; Wang, A.; Yuan, K. Improve Rate Capability of the Sulfur Cathode Using a Gelatin Binder. *J. Electrochem. Soc.* **2011**, *158*(6), A775–A779.

77. Peled, E. The Electrochemical Behavior of Alkali and Alkaline Earth Metals in Nonaqueous Battery Systems—The Solid Electrolyte Interphase Model. *J. Electrochem. Soc.* **1979**, *126*(12), 2047–2051.

78. Ji, X.; Liu, D.-Y.; Prendiville, D. G.; Zhang, Y.; Liu, X.; Stucky, G. D. Spatially Hetero-geneous Carbon-fiber Papers as Surface Dendrite-free Current Collectors for Lithium Deposition. *Nano Today* **2012,** *7*(1), 10–20.

79. Zhamu, A.; Chen, G.; Liu, C.; Neff, D.; Fang, Q.; Yu, Z.; Xiong, W.; Wang, Y.; Wang, X.; Jang, B. Z. Reviving Rechargeable Lithium Metal Batteries: Enabling Next-gener-ation High-energy and High-power Cells. *Energy Environ. Sci.* **2012,** *5*(2), 5701–5707.

80. Li, H.; Huang, X.; Chen, L.; Wu, Z.; Liang, Y. A High Capacity Nano Si Composite Anode Material for Lithium Rechargeable Batteries. *Electrochem. Solid-State Lett.* **1999,** *2*(11), 547–549.

81. Brückner, J.; Thieme, S.; Böttger-Hiller, F.; Bauer, I.; Grossmann, H. T.; Strubel, P.; Althues, H.; Spange, S.; Kaskel, S. Carbon-based Anodes for Lithium Sulfur Full Cells with High Cycle Stability. *Adv. Funct. Mater.* **2014,** *24*(9), 1284–1289.

82. Aurbach, D.; Markovsky, B.; Weissman, I.; Levi, E.; Ein-Eli, Y. On the Correlation Between Surface Chemistry and Performance of Graphite Negative Electrodes for Li Ion Batteries. *Electrochim. Acta* **1999,** *45*(1), 67–86.

Wei Seh, Z.; Li, W.; Cha, J. J.; Zheng, G.; Yang, Y.; McDowell, M. T.; Hsu, P.-C.; Cui, Y. Sulphur-TiO 2 Yolk-shell Nanoarchitecture with Internal Void Space for Long-cycle Lithium-Sulphur Batteries. *Nat. Commun.* **2013,** *4*(1).

83. Lithium–Sulfur Battery, Retrieved on 11-10-2017 from: https://en.wikipedia.org/wiki/Lithium%E2%80%93sulfur_battery.

84. Li, Z.; Wu, H. B.; Lou, X. W. D. Rational Designs and Engineering of Hollow Micro-/Nanostructures as Sulfur Hosts for Advanced Lithium–Sulfur Batteries. Energy Environ. Sci. **2016,** *9*(10), 3061–3070.

85. Sabu, T.; Thomas, R.; Zachariah, A.; Raghvendra, K. M. *Microscopy Methods in Nanomaterials Characterization,* 1st ed.; Sabu, M., Thomas, R. T., Zachariah, A., Raghvendra, K., Eds.; Elsevier: Amsterdam, 2017. DOI: https://doi.org/10.1016/B978-0-323-46141-2.01002-6.

86. Zhou, G.; Paek, E.; Hwang, G. S.; Manthiram, A. Long-life Li/Polysulphide Batteries with High Sulphur Loading Enabled by Lightweight Three-dimensional Nitrogen/Sulphur-codoped Graphene Sponge. *Nat. Commun.* **2015,** *6*, 7760.

87. Sabu, T.; Thomas, R.; Zachariah, A.; Raghvendra, M. *Spectroscopic Methods for Nanomaterials Characterization,* 1st ed.; Sabu, M., Thomas, R. T., Zachariah, A., Raghvendra, K., Eds.; Elsevier: Amsterdam, 2017. DOI: https://doi.org/10.1016/B978-0-323-46140-5.01002-5.

88. Zhang, Y.; Cairns, E. J.; Ji, L.; Rao, M. Graphene Oxide as a Sulfur Immobilizer in High Performance Lithium/Sulfur Cells. U.S. Patent 9,673,452, Issued June 6, 2017.

89. Cintil, J. C.; Abraham, J.; Mishra, R. K.; George, S. C.; Thomas, S. Instrumental Tech-niques for the Characterization of Nanoparticles. In *Thermal and Rheological Measure-ment Techniques for Nanomaterials Characterization,* 1st ed.; Sabu, T., Thomas, R., A. K. Z., R. K. M., Eds.; Elsevier: Amsterdam, 2017; pp 1–36. DOI: https://doi.org/10.1016/B978-0-323-46139-9.00001-3.

90. Xie, Y.; Meng, Z.; Cai, T.; Han, W.-Q. Effect of Boron-doping on the Graphene Aerogel Used as Cathode for the Lithium–Sulfur Battery. *ACS Appl. Mater. Interfaces* **2015,** *7*(45), 25202–25210.

91. Sabu, T.; Thomas, R.; Zachariah, A.; Raghvendra, M. *Thermal and Rheological Measurement Techniques for Nanomaterials Characterization,* 1st ed.; Sabu, M., Thomas, R. T., Zachariah, A., Raghvendra, K., Eds.; Elsevier: Amsterdam, 2017. DOI: https://doi.org/10.1016/B978-0-323-46139-9.01002-1.

92. Yuan, S.; Bao, J. L.; Wang, L.; Xia, Y.; Truhlar, D. G.; Wang, Y. Graphene-supported Nitrogen and Boron Rich Carbon Layer for Improved Performance of Lithium–Sulfur Batteries Due to Enhanced Chemisorption of Lithium Polysulfides. *Adv. Energy Mater.* **2016,** *6*(5).

93. Gu, X.; Tong, C.-J.; Lai, C.; Qiu, J.; Huang, X.; Yang, W.; Wen, B.; Liu, L.-M.; Hou, Y.; Zhang, S. A Porous Nitrogen and Phosphorous Dual Doped Graphene Blocking Layer for High Performance Li–S Batteries. *J. Mater. Chem. A* **2015,** *3*(32), 16670–16678.

94. Fan, X.; Sun, W.; Meng, F.; Xing, A.; Liu, J. Advanced Chemical Strategies for Lithium-Sulfur Batteries: A Review. *Green Energy Environ.* **2017.**

95. Ma, L.; Hendrickson, K. E.; Wei, S.; Archer, L. A. Nanomaterials: Science and Applications in the Lithium–Sulfur Battery. *Nano Today* **2015,** *10*(3), 315–338.

PART II
Innovations in Life Chemistry

CHAPTER 7

DENSITY CLUSTERING FOR AUTOMATIC LESION DETECTION IN DIABETIC RETINOPATHY FROM EYE BACKGROUND IMAGES

CAROLINA TRAVASSOS[1], RAQUEL MONTEIRO[1],
SARA PIRES DE OLIVEIRA[1], CARLA BAPTISTA[2],
FRANCISCO CARRILHO[2], and PEDRO FURTADO[1,*]

[1]Departamento de Engenharia Informática, Faculdade de Ciências e Tecnologia, Universidade de Coimbra, Pólo II Pinhal de Marrocos, 3030-290 Coimbra, Portugal

[2]Serviço de Endocrinologia, Diabetes E Metabolismo do Centro Hospitalar e Universitário de Coimbra, Praceta Prof. Mota Pinto, 3000-075 Coimbra, Portugal

*Corresponding author. E-mail: pnf@dei.uc.pt

CONTENTS

ABSTRACT

Since diabetic retinopathy (DR) is an ocular disease with a fast progression to blindness, prevention and early diagnosis are of utmost importance. Fast action in the early stages of the disease can be decisive in preserving the patient's vision. This disease is characterized by the appearance of a heterogeneous set of lesions, among which microaneurysms and hemorrhages are some of the most important ones. Detection and classification of these lesions is a part of DR classification. In this chapter, we present the approach and results of our experiments with automated detection of such lesions with a density clustering approach, by integrating preprocessing, superpixelation, and density clustering, and by targeting classification of individual lesions. The approach is based on preprocessing, simple linear iterative clustering (SLIC) and density-based spatial clustering (DBSCAN), filtering, and classification of the retina background images. After detailing the approach, we analyze it experimentally, in particular, we study its achievements in terms of accuracy, sensitivity, and specificity when different classifiers are employed. Finally, we reach important conclusions and delineate future work on the issue.

7.1 INTRODUCTION

Diabetic retinopathy (DR) is an affection related to microvascular changes in the retina, a result of diabetes and the leading cause of blindness in working age—between 20 and 60 years old. In Portugal, the number of people engaged in screening programs of this disease as well as those already identified for treatment have been increasing since 2009.[1,2]

DR is divided into two types: nonproliferative and proliferative. Nonproliferative DR (NP-DR) is characterized by the rupture of small capillaries, but without the formation of new blood vessels. The first manifestation of NP-DR is a set of microaneurysms, then hemorrhages and, finally, exudates. Proliferative DR (P-DR) is characterized by abnormal proliferative growth of new blood vessels (neovascularization) and is the most severe stage of the disease.[3,4]

Microaneurysms consist of dilated capillaries in the retina, similar to small red dots (red lesions). They are present from the earliest stages of the disease, allowing early diagnosis; the amount of these lesions is an indicator of disease progression (Fig. 7.1b).[4,5]

The exudates consist of proteins and lipids, which pass through the vessel walls and are deposited on the retina. These can be divided into two types: soft and hard exudates.[4] They are characterized by a yellowish color, rounded shape, and variable size/brightness, depending on the stage of the disease (Fig. 7.1b). The bleeding results from breaking of fragile capillary walls and is a characteristic of later stages of the disease (Fig. 7.1b):[4] the higher the number of hemorrhages in the retina, the greater propensity to damage.[5]

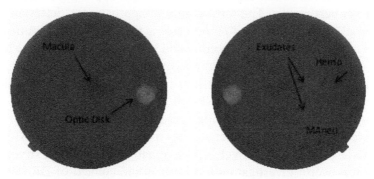

FIGURE 7.1   Schematic representation of constituents of the retina (a) some structures and (b) three main lesions that can be detected by automatic means (mas—microaneurysms; exs—exudates; and hms—hemorrhages).

Since DR is initially a silent disease, early diagnosis is critical for the prevention of blindness. Analysis of the fundus images is one of the simplest forms of diagnosis, since it is a noninvasive technique, easy to perform, and, above all, allows the visualization of all types of lesions.[1,3] These images are used to extract features (area of lesions, perimeter, compactness, and color, among others) and thus to detect lesions and determine its location, size, and degree of severity.[6]

Automatic detection through computer analysis can be advantageous, since it reduces costs associated with the evaluation by experts and analysis time. However, automatic detection of lesions in the retina, and in particular of hemorrhages, poses a series of obstacles. The most relevant challenges include:

- Segmenting small lesions in low-contrast regions,
- Similarity between lesions and some blood vessels segments,
- Proximity of some lesions from blood vessels,

- Very small dimension of some lesions, which are easily confounded with noise,
- Disparity in form, size, and color of lesions forma (both inter- and intraimages).

It is worth to investigate all sophisticated state-of-the-art approaches to try to handle those difficulties and return accurate classification. There is probably not one single approach or algorithm that is expected to solve all challenges, but rather one should adapt different approaches and combine them to get best results. The most common approach for detection of DR from retinal images consists of preprocessing the image (target optical disc, vascular and fovea network), detecting and segmenting of candidate lesions, feature extractions from these, and classification as DR or non-DR, based on the identified lesions.[7] The classifier can also be made to determine the degree of severity of the disease (NP-DR, P-DR).

Density-based clustering methods preceded by determination of super-pixels are a promising approach to be used in this context. SLIC and DBSCAN[34,36] can very effectively segment and detect relevant structures in images. It is worth investigating their use in the context of DR detection. However, these methods need to be carefully integrated and accompanied by a set of pre- and post-processing steps. Preprocessing prepares, normalizes, and cleans the data. After segmentation, post-processing steps clean further and create a set of features from the remaining segments. The final step feeds segments properties to a classifier that can be designed in different ways. In this chapter, we propose the automated approach to detection using SLIC-DBSCAN clustering as segmentation step, and integration of those in a whole automated processing pipeline that is able to automatically detect lesions. The approach actually individualizes candidate lesions using the density clustering method, and uses classification of those clusters as part of the process. We describe the pre + segment + post processing pipeline, and investigate details to obtain a good detection model using the techniques. Extensive experimentation follows to test classification alternatives, and we conclude about the quality of the approach. Finally, we discuss further challenges and future work.

7.2 RELATED WORK

The topic of automated or semiautomated detection of DR from eye fundus images has been investigated extensively in the past, some of the related works being listed in this study.[3,4,6,8,10–13,15,17,19–23,25,27,28,30,44,45] Most works rely on pattern recognition principles[45] a pipeline of transformations, starting with preprocessing, detection of lesion candidates, extraction of features, and classification. We discuss the typical approaches used in each step.

7.2.1 INITIAL PREPROCESSING

Preprocessing eliminates unnecessary information, and improves the data to be used for further analysis. According to Rocha et al.,[7] it mitigates the variation by normalizing images relative to a reference model (an image chosen by an expert), so that a more accurate and efficient analysis becomes possible. The usual approach to handle variability distinguishes intravariability (differences in light scattering, presence of abnormalities and/or variation of reflectance, and thickness of the retina) and intervariability (different means/angles, acquisition, illumination, and retinal pigment).[9] This step involves, among others, brightness correction, contour detection, intensity adjustment, contrast, and histogram equalization.[7]

Shadow correction is a preprocessing step based on a mean/median filter method. The image background (which contains all the lighting variations) is obtained by smoothing the original image, with the mean or median filter configured with a size that should be larger than the largest retinal structure. Shadow correction reduces noise in the image (shot noise and salt and pepper noise).[8,17]

Morphological transformations, dilation, and erosion (based on addition and subtraction operations, respectively), are applied to binary images, relying on predefined structural elements. Structural elements interact with geometric shapes in the image, modify them, thus obtain relevant information.[5] To highlight fine and well-defined contours, the ideal is to use a small structuring element, since major structural elements do not detect objects smaller than the size of the structuring element itself.[14,18]

Histogram equalization redistributes the histograms for each color channel of the original image, so that the processed image contains a more uniform distribution of the pixel values. Alternatively, specification of the histogram allows normalization of colors by interpolation of the pixel values in each channel. This method makes structures more distinguishable from each other and from the background.[9] Adaptive histogram equalization is used to improve local contrast.[11]

The Radon transform is applied using a small window centered on the lesion candidate (more specifically, the microaneurysm), providing information for obtaining features to be used in pattern classification techniques.[4,19]

Adaptive thresholding takes into account brightness variations over the image and changes the threshold value from pixel to pixel. It consists of two sequential steps: image division into homogeneous regions and targeting injury candidates with border detection and region-growing techniques.[6]

7.2.2 DETECTION OF MICROANEURYSMS

For detection of microaneurysms, it is sensible to first remove blood vessels from the image, in order to minimize false positives (FPs).[4] This can be done by simply removing large objects using the Radon transform,[19] morphological operators (top-hat transform),[8] double-ring filters,[10] or pixel classification (supervised method).[20] It is also possible to use segmentation techniques, similar to those used for microaneurysm extraction.[4] Since microaneurysms are characterized by darker lesions than the other regions of the retina, it is expected that they can also be detected using a double filter ring. This filter makes the comparison of the value of a given target pixel with the values of neighboring pixels. The filter consists of an internal and an external circle; if the pixel value inside the small circle is smaller than the pixel values of the outer circle, the candidate target pixel becomes the microaneurysm.[10]

Region-growing algorithms are used to estimate the shape of a potential injury and to assemble the pixels in regions until they find a discontinuity. It is based on the average of the pixels in the agglomerates, to assess whether neighboring pixels should be included or not. The algorithm ends when no pixel meets the inclusion criteria.[17,21,22]

7.2.3 DETECTION OF EXUDATES

After preprocessing, since the optical disc has similar characteristics to exudates (brightness and contour), the optical disc needs to be identified, so that it does not interfere with detection of exudates. This structure can be distinguished from the others by its smooth texture and removed by morphological operators, namely the watershed transformation and the Otsu algorithm.[11,12] The watershed transformation detects the boundary of the optical disc based on the variation of the image gray levels, since this structure is a bright area and blood vessels that emerge from it are dark, with a very intense variation.[5,8] The level set method is a method for targeting the optical disc that incorporates an initial curve—level set of level 0—of a higher dimensional surface, and evolves this surface so that the level set of level 0 converges to the optical disc.[23]

Morphological filtering is used to isolate regions corresponding to exudates, based on a linear classifier constructed from color properties and contours of lesions. Structures that satisfy both brightness and boundary conditions are classified as exudates.[24]

7.2.4 DETECTION OF HEMORRHAGES

There are two general methods in the literature for the detection of hemorrhages in images of the fundus, detection of blood vessels, and detection of blood vessels with hemorrhages. Both top-down and bottom-up methods were designed to be robust to illumination changes in the image. They use contrast enhancement algorithms with region-growing approaches to delineate candidate regions.[5] The Otsu algorithm is used, in general, to threshold an image. The values of the thresholds differ from image to image, depending on the selection made automatically by the algorithm. As a result, high-intensity pixels are reconstructed, while the remaining ones are removed. Thresholding can also be used for elimination of the optical disc, by applying a closing grayscale operator, in order to eliminate blood vessels from the region.[12]

7.2.5 CLASSIFICATION

The choice of the most appropriate method for classifying each type of lesion should be based on two important aspects: robustness, in case of existence of outliers in the training set, and robustness, in cases where the distribution of features is unknown.[24]

Before classification, correlation techniques are frequently used to select the subset of most significant features.[24] It consists on iteratively correlating features two-by-two, and eliminating one of the two mostly correlated ones (elimination of highly redundant information).

Several authors consider k-nearest neighbor (k-NN) to be a good choice in comparison with other classification methods, since it has the advantage of robustness to outliers and it is nonparametric. However, all the neighbors have the same weight in the decision, regardless of its distance from the candidate to rank, which can be disadvantageous. Furthermore, if the training set is very asymmetrical, the method may be ineffective, particularly for high values of K.[24]

Classification using naive Bayes plus support vector machines (SVM) is a frequent option in many works. Feature selection is done using a naive Bayes classifier, to reduce the number of features iteratively. A final set of features is then fed into an SVM classifier. The SVM classifier tends to outline the contours of exudates with improved accuracy, frequently obtaining a small number of FPs.[11,19]

K-means clustering is used frequently to determine two groups: exudates and nonexudates. The first is characterized by high-intensity levels, unlike the second. Initially, the maximum and minimum intensities are calculated and used as centers of clusters, which are updated after every iteration of the algorithm.[24,26]

Fuzzy C-means clustering is another well-known technique for image segmentation, having been used also with retinal images. Sopharak et al.[11] proposes a method based on fuzzy clustering and data analysis algorithms on a set of features carefully chosen by ophthalmologists: pixel intensity after preprocessing (pixels of exudates can be distinguished from normal pixels by the intensity level), standard deviation of intensity, hue, and the number of edge pixels (because exudates are usually grouped into small clusters, and therefore present many edge pixels).[12]

Several other authors use neural networks with sigmoidal or linear activation functions.[27] In this case, back propagation is used as a convergence method (e.g., the Levenberg–Marquardt algorithm).[27] Diffuse neural systems can also be applied,[28] and more recently convolution networks and deep learning approaches are preferred, with its series of convolutional operations and neural computations.[47,48]

Another simpler classification method for the detection of two kinds of exudates uses histogram data. The region of the histograms between 0.8 and 0.85 is considered as soft exudates, while the rest corresponds to hard exudates.[12]

7.3 DENSITY CLUSTERING-BASED RETINOPATHY DETECTION

Density-based clustering methods can be quite effective in image segmentation, being used as part of pattern recognition approaches. Recently, the combination of the computation of superpixels (SLIC),[34] and density clustering (DBSCAN)[49,36] was proposed as an effective mechanism to segment images. SLIC preprocesses the dataset to obtain superpixels, from which

DBSCAN obtains clean segments. SLIC determines small homogeneous sets of neighboring pixels as superpixels, based on the maximum number of superpixels. DBSCAN then segments the whole picture or region of interest by finding dense homogeneous regions out of the superpixels computed by SLIC. The use of the CIElab color space also contributes to a good segmentation obtained by the sequence SLIC-DBSCAN.

We have devised a method for fully automated detection of lesions in eye fundus images based on applying preprocessing, density clustering, postprocessing with removal of blood vessels, feature extraction, normalization, and finally classification. In this section, we discuss the method.

7.3.1   PREPROCESSING AND EXTRACTION OF VASCULAR TREE

Due to the similarity between the color of microaneurysms, retinal hemorrhages, and blot hemorrhages on one side, and blood vessels on the other, it is important to be able to detect the vascular tree as a preprocessing step, so as to remove it during detection of those kinds of lesions, thus improving the results obtained. The vascular tree is also useful in determining other lesions on the tree itself that are characteristic of more advanced stages of the disease, such as venous beading, venous looping, and new vessels.

Figure 7.2 shows the steps followed to arrive at a blood vessels mask, adapted from Ref. 32. Figure 7.2a shows an example of an input image, and Figure 7.2b does a PCA over that image and Figure 7.2c selects the b channel of the image (since the b channel presents highest contrast). Next, a grayscale image is computed from the output of the step in Figure 7.2b, by normalizing values into the interval [0,1] (Fig 7.2d). The next step does contrast enhancement of the gray image by applying adaptive equalization of the histogram (CLAHE: Contrast-Limited Adaptive Histogram Equalization). This results in the contrast enhancement shown in Figure 7.2e. Intensity variations due to nonuniform illumination were removed using a mean filter with high kernel[33] (Figure 7.2f). Figure 7.2g is the difference between (f) and (e). The next step, returning Figure 7.2h, computes a black and white mask from (g). This is done by first thresholding based on the isodata method (which determines the background/foreground separator iteratively), then by applying a conversion to binary mask based on the computed threshold. Figure 7.2j cleans noise by removing very small regions from the black and white mask, and is the complement of Figure 7.2h. The final step of Figure 7.2i overlays the obtained black and white mask on the initial image, resulting in the vascular tree shown in the image. This resulting image isolates the vascular tree as desired.

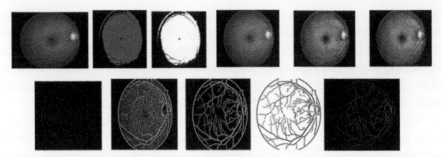

FIGURE 7.2 Preprocessing: blood vessels segmentation steps: (a) original, (b) pca ,(c) 1 channel, (d) gray, (e) enhance, (f) avg filter, (g) diff to avg, (h) bin by thresh, (i) remove small, (j) complement, and (k) overlay image.

The optical disc can also be identified and extracted, so that it does not interfere with detection of exudates. This structure can be distinguished from the others by its smooth texture and removed by morphological operators, namely the watershed transformation and the Otsu algorithm,[11,12] or the level set method.[23]

7.3.2 AUTOMATIC SEGMENTATION BY DENSITY CLUSTERING

Density clustering is a concept very closely related to our own perceptual mechanisms when identifying structures in images. We recognize clusters of some homogeneity that may correspond to structures in images as being regions for which the density of somehow homogeneous pixels is considerably higher than outside the cluster. Furthermore, the density within the areas of noise is lower than the density in any of the clusters. The key idea is that, for each point/pixel of a cluster/region, the neighborhood of a given radius has to contain at least a minimum number of points/pixels with similar perceptual characteristics (i.e., perceptual distance within a certain range). In other words, the density in the neighborhood has to exceed some threshold. DBSCAN was proposed in Ester et al.[49] We apply a spatial version of DBSCAN as density algorithm.

Bad quality and definition of images can affect the quality of the outcome of segmentation algorithms significantly. All digital images have a certain amount of electronic noise, which is random (not present in the object imaged) variation of brightness or color information. Noise can be produced by the sensor and circuitry of the image acquisition and processing devices. It is an undesirable by-product of image capture that adds spurious and extraneous information.

Superpixel algorithms can help the density clustering algorithm to reduce its difficulties with noise and bad quality, by pregrouping pixels into perceptually meaningful atomic regions that replace the rigid structure of the pixel grid. The name, superpixels, derives from the fact that they group together neighboring pixels into homogeneous larger (super) pixels. Superpixel algorithms capture image redundancy, provide a convenient primitive from which to compute image features, and greatly reduce the complexity of subsequent image processing tasks. We apply SLIC as a preprocessing step prior to applying DBSCAN.[34,36]

The SLIC method is based on the k-means clustering algorithm, differing only in two important points: the number of distance calculations upon optimization is much smaller (since the area taken into account is proportional to the superpixel size) and the distance is computed by combining spatial and color properties. SLIC generates superpixels by grouping pixels based on the color similarity and spatial proximity. The implementation of the method is done in lab xy space in which, lab is the pixel color in the CIELAB color space (considered as determining perceptually uniform color for small distances), and xy is the position of the pixel.[35] The resulting image is shown in Figure 7.3a.

The superpixelized image in Figure 7.3a is then ready for density-based clustering into segments. Figure 7.3b shows the result of applying DBSCAN. Given a set of superpixels in space, DBSCAN groups those that are closest in color and spatial properties leave outliers isolated in low-density regions.

The clustering process is based on the classification of points in three different groups: core points, border points, and noise points, and based on the relations between density points (directly density-reachable, density-reachable, density-connected).

A pixel is classified as a core point if it has at least a specified number of points (minPts—user defined) at a distance less than r (radius—user defined). These points are called 'directly density-reachable' from p. A pixel q is accessible from p if there is a path $p_1 \ldots p_n$ accessible from p with $p_1 = p$ and $p_n = q$ in which each pixel p_{i+1} is directly accessible from p_i. This way, if p is a core pixel, then it forms a cluster with all pixels (core and noncore) that are accessible from it. Every cluster has at least one core point, and each pixel in the cluster must have a neighborhood (defined by a radius) with a minimum number of pixels with similar characteristics above a user-defined threshold.[37]

DBSCAN is particularly useful on datasets with noise and in which the form of clusters can be arbitrary, which makes it well-adapted to segmenting images and image structures. The fact that each pixel needs to be analyzed more than once to evaluate to which cluster it might belong makes the algorithm run slow. However, since we apply it on superpixels, the process runs sufficiently fast to be well-suited to automated, real-time segmentation.

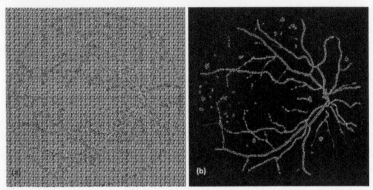

FIGURE 7.3   Eye vascular tree image after SLIC and DBSCAN.

7.3.3   REMOVING BLOOD VESSELS FOR DETECTION OF HEMORRHAGES

The next step after density-based clustering for detecting hemorrhages is to remove the vascular tree and to keep only clusters corresponding to candidate lesions (Fig. 7.4). To obtain lesion clusters automatically, four parameters are tuned for filtering in valid candidates and for filtering out blood vessels: the threshold to be used in DBSCAN, the minimum and maximum numbers of pixels used in lesion candidate clusters, and the maximum cluster eccentricity.

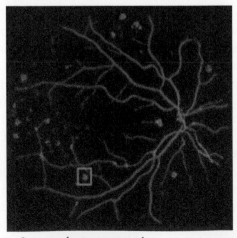

FIGURE 7.4   Clusters after vascular tree removed.

The four parameters were tuned based on manual labeling of lesions and vascular tree clusters on a training set, from which cutoff characteristics of potential blood lesions were extracted [npixels in (40,500), excentricity <0.91]. Figure 7.5 shows the algorithm.

Algorithm: remove vascular tree

Inputs: clusters (label Ci, pixels (x,y)), image pixel mask with cluster label

For each cluster
    Compute number of pixels npixels;
    Compute eccentricity;

If eccentricity< e and npixels within lesion limits
    tag cluster as potential lesion;

For each cluster Ci
    If Ci is not tagged as potential lesion
        Switch pixel mask values labeled as Ci to black (background, non-lesion)

FIGURE 7.5   Vascular tree removal algorithm.

7.3.4 FEATURES EXTRACTION, NORMALIZATION, AND WEIGHTING

After vascular tree extraction, the algorithm automatically extracts features from remaining clusters. The selected features include data cluster area, major and minor axis, eccentricity, diameter, perimeter, compactness, average color of the cluster, and the ratio between the average color of the cluster and the average color of the image. Feature compactness corresponds to the ratio between area and the square of the perimeter of the cluster.

Normalization maps the values of each feature to the interval [0,1]. The objective is to avoid different numeric ranges of different features to weight differently. Instead of being defined artificially by different natural ranges of attributes, if we want to assign different weights to different features, reflecting different degrees of importance of those features, it should be done by using explicit weights.

7.4 CLASSIFICATION

The classifier is the last step in the whole automated pipeline that detects the lesions. In our approach, the classifier outputs whether segments should be considered lesions or not (since the described approach is restricted to detection of microaneurysms and retinal hemorrhages). In this section, we first describe how we have built the classifier (model) based on the features.

Although in this work we focus only on the detection of microaneurysms and retinal hemorrhages, we also discuss in this section how the approach is integrated in a full-DR detection pipeline, which incorporates the approach described in this chapter, and also detection of other types of lesions, then finally detection of the degree of DR based on the outcomes of those parts.

7.4.1 CLASSIFIER FOR U-ANEURYSMS AND RETINAL HEMORRHAGES

The classifier for U-aneurysms and retinal hemorrhage lesions is dedicated to output whether each cluster detected in the previous steps of the approach as described in this chapter is classified as a lesion or not. The number properties of such lesions are then used in DR classification.

A classifier is a standard block in classification tasks, and a set of alternative classification algorithms is employed. For this step, we tested different algorithms, to determine empirically which would be most appropriate (i.e., the one with best sensitivity, specificity or accuracy). The classifier was built from a set of training cases, and its accuracy was tested using a set of test cases.

The dataset from which we extracted train and test sets was made of all candidate lesions that were extracted from a database of eye fundus images with and without lesions, each one labeled as "lesion" and "not lesion", according to the class the eye fundus image belongs to (as classified by a human expert).

7.4.2 TYPES OF CLASSIFIERS TESTED

We wanted to have the best possible results for the lesion detection task; therefore, a set of alternative classifier algorithms were used as candidates before we chose the one with best performance. In this subsection, we discuss the types of classifiers tested.

7.4.2.1 FISHER LINEAR DISCRIMINANT

Most classification algorithms can be seen as projecting a higher dimensional space of features into a unidimensional axis of classes (either binary as in our case or multiple choices). The Fisher linear discriminant (FLD) aims to find the linear combination of features that makes the best separation of two or more classes in one dimension. In general, projection data for a dimension leads to a considerable loss of information, and classes that should be well separated in the original space can be considerably overlapped in the unidimensional space. By adjusting the components of the vector w which performs the transformation $y = wTx$, the algorithm can select a projection that maximizes the separation between classes. In detail, the idea proposed by Fisher is to maximize the function which would give a large separation between the mean projected classes and at the same time maintain a reduced intraclass variance, which minimizes overlapping of classes. The intraclass variance of the transformed data C_k class is given by:

$$s_k^2 = \sum_{n \ in \ C_k} \left(y_n - m_k \right)^2 \qquad (7.1)$$

$$y_n = w^T x_n. \qquad (7.2)$$

The total intraclass variance is simply $S_1^2 + S_2^2$. The Fisher criterion is defined as the quotient of the interclass variance and the total variance intraclass, namely:

$$J(w) = \frac{\left(m_2 - m_1 \right)^2}{s_1^2 + s_2^2} = \frac{w^T S_B w}{w^T S_w w}, \qquad (7.3)$$

where S_B is the interclass covariance matrix and S_W is full intraclass covariance matrix. By maximizing (w) we get:

$$w \times S_w^{-1} \left(m_2 - m_1 \right) \qquad (7.4)$$

This is known as Fisher's discriminant.[40]

7.4.2.2 BAYESIAN CLASSIFIER

The Bayesian classifier[38] is a statistical classifier that makes a probabilistic forecast of belonging to one of the classes. It is based on Bayes theorem and is particularly useful when the dimensionality of the data is high. Despite its simplicity, this classifier can frequently exhibit better performance than other more complex ones. Given a set of variables $X = \{x_1, x_2, ..., x_n\}$, and assuming that there are J classes $C = \{C_1, C_2, ..., C_j\}$, the objective is to find the maximum a posteriori probability, which is obtained directly from Bayes' theorem:

$$p\left(C_j \mid x_1, x_2, ..., x_d\right) \alpha\, p\left(x_1, x_2, ..., x_d \mid C_j\right) p\left(C_j\right),$$

where $p\left(C_j \mid x_1, x_2, ..., x_d\right)$ is the probability of C_j belonging to X. As it is assumed that conditional probabilities of independent variables are statistically independent, it can be decomposed into a product of the probability terms:

$$p\left(X \mid C_j\right) \alpha\, \Pi\, p\left(x_k \mid C_j\right)$$

The a posterior probability is;

$$p\left(C_j \mid X\right) \alpha\, p\left(C_j\right) \Pi\, p\left(x_k \mid C_j\right).$$

Using the above rule, a new case X is identified as belonging to the class C_j by identifying its highest a posteriori probability.[36]

7.4.2.3   SUPPORT VECTOR MACHINES

The Support Vector Machines classifier[46] (SVM) is a supervised learning method that analyzes the data for classification and regression analysis; given a training set, the method finds a hyperplane that best separates the classes. The best hyperplane is the one that achieves greatest margin between the classes, the maximum width of the region parallel to the hyperplane that has no points inside of it. The points that are closest to the separating hyperplane are called support vectors (Fig. 7.6).

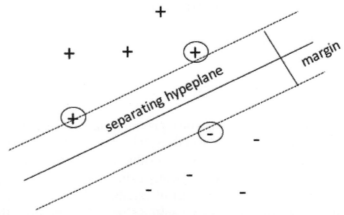

FIGURE 7.6   SVM—separating hyperplanes, margin, and support vectors.

The classification is made by mapping instances to classify in space and predicting the class they belong to based on the location of the hyperplane.

We test both LIBSVM[39] and LIBLINEAR.[40,42] The difference between the two methods lies in the optimization function that each uses.

LIBSVM: Given a set of event-label pairs,

$$(x_i, y_i), (i = 1, \ldots l), x_i \in R^n \; e \; y_i \in \{-1, 1\} \tag{7.5}$$

the algorithm solves the following optimization problem:

$$min_{w,b,e} \left(\frac{1}{2} w^T + C \sum_1^l e_i \right) \tag{7.6}$$

subject to

$$y_i \left(w^T \varnothing (x_i) + b \right) \leq 1 - e_i \tag{7.7}$$

The x_1 training vectors are mapped into a higher dimension space by the Φ function. $C > 0$ is a penalty parameter, and a radial-type basis kernel function $(x_i, x_j) = \Phi(x_i)T \; \Phi(x_j)$ is used,

$$K\left(x_i, x_j \right) = e^{-\gamma x_i - x_j^2}, \gamma > 0 \tag{7.8}$$

LIBLINEAR: Similarly to LIBSVM, given a set of event-label pairs,

$$(x_i, y_i), (i = 1, \ldots, l), x_i \in R^n \; e \; y_i \in \{-1, 1\} \tag{7.9}$$

the algorithm solves the following optimization problem:

$$min_w (\frac{1}{2} w^T w + C \sum_{i=1}^l \max \left(0.1 - y_i w^T x_i \right)^2 \tag{7.10}$$

This operation is called L2-regularized L2 loss.

7.4.2.4 *k-NEAREST NEIGHBORS CLASSIFIER*

Given a dataset (training set), and defining a metric to compute the distance (usually Euclidean) and the value of k (number of nearest neighbors that will be considered by the algorithm), k-NN is based on the idea that nearest points in the features space are more likely to belong to the same class.

Considering a new event, the algorithm calculates the distance between this and the examples of the training set, and identifies the k-NNs. The new event will be classified as belonging to the majority class of the set of neighbors.

This is a very simple and easy to implement method, and it frequently returns good results. On the down side, it can be a computationally complex and time-consuming process (comparison with all training cases).

7.4.2.5 MINIMUM DISTANCE CLASSIFIER (EUCLIDIAN)

The minimum distance algorithm[41] is a supervised learning method, where a new event is classified as belonging to the class whose distance to the prototype (midpoint of each class of events) is smallest (e.g., using the Euclidean distance). This method is particularly fast for the classification of new events.

7.4.3 CLASSIFICATION OF DR AND DR DEGREE

The above-described approach that we propose in this chapter covers automated detection of blood hemorrhages (microaneurysms, dot and blot hemorrhages) in eye fundus images. Microaneurysms are one of the earliest clinical signs of DR. They appear as small, red dots in the superficial retinal layers. Dot and blot hemorrhages are another common sign of DR. They appear similar to microaneurysms if they are small, and usually occur when microaneurysms rupture in the deeper layers of the retina, such as the inner nuclear and outer plexiform layers.

In this approach, the output of the classifier indicates whether each cluster found by density clustering is a lesion (microaneurysm or hemorrhage) or not. Based on that classification for all clusters of an eye fundus image, information is given concerning the number and characteristics of the lesions that were detected. That result should then be mixed with the ratings for other lesions (e.g., exudates, venous beading and venous loops, new vessels) to arrive at a final classification of the degree of evolution of the disease.

The degree of the DR disease is determined based on a broad set of characteristics. For instance, according to one approach, NP-DR might be classified as mild, based on the presence of at least one microaneurysm, moderate, when hemorrhages, microaneurysms, and hard exudates are present, or severe, based on a (4–2–1) rule: hemorrhages and microaneurysms in four quadrants, venous beading in at least two quadrants, and intraretinal microvascular abnormalities in at least one quadrant.

Since the approach we propose covers only automated detection of blood hemorrhages, it must be mixed with detection of the other kinds of lesions to fully detect DR degree, and it is necessary to add DR classification-dependent details. For instance, according to the (4–2–1) rule above, we should count the number of hemorrhages per quadrant, which is actually easy to do because we have the location of each cluster that was classified as lesion (positive). Detection of all types of lesions and full determination of the degree of DR is out of scope of this chapter, since we focused on how to apply density clustering to detect and classify blood hemorrhages. As part of our own future work on the subject, we intend, on one hand, to apply approaches discussed in the related work for detection of the other types of lesions and on the other hand, to experiment with density clustering approaches for those classifications as well. Finally, we also intend to bring the parts together for fully classifying the degree of DR.

7.5   EXPERIMENTAL SETUP

Our experimental dataset was taken from the images of the Messidor database (methods for evaluating segmentation and indexing techniques dedicated to retinal ophthalmology), made available by the Analysis Research Center of Caen—Advanced Concepts in Imaging Software (ADCIS).[29] This database contains 1200 images of the eye fundus, with and without pupil dilation, acquired in three ophthalmology departments using a 3CCD camera in a retinographer Topcon TRC NW6. These images are in TIFF format, with 1440 × 960 pixels and are classified according to the degree of DR as shown in Table 7.1.

TABLE 7.1   Classification of Messidor Database Images.

DR degree	Nr. of micro-aneurisms	Nr. of hemorrhages	Neo-vascularization
0	0	0	–
1	0–5	0	–
2	mai/15	0–5	0
3	<15	>5	1

Source: Adapted from ADCIS (Ref. 29)" in footnote of Table 7.1.

7.5.1   TRAINING AND TESTING DATASETS

The dataset extracted randomly and used to build the classifier using our approach (training and testing data) contained 1605 instances classified into nine features, as we discuss in the approach, plus a column identifying the cluster number and another column identifying the class (lesion = 1, nonlesion = 0). The dataset is composed of 77% of lesion instances and 23% of nonlesion instances. To obtain the best set of parameters (specific for each classifier), we used cross-validation. According to this method, the data were divided into k mutually exclusive sets of equal size, where k-1 sets are used for training and the remaining assembly is used for testing. This process is performed k times, using in each iteration different training and test sets.

7.5.2 METRICS

True positives (TRs) and true negatives (TNs) correspond, respectively, to the positive and negative cases correctly classified as such. FP and false negatives (FNs) correspond respectively to the positive and negative events that were misclassified.

Sensitivity is also known as recall or TR rate. It measures the proportion of positives that are correctly identified as such (in our case, it is the percentage of lesions that are correctly identified as such),

$$\text{Sensitivity} = TP/(TP + FN).$$

Specificity or TN rate measures the proportion of negatives that are correctly identified as such (in our case, that is the fraction of nonlesions that are correctly identified as such).

$$\text{Specificity} = TN/(TN + FP).$$

Accuracy is given by the fraction of correctly classified instances (TP and TN),

$$\text{Accuracy} = (TP + TN)/(TP + FN + TN + FP).$$

In our experimental analysis, we value these three performance indicators alike. Several authors focus on getting high levels of sensitivity without caring for specificity, which makes those methods sensitive to various image structures (not only lesions) that should not be classified as lesions. Having a high value of sensitivity but low level of specificity means that the model detects the majority of lesions (TR), but it also mistakenly detects other structures as being lesions (FP). On the other extreme, a high value of specificity and low sensitivity value means that the model correctly detects areas without lesions (TNs), but it is not able to identify many actual lesions, classifying them as nonlesions (FNs).[44] Therefore, it is considered that an accurate classifier must present high and as homogeneous as possible values of accuracy, sensitivity, and specificity.

We also show the accuracy, sensitivity, and specificity of the approach classifying test images according to the DR degrees 0–3 of Figure 7.7.

Classifier	Sensitivity	Specificity	Accuracy
Bayesian Classifier	51.79%	80.58%	60.68%
Fisher Linear Discriminant	48.33%	86.48%	70.04%
SVM (linsvm)	12.50%	96.39%	76.31%
SVM (libsvm)	25.00%	94.10%	77.56%
k-NN	52.83%	80.46%	60.48%
Min Distance Classifier	51.22%	83.07%	65.32%

Classifier	Sensitivity	Specificity	Accuracy
Bayesian Classifier	66.67%	80.22%	61.82%
Fisher Linear Discriminant	52.63%	86.41%	71.36%
SVM (linsvm)	20.20%	96.36%	77.56%
SVM (libsvm)	26.26%	97.02%	79.55%
k-NN	76.92%	80.94%	63.66%
Min Distance Classifier	48.51%	84.67%	67.29%

FIGURE 7.7 Classifier results (U-aneurysms and retinal hemorrhages). (a) Nonnormalized and (b) normalized.

7.5.3 METHODS COMPARED

In this work, we were focused on maximizing the accuracy, sensitivity, and specificity of the approach. In particular, we identified the type of classifier as the first relevant factor that we should try to optimize. For that reason, our experiments compare results using different classifier alternatives. We were also particularly concerned with evaluating whether normalizing the dataset would contribute significantly to improve metric results, therefore we compare normalized versus nonnormalized classification.

7.6 EXPERIMENTAL RESULTS

For the experiments, the original eye fundus dataset, consisting of Messidor images,[29] was ran through the processing pipeline described in this chapter. The features extracted from the processing pipeline were submitted to different classifier algorithms and accuracy, sensitivity, and specificity were measured. Figure 7.7 shows the results.

Figure 7.8 shows the experimental detail of how accuracy varied with parameters C in LIBSVM and k in k-NN. We used this procedure to tune the values of those parameters to obtain most accurate classifications, comparing also the evolution when the dataset is normalized (b) versus when it is not normalized (a). It is quite noticeable from these results that normalization influences significantly both the accuracy and the best tuning values for the parameters. Figure 7.9 shows the results for tuning parameter k of k-NN.

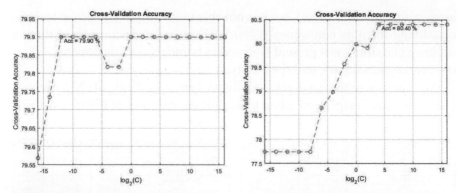

FIGURE 7.8 Tuning of LIBSVM parameter C (MATLAB). (a) Nonnormalized and (b) normalized.

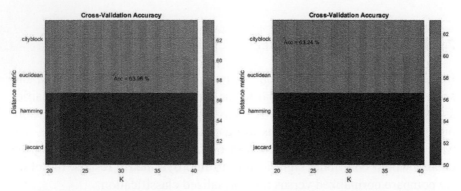

FIGURE 7.9 Tuning parameter K (number of neighbors) in K-NN (MATLAB). (a) Nonnormalized and (b) normalized.

7.7 ANALYSIS, CONCLUSIONS, AND CHALLENGES

Classifier results shown in Figure 7.7a and b concern classification of individual clusters as either lesions or nonlesions. These classifiers are at the end of a chain of processing that started with preprocessing of the image to get the vascular tree, removal of blood vessels based on that vascular tree, segmentation using SLIC-DBSCAN, some post-processing to further filter candidate lesions, and finally feature extraction. Results show that accuracy varied roughly between 60% and 80%, depending on the classifier used, sensitivity varied between 50% and 80%, and specificity varied between 80% and 97%. One interesting fact we extract from the results is that SVMs seem to yield the best accuracy, because they excel in specificity, while in fact they are extremely bad in terms of sensitivity (about 12–26% versus 50–80% for other classifiers). This means that SVMs were good at detecting that a certain lesion candidate is not actually a lesion, however this was achieved at the expense of bad capacity to detect that actual lesions were lesions indeed. High values of specificity combined with low values of sensitivity means that the model correctly detects areas without lesions (TNs) but is not able to identify all lesions, classifying them as noninjury (FNs).[44] Although, many authors focus on getting high levels of sensitivity without striving for specificity, which makes their methods sensitive to various image structures (not only injuries). Those models detect the majority of lesions (TRs), but also mistakenly classify other structures as being lesions (FPs). Since a good classifier should present good values for both sensitivity and specificity, and based on the experimental results, we considered the most appropriate classifiers for our case to be Bayes and k-NN.

These experimental results have also shown that data normalization improved accuracy, sensitivity, and specificity for most classifiers.

As part of our experimental work, we had to deal with other parts of the pipeline besides classification, and we had to tune parameters and to handle images with very different qualities and characteristics. In the next, we describe conclusions and major challenges that we think must still be handled to improve our approach in each part of the processing.

Preprocessing: The main purpose of this step was the removal of pixels corresponding to the background of the retina and to the vascular tree. Due to the heterogeneity of the images (luminance, color, intersubject variability amongst others), the preprocessed images show dissimilarities in the quality of preprocessing. The major difficulties were related to the removal of structures and noise (e.g., fovea, optic disc, and isolated pixels).

Segmentation and Clustering: The purpose of this step was to segment the preprocessed image based on the density and color of the pixels in a given area. To this end, we applied superpixelization and density clustering (SLIC and DBSCAN) algorithms. In this part of the processing, we faced two main difficulties: optimization of parameters for each image and processing time (mainly for the SLIC algorithm). In order to reduce this time, it was decided to reduce the size of images to half and test a few combinations of parameters. Consequently, and due to the subjectivity of visual analysis of the resulting images, the results may not have been the best. Nevertheless, segmentation results were still satisfactory.

Removal of blood vessels: In one of the last preprocessing steps, we removed blood vessels, in order to obtain an image with only lesion candidates. This process has involved a great effort in optimizing parameters (again due to inter- and intrapicture heterogeneities) that discriminate lesions and nonlesions. However, after removal of the clusters relating to blood vessels, there still remained nonlesion clusters that could not be removed since they were very similar to lesions (in size and eccentricity).

Features extraction—we have chosen an aforementioned set of features that, according to the literature, have relevant discriminative potential. For a more detailed analysis, we could have used a greater number of features. We could have tried to add features of color in the CIELAB space, since those would more closely correlate color values with visual perception (i.e., a variation in color value produces the same variation in visual perception).

7.8   CONCLUSION

Since DR is one of the most frequent and severe consequences of diabetes, it is extremely important to develop methods for automatic detection of lesions of the retina, which facilitate and make faster the process of identification of the disease and its degree. Without these, lesions have to be exclusively detected manually by skilled professionals. In spite of their advantage as a fast and complementary mechanism, these algorithms are extremely difficult to develop due to heterogeneity of both inter- and intrapicture. For these reasons, this is one of the issues that remains as a research challenge, and dozens of research groups around the world continue to propose innovative approaches.

In this work, we focused on the detection of bleeding in the retina. Despite it being an exploratory work, the results are encouraging, as we were able to detect individual lesions with high accuracy, sensitivity, and specificity. We also discussed the main challenges that still need to be faced. Our future work should focus on improving the quality of the preprocessing steps, on adding similar approaches to detect the remaining characteristic affections of DR, and on fully classifying the degree of the illness based on detection results.

ACKNOWLEDGMENTS

The Messidor dataset was kindly provided by the Messidor program partners.

KEYWORDS

- computer vision
- image processing
- segmentation
- clustering
- classification
- machine learning

REFERENCES

1. Figueira, J; Nascimento, J; Henriques, J; Gonçalves, L; Rosa, P; Silva, R., Eds. *Retinopatia Diabética Guidelines* 1st ed.; Sociedade Portuguesa de Oftalmologia: Lisboa, Portugal, 2009.
2. Diabetes: Factos e Números 2014 – Relatório Anual do Observatório Nacional da Diabetes 11/2014 Sociedade Portuguesa de Diabetologia, 11/2014. https://www.dgs.pt/estatisticas-de-saude/estatisticas-de-saude/publicacoes/diabetes-factos-e-numeros-2014-pdf.aspx.
3. Walter, T.; Massin, P.; Erginay, A.; Ordonez, R.; Jeulin, C.; Klein, J.-C. Automatic Detection of Microaneurysms in Color Fundus Images. *Med. Image Anal.* **2007,** *11*(6), 555–566.
4. Oliveira, J. Estudo e Desenvolvimento de Técnicas de Processamento de Imagem Para a Identificação de Patologias em Imagem de Fundo do Olho. tese de Dissertação de Mestrado Integrado em Engenharia Biomédica, University of Minho, 2012. Accessible Online in http://repositorium.sdum.uminho.pt /handle/1822/23422.
5. Guerra, R. Identificação Automática do Disco Óptico em Imagens Coloridas da Retina. tese de Dissertação de Mestrado Integrado em Engenharia Biomédica, Faculdade de Engenharia da Universidade do Porto, 2008.
6. Jaafar, H. F.; Nandi, A. K.; Al-Nuaimy, W. In *Automated Detection and Grading of Hard Exudates from Retinal Fundus Images*, European Signal Processing Conference, Eusipco, 2011, pp 66–70.
7. Rocha, J. W.; Carvalho, T; Goldenstein, S; Wainer, J. Points of Interest and Visual Dictionary for Retina Pathology Detection. *IEEE Trans. Biomed. Eng.* **2012,** *59*(8):2244–2253.
8. Walter, T.; Klein, J. C.; Massin, P., Erginay, A. A Contribution of Image Processing to the Diagnosis of Diabetic Retinopathy—Detection of Exudates in Color Fundus Images of the Human Retina. *IEEE Trans. Med. Imaging.* **2002,** *21*(10), 1236–1243.
9. Prentasic, P.; Loncaric, S.; Vatavuk, Z.; Bencic, G.; Subasic, M., Petkovic, T.; Dujmovic, L.; Malenica-Ravlic, M.; Budimlija, N.; Tadic, R. In *Diabetic Retinopathy Image Database(DRiDB):A New Database for Diabetic Retinopathy Screening Programs*

Research, 8th International Symposium on Image and Signal Processing and Analysis (ISPA 2013), 2013.

10. Mizutani, A.; Muramatsu, C.; Hatanaka, Y.; Suemori, S.; Hara, T.; Fujita, H. In *Automated Microaneurysm Detection Method Based on Double Ring Filter in Retinal Fundus Images,* Medical Imaging Computer-aided. Diagnosis, 2009, Vol. 7260; pp 72601N–72601N–8.

11. Sopharak, A.; Dailey, M. N.; Uyyanonvara, B.; Barman, S.; Williamson, T.; New, K. T.; Moe, Y. A. Machine Learning Approach to Automatic Exudate Detection in Retinal Images from Diabetic Patients. *J. Modern Optics*. **2010,** *57*(2), 124–135.

12. Sopharak, A.; Uyyanonvara, B.; Barman, S.; Williamson, T. H. Automatic Detection of Diabetic Retinopathy Exudates from Non-dilated Retinal Images Using Mathematical Morphology Methods. *Comput. Med. Imaging Graph*. **2008,** *32*(8), 720–727.

13. Sopharak, A.; Uyyanonvara, B.; Barman, S. Automatic Exudate Detection from Non-dilated Diabetic Retinopathy Retinal Images Using Fuzzy C-means Clustering. *Sensors* **2009,** *9*(3), 2148–2161.

14. Niemeijer, M.; Van Ginneken, B.; Staal, J.; Suttorp-Schulten, M. S.; Abràmoff, M. D. Automatic Detection of Red Lesions in Digital Color Fundus Photographs. *IEEE Trans. Med. Imaging*. **2005,** *24*(5), 584–592.

15. Fleming, A. D.; Philip, S.; Goatman, K. A.; Olson, J. A.; Sharp, P. F. Automated Micro-aneurysm Detection Using Local Contrast Normalization and Local Vessel Detection. *IEEE Trans. Med. Imaging*. **2006,** *25*(9), 1223–1232.

16. Narasimha-Iyer, H.; Can, A.; Roysam, B.; Tanenbaum, H. L.; Majerovics, A. Integrated Analysis of Vascular and Nonvascular Changes from Color Retinal Fundus Image Sequences. *IEEE Trans. Biomed. Eng*. **2007,** *54*(8), 1436–1445.

17. Ege, B. M.; Hejlesen, O. K.; Larsen, O. V; Møller, K.; Jennings, B.; Kerr, D.; Cavan, D. A. Screening for Diabetic Retinopathy using Computer Based Image Analysis and Statistical Classification. *Comput. Methods Programs Biomed*. **2000,** *62*(3), 165–175.

18. Quellec, G.; Lamard, M.; Josselin, P. M.; Cazuguel, G.; Cochener, B.; Roux, C. Optimal Wavelet Transform for the Detection of Microaneurysms in Retina Photographs. *IEEE Trans. Med. Imaging* **2008,** *27*(9), 1230–1241.

19. Giancardo, L.; Mériaudeau, F.; Karnowski, T. P.; Tobin, K. W.; Li, Y.; Chaum, E. In *Microaneurysms Detection with the Radon Cliff Operator in Retinal Fundus Images*, Medical Imaging 2010: Image Processing, May, 2010, Vol. 7623; p 76 (TransTrans230U).

20. Niemeijer, M.; Van Ginneken, B.; Cree, M. J.; Mizutani, A.; Quellec, G.; Sanchez, C. I.; Zhang, B.; Hornero, R.; Lamard, M.; Muramatsu, C.; Wu, X.; Cazuguel, G.; You, J.; Mayo, A.; Li, Q.; Hatanaka, Y., Cochener, B. E.; Roux, C.; Karray, F.; Garcia, M.; Fujita, H.; Abramoff, M. D. Retinopathy Online Challenge: Automatic Detection of Microaneurysms in Digital Color Fundus Photographs. *IEEE Trans. Med. Imaging* **2010,** *29*(1), 185–195.

21. Yang, G.; Gagnon, L.; Wang, S.; Boucher, M.-C. In *Algorithm for Detecting Micro-aneurysms in Low-resolution Color Retinal Images*, Proceedings of Vision Interface, Ottawa, Canada, 2001; pp 265–271.

22. Zhang, B.; Zhang, L.; You, J.; Karray, F. In *Microaneurysm (MA) Detection via Sparse Representation Classifier with MA and Non-MA Dictionary Learning*, 20th International Conference on Pattern Recognition, Istanbul, Turkey, 2010; pp 277–280.

23. Sae-Tang, W.; Chiracharit, W.; Kumwilaisak, W. In *Exudates Detection in Fundus Image using Non-uniform Illumination Background Subtraction*, IEEE Region 10

Annual International Conference, Proceedings/TENCON, Fukuota, Japan, 2010; pp 204–209.

24. Narasimhan, K.; Neha, V. C.; Vjayarekha, K. A Review of Automated Diabetic Retinopathy Diagnosis from Fundus Image. *J. Theor. Appl. Inf. Technol.* **2012,** *39*(2), 225–238.

25. Siddalingaswamy, P. C.; Prabhu, K. G. In *Automated Detection of Optic Disc and Exudates in Retinal Images*, IFMBE Proceedings, Springer: Berlin, Heidelberg, 2009; Vol. 23, pp 277–279.

26. Ram, K.; Joshi, G. D.; Sivaswamy, J. A Successive Clutter-rejection-based Approach for Early Detection of Diabetic Retinopathy. *IEEE Trans. Biomed. Eng.* **2011,** *58*(3), 664–673 (PART 1).

27. Cree, M. J.; Olson, J. A.; McHardy, K. C.; Sharp, P. F.; Forrester, J. V. A Fully Automated Comparative Microaneurysm Digital Detection System. *Eye* (Lond) **1997,** *11*(5), 622–628.

28. Jayakumari, C.; Santhanam, T. Detection of Hard Exudates for Diabetic Retinopathy Using Contextual Clustering and Fuzzy Art Neural Network. *Asian J. Inform. Technol.* **2007,** *6*(8), 842–846.

29. ADCIS, Methods to Evaluate Segmentation and Indexing Techniques in the Field of Retinal Ophthalmology (MESSIDOR), 2010. http://www.adcis.net/en/Download-Third-Party/Messidor.htmlindex-en.php (accessed Jan 20, 2016).

30. Bharali, P.; Medhi J. P.; Nirmala, S. R. In *Detection of Hemorrhages in Diabetic Retinopathy Analysis Using Color Fundus Images*, IEEE International Conference on Recent Trends Information System, Kolkata, India, 2015; pp 237–242.

31. Gulati, S.; Kleawsirikul, N.; Uyyanonvara, B. A Review on Hemorrhage Detection Methods for Diabetic Retinopathy Using Fundus Images. *Int. J. Biol. Ecol. Environ. Sci.* **2012,** *1*(6).

32. Coye, T. L. A Novel Retinal Blood Vessel Segmentation Algorithm for Fundus Images, 2015. http://www.mathworks.com/matlabcentral/fileexchange/50839-a-novel-retinal-blood-vessel-segmentation-algorithm-for-fundus-images (accessed Nov 01, 2015).

33. Marín, D.; Aquino, A.; Gegúndez-Arias, M. E.; Bravo, J. M. A New Supervised Method for Blood Vessel Segmentation in Retinal Images by Using Gray-level and Moment Invariants-based Features. *IEEE Trans. Med. Imaging* **2011,** *30*(1), 146–158.

34. Achanta, R.; Shaji, A.; Smith, K.; Lucchi, A.; Fua, P.; Süsstrunk, S. SLIC Superpixels Compared to State-of-the-Art Superpixel Methods. *IEEE Trans. PAMI* **2012,** *34*(11), 2274–2282.

35. Alonso, O.; Linares, C. Segmentação de Imagens de Alta Dimensão por meio de Algoritmos de Detecção de Comunidades e Super Pixels. Dissertação apresentada ao Instituto de Ciências Matemáticas e de Computação - ICMC-USP, como parte dos requisitos para obtenção do título de Mestre em Ciências—Ciências de Computação e Matemática Computacional, USP-São Carlos, 2013.

36. Kovesi, P. Image Segmentation using SLIC SuperPixels and DBSCAN Clustering, 2013. http://www.peterkovesi.com/projects/segmentation/ (accessed Jan 20, 2016).

37. Moreira, A.; Santos, M. Y.; Carneiro, S. Density-based Clustering Algorithms – DBSCAN and SNN, 2005. http://ubicomp.algoritmi.uminho.pt/local/download/SNN&DBSCAN.pdf (accessed Jan 20, 2016).

38. DELL Software, Naive Bayes Classifier. http://www.statsoft.com/Textbook/Naive-Bayes-Classifier, (accessed Jan 20, 2016).

39. Franc, V. Statistical Pattern Recognition Toolbox for Matlab, 2008. http://cmp.felk.cvut.cz/cmp/software/stprtool/ (accessed Jan 20, 2016).

40. Bishop, C. M. *Pattern Recognition and Machine Learning*; Springer, 2006; Vol. 4, No. 4, ISBN-13: 978-0387310732.

41. Chang, C.-C.; Lin C.-J. LIBSVM: A Library for Support Vector Machines. *ACM Trans. Intell. Syst. Technol.* **2013**, *2*, 1–39.

42. Fan, R.-E.; Chang, K.-W.; Hsieh, C.-J.; Wang, X.-R.; Lin, C.-J. LIBLINEAR: A Library for Large 29 Linear Classification. *J. Mach. Learn.* **2008**, *9*(2008),1871–1874.

43. Marques, J. P. de Sá. *Pattern Recognition Concepts, Methods and Applications*; Springer Science & Business Media: Berlin, 2001; p 318.

44. Júnior, S. B.; Welfer, D. Automatic Detection of Microaneurysms and Hemorrhages in Color Eye Fundus Images. *Int. J. Comput. Sci. Inf. Technol.* **2013**, *5*, 21.

45. Usman Akram, M.; Khalid, S.; Tariq, A.; Khan, S. A.; Azam, F. Detection and Classification of Retinal Lesions for Grading of Diabetic Retinopathy. *Comput. Biol. Med.* **2014**, *45*, 161–171.

46. Mathworks, Support Vector Machines (SVM), 2015. http://www.mathworks.com/help/stats/support-vector-machines-svm.html (accessed Jan 21, 2016).

47. De Fauw, J. Detecting Diabetic Retinopathy in Eye Images, July 28, 2015. http://jeffreydf.github.io/diabetic-retinopathy-detection/.

48. Hou, L.; Samaras, D.; Kurc, T. M.; Gao, Y.; Davis, J. E.; Saltz, J. H. In *Patch-based Convolutional Neural Network for Whole Slide Tissue Image Classification*, Computer Vision and Pattern Recognition (J.3, I.4, I.5).

49. Ester, M.; Kriegel, H.-P.; Sander, J.; Xu, X. In *A Density-based Algorithm for Discovering Clusters in Large Spatial Databases with Noise*, Proceedings of the Second International Conference on Knowledge Discovery and Data Mining (KDD-96), AAAI Press, 1996; pp 226–231.

CHAPTER 8

REFLECTIONS ON ARTEMISININ: PROPOSED MOLECULAR MECHANISM OF BIOACTIVITY AND RESISTANCE

FRANCISCO TORRENS[1,*], LUCÍA REDONDO[2], and GLORIA CASTELLANO[2]

[1]Institut Universitari de Ciència Molecular, Universitat de València, Edifici d'Instituts de Paterna, P.O. Box 22085, E-46071 València, Spain

[2]Departamento de Ciencias Experimentales y Matemáticas, Facultad de Veterinaria y Ciencias Experimentales, Universidad Católica de Valencia San Vicente Mártir, Guillem de Castro-94, E-46001 València, Spain

*Corresponding author. E-mail: torrens@uv.es

CONTENTS

Originally published as Torrens, F.; Castellano, G. Artemisinin: Tentative Mechanism of Action and Resistance. In Proceedings of the 2nd Int. Electron. Conf. Med. Chem., 1–30 November 2016; Sciforum Electronic Conference Series, Vol. 2, 2016 , A008; doi:10.3390/ecmc2A008. Modified by the authors and used with permission.

ABSTRACT

Every year, 1–2 million people (mainly children) living in the tropics and subtropics die of malaria. Artemisinin is the most effective treatment against malaria, the most infectious disease in the world today. Artemisinins are derived from extracts of sweet wormwood (*Artemisia annua*) and are well established for the treatment of malaria, for example, highly drug-resistant strains. They resulted in one of the most significant advances in the treatment of malaria since the discovery and first use of quinine (1820). Their efficacy also extends to phylogenetically unrelated parasitic infections, for example, schistosomiasis. They showed potent and broad anticancer properties in cell lines and animal models. What hope does the drug offer for the future?

Malaria is caused by the bite of a female mosquito (*Anopheles*), resulting in the malaria parasite, *Plasmodium*, entering the human blood stream. Once an erythrocyte becomes invaded with the parasite, several rounds of asexual reproduction ensue, leading to its eventual rupture. The cyclic release of parasites from the erythrocytes causes the intermittent symptoms of fever, shivering, and anemia, which are characteristics of malaria. Of the four species of *Plasmodium* that infect humans, the most dangerous is *Plasmodium falciparum*, which accumulates in the capillaries of vital organs (e.g., brain, kidney, intestine, lungs). Cerebral malaria, *P. falciparum* accumulation in the brain, is responsible for most deaths associated with malaria. During the erythrocyte cycle, the parasite uses the host's hemoglobin (Hb) as food. Protein Hb is broken down by enzymes and the parasite assimilates the amino acids that are released. The digestive process liberates hem, Fe-containing porphyrin, which is normally buried within Hb molecule. The free hem molecule, bearing an exposed Fe^{II} atom at its center, provides the unique line of attack for the drug peroxidic-sesquiterpene-lactone (STL) artemisinin (ART, *qinghaosu*, QHS, cf. Fig. 8.1a). The remarkable story of the discovery of ART and establishment of its antimalarial activity by Chinese scientists represents one of the great discoveries in medicine in the latter half of the 20th century. The ART derivatives (ARTDs) present structures different from the classical quinoline.

The STLs constitute a large class of secondary plant metabolites, which carry α-methylene-γ-lactone (ML) groups as common structural element and display a number of bioactivities (e.g., cytotoxic, antineoplastic, cardiovascular, antimicrobial). What did make STLs reach cancer clinical trials?[1]

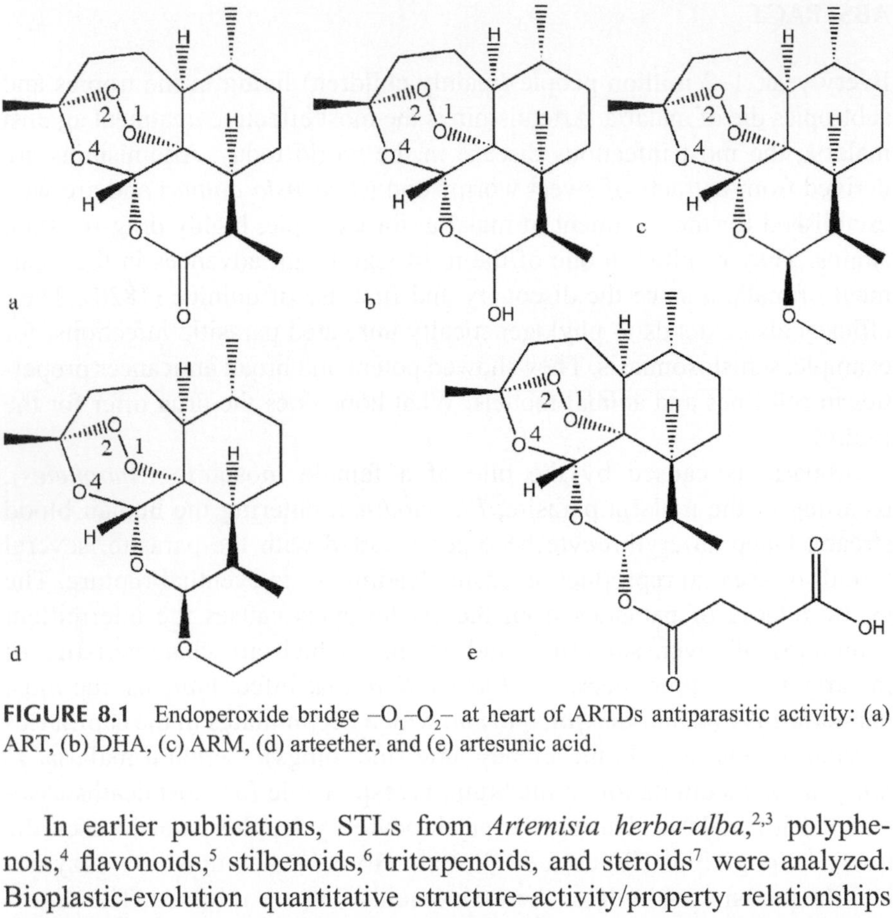

FIGURE 8.1 Endoperoxide bridge $-O_1-O_2-$ at heart of ARTDs antiparasitic activity: (a) ART, (b) DHA, (c) ARM, (d) arteether, and (e) artesunic acid.

In earlier publications, STLs from *Artemisia herba-alba*,[2,3] polyphenols,[4] flavonoids,[5] stilbenoids,[6] triterpenoids, and steroids[7] were analyzed. Bioplastic-evolution quantitative structure–activity/property relationships were applied to phenylalcohols, 4-alkylanilines,[8] aromatics,[9] phenyl urea herbicides,[10] pesticides,[11,12] isoflavonoids,[13] methylxanthines, cotinine,[14] chlorogenic acids,[15] natural STLs,[16] and ARTs.[17] The aim of the present chapter is to review ART proposed molecular mechanism of bioactivity and resistance. The main objective is to initiate a debate by suggesting a number of questions (Qs) that can arise when treating malaria with ARTDs and providing, when possible, answers (As) and/or hypotheses (Hs).

8.1 THE DISCOVERY OF ARTEMISININ

In 1967, versus a scene of rising chloroquine (cf. Fig. 8.2b) resistance, and on-going wars in neighboring Cambodia and Vietnam, Chinese Government Project May 23 (523), began a systematic examination of plants used in traditional Chinese medicine (TCM) with the aim of finding a new antimalarial.[18-21]

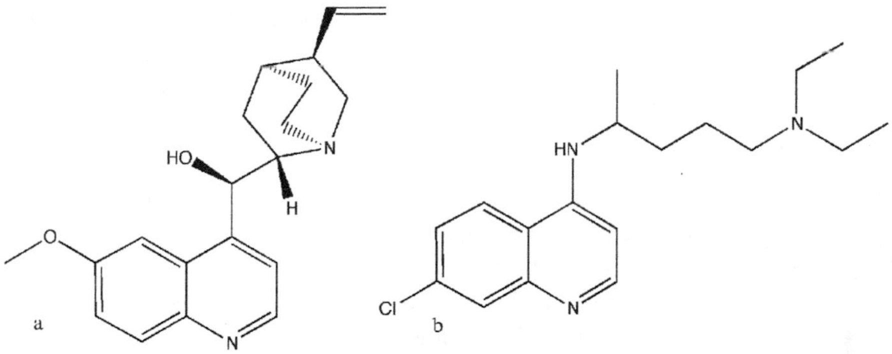

FIGURE 8.2 Molecular structures of quinine and its derivative: (a) quinine and (2) chloroquine.

The herb *Artemisia annua* L. (sweet/annual wormwood, *qinghao*; Asteraceae) was one of several hundred species listed in TCM *Pharmacopoeia* to be considered. According to its listing, *A. annua* is *cool* in nature (yin) and can be used to treat internal *heat* conditions (yang). The first report of the use of *A. annua* in this context appeared in 1596 (in TCM materia medica *Ben Cao Gang Mu*) for treating *hot and cold because of intermittent fever illness*. According to the old pharmacopeias, one bunch of the leaves, collected in spring or summer, was taken with two *sheng* (0.4 L) of water and pounded with a pestle and mortar to extract the *juice*. This procedure was, perhaps, intended to improve the recovery of essential oils (EOs) from the leaves, in which the active principal ART is now known to be most concentrated. However, when the Chinese scientists made hot water extracts of *A. annua* according to the ancient texts, they observed no activity versus mice infected with the murine malarial parasite *Plasmodium berghei*. They tested cold ethereal extracts of *A. annua*, which showed encouraging activity, and, in 1973, ART (a stable and easily crystallizable compound) was isolated and characterized (chemical structure, 1975). Zhang proposed Qs, Hs, and As on Project 523.[22]

Q1. How could the Chinese, in such a short time, have achieved such high-quality results?

H1. High temperature used in preparing *A. annua* destroyed plant active ingredient, affecting antimalarial properties.

H2. (Li, 1974). Mixing crushed ART tablets with water, drug is administered to comatose cerebral malaria patients by a feeding tube via nose.

H3. (Li, 1974). Parasites are destroyed faster in this way than via intravenous quinine (Fig. 8.2a).

H4. Preliminary clinical trial H: ART is a fast-acting antimalarial drug with rapid control of malaria parasite development achieved with a first 0.3–0.5 g dose.

H5. Untrue H (after quinine): A nitrogen heteroring is necessary for an effective antimalarial.

Q2. Who discovered what first?

Q3. Did the extracted crystals contain active antimalarial substance?

Q4. Who discovered what and when?

Q5. Was ART treatment effective?

H6. (Project 523, 1975). ARTDs development/new compounds in ART combination therapy (ACT) for other diseases.

Q6. What is the action of the peroxide group $-O_1-O_2-$ in ART?

Q7. If peroxide group $-O_1-O_2-$ was effective versus malaria, would other peroxides also be efficient?

H7. In creating an effective ARTD, it was necessary to keep ART peroxide group/parent nucleus.

H8. (Project 523, 1977). Increasing the solubility of the water-soluble ARTDs.

Q8. How would one know the chemical structure of ARTD dihydroartemisinin (DHA) (Fig. 8.1b)?

Q9. *Who does know which glass is sterile?*

Q10. Why did not the World Health Organization support more effective/clinically tested artemether (ARM, Fig. 8.1c) rather than less efficient arteether (Fig. 8.1d)?

A10. Unfounded rumors existed that ARM metabolized releasing methanol, increasing toxicity.

Q11. How could decision to drop ARM collaborative research and develop arteether happen?

Q12. Why could Chinese discover/develop ART/ARTDs and not they produce a final drug product meeting world-class standards from manufacturing capabilities?

A12. China's isolation, not lack of ability or intelligence, was at fault.

Q13. How to develop international collaboration to create a pathway for Chinese drugs?

Q14. Are the methods accurate and reliable for measuring blood levels of ACT drugs?

Q15. Is the therapeutic efficacy of ACT credible?

Q16. Did ARTDs affect the development of the babies born to treated pregnant women?

A16. ARTDs were appropriate in treating uncomplicated *falciparum* severe malaria in women in pregnancy 2/3rd trimester?

H9. Untrue H (Project 523, 1978): ART had no effect on gametocytes in *Anopheles* mosquitoes.

H10. An alternative method exists for controlling malaria spread eliminating gametocytes infectivity.

Q17. Why did Li team decide to devote completely their research to new ACTs?

H11. (Li, 1991). He indicated immediate artesunate (ARS, Fig. 8.1e) use in all epidemic/endemic Vietnam areas.

H12. (Li). Use of large amounts of antimalarial drugs during Indochina War (1960/1970s) rendered Vietnam most severe multidrug-resistant *falciparum* malaria endemic area.

H13. Artequick ACT is an alternative method to reverse severity of/bring under control malaria epidemics.

Q18. How, during *Cultural Revolution* with repressed science/intellectual endeavors and limited technical resources/equipment, could China made such important scientific progress?

Q19. Key Q: How to solve the drug resistance problem?

H14. To solve such problem, it is necessary to find new drugs with new chemical structures.

H15. Project 523 units have the ability to accomplish assignment, as in a major military campaign.

8.2 TENTATIVE MECHANISM OF ACTION OF ARTEMISININ

The unusual ART structural feature, the 1,2,4-trioxane ring, is the basis for the unique antimalarial action of the drug. The pharmacophoric peroxide linkage $-O_1-O_2-$ in the endoperoxide ring *triggers* ART to *explode* (but only in the vicinity of the *Plasmodium* parasite). Bond $-O_1-O_2-$ is cleaved when it comes into contact with Fe^{II}, releasing reactive radicals, which ultimately destroy the parasite (cf. Fig. 8.3). Substantial quantities of reactive Fe^{II} accumulate inside an infected erythrocyte as a result of the liberation of the hem group, which is a by-product of the digestion of Hb. However, the effective *delivery* of ART *bomb* to the infected erythrocytes, where it can be detonated, suffers from one fundamental problem, which the Chinese scientists encountered when attempting to prepare their herbal teas from *A. annua*:

ART is poorly soluble in water (and oil). When creating the first generation of ARTD drugs in 1970s, the overriding goal was to improve its solubility characteristics, so that ARTDs be more easily formulated and efficiently delivered.

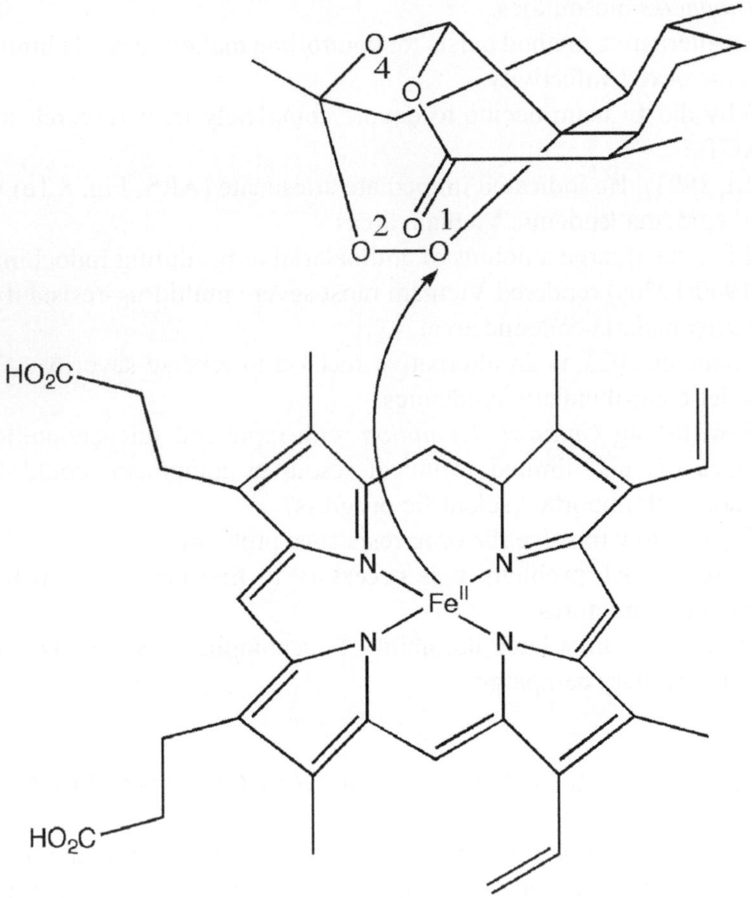

FIGURE 8.3 Cleavage of the endoperoxide bond $-O_1-O_2-$ in ART by Fe^{II} in a hem group.

8.3 ARTEMISININ (*QINGHAOSU*): AN ANTIMALARIAL DRUG FROM CHINA

The herb *A. annua* was used for many centuries in TCM as a treatment for fever and malaria.[23] In 1971, Chinese chemists isolated, from the leafy portions of the plant, the substance responsible for its reputed medicinal action. The compound, called ART, is an STL that bears a peroxide group $-O_1-O_2-$ and, unlike most other antimalarials, lacks an N-containing hetero-cyclic ring system. The ART was used successfully in several thousand malaria patients in China, for example, with both chloroquine-sensitive and -resistant strains of *P. falciparum*. The ARTDs (e.g., DHA, ARM, water-soluble Na-ARS) appear to be more potent than ART. Na-ARS acts rapidly in restoring to consciousness comatose patients with cerebral malaria. The ART/ARTDs offer promise as a totally new class of antimalarials. Klayman proposed the following H.

H1. (Qinghaosu Antimalaria Coordination Research Group, 1979). ART has no effect on early/persistent exoerythrocytic tissue stage.

8.4 ARTEMISININ AND A NEW GENERATION OF ANTIMALARIALS: RESISTANCE

The following questions were raised on ART-resistant malaria, challenges and public health.

Q1. Why ARTD compounds with their elaborate functionality are so effective?

Q2. How are parasites becoming more tolerant to ARTDs?
Brown proposed questions and Hs on ART new generation of antima-larial drugs.[24]

Q3. What hope does ART drug offer for the future?

H1. Pestle-and-mortar procedure betters EOs recovery from leaves, in which ART is concentrated.

Q4. Could the 1,2,4-trioxane ring embedded in the structure of ART be stable?

Q5. Could ART ever really be useful as a drug?

Q6. How long can the situation of no documented clinical case of resis-tance to ARTDs last?

H2. (Jambou, 2005). Resistance is arising in areas with uncontrolled use of ARTDs.
Krishna reviewed and proposed H/Q on ARTDs growing importance in medicine.[25,26]

H3. Safe and cheap ARTD drug class saving lives at risk from malaria is important in oncology.

H4. The ARTDs rival acetylsalicylic acid (ASA) in the breadth of their antidisease properties.

Q7. Why have ARTDs the potential to rival ASA in the breadth of their antidisease properties?

Q8. What are the mechanisms of antimalarial action of ARTDs?

Q9. (Haynes, 2007). What is the role of Fe^{II} species in the antimalarial actions of ARTDs?

Q10. Is there a single important target for ARTDs in *Plasmodium* spp. or are multiple targets?

Q11. (Krishna, 2006). How do ARTDs work in light of emerging evidence of in vitro resistance?

Q12. How might structurally related drugs, for example, the fully synthetic trioxolanes, work?

Q13. What is the basis for the development of resistance by parasites to this class of antimalarial?

H5. (Peters and Richards, 1984). Murine malarias are models for understanding resistance mechanisms to different antimalarial classes.
They proposed Hs on ARTDs mechanisms of action and molecular targets in *food vacuole*.

H6. Hem association forming insoluble hemozoin (Hz) is aided by histidine (His)-rich protein II.

H7. Hem detoxification pathway: ARTDs act on parasite Hb-digestion processes in *food vacuole*.

H8. SERCA H: Ca^{2+} pump localized to endoplasmic reticulum (ER) PfATP6, *P. falciparum* SERCA.

H9. ART disrupts mitochondrial membrane potential when grown in nonfermentable conditions.

H10. The ARTDs target the translationally controlled tumor protein ortholog.

H11. The shared-target H for mechanisms of action.
They proposed additional Qs/Hs on ARTDs and their growing importance in medicine.

Q14. Should ARTDs remain relegated to compounds category with in vitro properties versus cancers?

H12. The ARTDs do not remain relegated to such large category of compounds.

Q15. How are the dosing regimes?

Q16. How is the safety of long-term use?

Q17. How might interactions with existing therapies/toxicities be related to tumors treatment?

H13. Interactions exist with existing therapies/toxicities that might be related to tumors treatment.

8.5   ART-INDUCED DORMANCY IN *P. FALCIPARUM*, DURATION, RECOVERY, AND FAILURE

Cheng group proposed H/Qs on ART-induced dormancy in *P. falciparum*, recovery and failure.[27]

H1.   Temporary growth arrest of ring-stage parasites (*dormancy*) after ARTs exposure explains recrudescence.

H2.   ART-induced growth arrest occurs being a key factor in *P. falciparum* malaria treatment failure.

H3.   The failures are not because of the development of parasite resistance.

Q1.   (Giao, 2001). Is there a place for ART monotherapy for treatment of uncomplicated *falciparum* malaria?

H4.   (White, 1997; Giao, 2001). Recrudescence is because short ART half-life results in plasma–drug concentrations not remaining above minimum inhibitory concentration (IC).

Q2.   How may parasite dormancy effect ART treatment failure?

Q3.   How long do parasites remain dormant after ART exposure?

Q4.   What proportion of dormant parasites does recover?

Q5.   How is the dynamics of the recovery?

Q6.   Are companion drugs used in ACT formulations effective versus dormant parasites?

H5.   Dormancy facilitates the development of resistance to ART drugs.

H6.   A better understanding of dormancy helps to reveal the mechanisms of ART resistance.

H7.   Some W2 parasites recover from dormancy after day 20.

H8.   Dormancy theory (DT, Kyle, Webster, 1996): Dormancy is specific to ARTDs, given that it could not be induced by other antimalarials, for example, quinine.

Q7.   How long do dormant parasites survive and recover?

H9.   The recovery of dormant parasites was continuing at the end of the experimental period.

H10.   Repeated-treatment continuation over a longer period results in a reduction of recovery rate.

H11.   (White, 1999). Antimalarials should no longer be used alone to protect from resistance emergence.

H12.   Delay/decay in recovery was caused by mefloquine/DHA-killing parasites that continued growing after first treatment and dormant parasites recovering early after treatment.

H13. (Lewis, 2007). Dormancy facilitates ART-resistance development giving that persister cells of various microbes were associated with resistant mutant emergence.

H14. Dormancy is a mechanism used by *P. falciparum* parasites to survive treatment with ART.

H15. The link between dormancy and the development of resistance requires urgent attention.

Nosten proposed Hs/Qs on *waking the sleeping beauty* and ART-induced dormancy.[28]

H16. DT (Kyle, Webster, 1996): Some parasites exposed at ring stage in vitro become metabolically inactive and resume growth after drug removal.

H17. Soon dormant-parasites recovery explains some recurrences in patients during weeks after treatments.

H18. The proportion of parasites recovering from the dormancy stage is dose dependent.

Q8. Is dormancy observed in nature?

Q9. How may it contribute to recurrences of malaria in patients treated with ART drugs?

Q10. Have different ARTDs or other (synthetic) peroxides –O–O– different effects on dormancy?

Q11. Have different partner drugs unlike effects on proportion of parasites recovering from dormant stages?

Q12. Can this observation be related to resistance to ARTs?

H19. With a correct prescription, ARTDs were universally effective versus *P. falciparum*.

Q13. Could *P. falciparum* Western Cambodia new phenotype be explained by DT?

H20. Decline in parasitemia levels after treatment in Cambodia does not indicate recovery of a transiently dormant parasites minority.

Q14. Are both observations (dormancy phenomenon, slow-clearance phenotype) totally unrelated?

Q15. Is observation that parasites must be exposed to DHA *at ring stage* random or are some parasites genetically primed to become dormant?

Q16. Are parasites that recover from dormancy more *tolerant* to DHA?

Q17. Have they adapted?

Q18. Could these effects on the ring stage explain the delayed clearance?

8.6 ART DISCOVERY FROM CHINESE GARDEN AND DIFFERENTIAL SENSITIVITY

Miller and Su proposed questions and answer on ART discovery from Chinese herbal garden.[29]

Q1. Without a publication record, who should be credited with the discovery of ART?

A1. Major credit must go to Tu (Inst. Chinese Materia Medica, China Acad. Chinese Medical Sci.).

Q2. Important unanswerable Q. How effective ART or ARTDs will be in the future?

Q3. What drug to adopt next if resistance to ART becomes a problem?

Q4. However, who is going to develop these new drugs?

Q5. How effective is any treatment or control measure in reducing mortality?

Tilley group proposed Hs, paradox (Pa), and Qs on ART activity versus *P. falciparum*.[30]

H1. (O'Neill and Posner, 2004). Hb parasite uptake and hydrolysis are necessary for ART activity.

H2. (Haynes, 2010). Activation of ARTs is metal independent.

H3. (Eckstein-Ludwig, 2003; Krishna, 2010). ART inhibits Ca^{2+} ATPase (PfATP6).

H4. The Hb digestion plays a critical role in the mechanism of action of ART drug class.

Pa1. One of the effects of ART is to inhibit uptake of Hb, the source of the ART activator.

H5. An Hb degradation product is needed for the antimalarial activity of ART.

H6. (Becker, 2004). The detoxification processes are barely sufficient to prevent parasite damage.

H7. If parasites evolve mechanisms to delay Hb uptake or hydrolysis, they circumvent ART action.

H8. This effect is particularly relevant clinically because of short half-lives of ART antimalarials.

H9. The ART-mediated inhibition of Hb uptake drives some parasites toward a quiescent state.

H10. Slowed growth is a general stress response.

H11. Such ART-mediated dormant state allows parasite to tolerate afforded short drug exposure.

H12. Growth stalling at a stage before Hb uptake is initiated or short-term inhibition of Hb uptake/degradation is sufficient to allow parasite survival until ARTs serum levels fall below ICs.

H13. (Noedl, 2008; Dondorp, 2009; Witkowski, 2010). Delayed Hb uptake explains postponed parasites clinical ARTs' clearance in Southeast Asia.

H14. Inhibitors of Hb hydrolysis are not suitable for use in combination with ARTs.

H15. Unique chemotherapy modes facilitating Hb uptake/hydrolysis/intracellular oxidative stress enhance ARTs action/circumvent evolving tolerance mechanisms.
They proposed Hs/Q on altered temporal response and differential sensitivity to ART.[31]

H16. The parasite experiences a saturable effective drug dose.

H17. Young ring-stage parasites exhibiting no Hb uptake show low sensitivity to ARTs.

H18. Youngest rings exhibited ARTs hypersensitivity due to stress from synchronization protocol rather than intrinsic sensitivity difference.

H19. Standard 3-day drug assays do not reveal differences in short-lived ART parasite strains sensitivity.

H20. Long exposure times conceal what results clinically relevant differences in drug sensitivities.

H21. Population average sensitivity is not sufficient to classify a strain as resistant or sensitive.

H22. Parameter t_{50}^e, a parasite-population average property, is not a predictor of fraction of parasite population surviving drug.

H23. Parameter t_{50}^e is only poorly correlated with delayed parasite clearance.

H24. There is a tightly synchronized infection of mid-ring-stage parasites at treatment beginning.

H25. The parasites remain tightly synchronized during treatment.

H26. Treatment of an infection with D10 strain reduces parasite load to 0.004% after 48 h.

H27. Assays via short pulses in vitro, combined with simple analytical model, provide correlation with parasite clearance times in field.

H28. (Ter Kuile, 1993; Skinner, 1996). Hypersensitive rings (hRings) presence in culture with ring-stage parasites is responsible for variability in ring-stage parasites drug-sensitivity reports.

H29. Clinically relevant differences in the responses of strains are not picked up in standard assays.

H30. To measure one parameter, for example, 50% lethal dose LD_{50} (t^e), representing a parasite-population average property is not sufficient.

Q6. (Teuscher, 2010; Witkowski, 2010; Tucker, 2012). Are dormant/ tolerant parasites subpopulations involved in ART resistance?

H31. (Mok, 2011). *P. falciparum* resistance is associated with altered transcription temporal pattern.

H32. (Cheeseman, 2012). Resistance in *P. falciparum* is linked to a region of chromosome 13.

8.7 A MOLECULAR MECHANISM OF ART RESISTANCE IN *P. FALCIPARUM* MALARIA

Haldar group proposed H/Q on ER phosphatidylinositol 3-phosphate (PI3P) lipid-binding targeting malaria proteins to host cell.[32]

H1. (Boddey, 2010; Russo, 2010). Cleavage is because of a resident ER protease plasmepsin V.

H2. (Boddey, 2010; Russo, 2010). Cleavage by plasmepsin V is mechanism for host targeting (HT).

H3. Lipid binding and protein export are linked.

H4. The HT signals of multiple, important malarial effector proteins bind PI3P.

H5. PI3P binding is generalized property of range of malaria secretome effectors exported to erythrocyte.

H6. The PI3P is present in highly localized secretory regions within the parasite.

H7. PI3P-enriched regions are present early in secretory pathway, in ER, but not concentrated in Golgi/periphery.

Q1. Is HT signal on the endogenous ER form of an effector protein associated with PI3P?

H8. Most N-terminal-peptide analysis indicated AAAA-site cleavage for wild type/mutant.

H9. The PI3P binding is the mechanism of export to the host cell.

H10. Green fluorescent protein is recognized by signal peptidase/released from membrane.

H11. Lipid binding plays a major role in targeting malaria parasite proteins to the host erythrocyte.

H12. Nanomolar affinity displayed by HT signal of virulence determinants containing R/KxLxE motif indicates new PI3P-binding mode.

H13. The HT-signal lipid interactions occur early in ER.

H14. At steady state, detectable amount of endogenous precursor carrying HT signal was not plamepsin-V processed.

H15. Export mechanism is efficient, early sorting event in ER dependent on high-affinity binding to PI3P in ER lumen.

H16. Association by PI3P facilitates protease cleavage by plamepsin V.

H17. Cleavage per se does not provide HT specificity rather it releases protein from ER membrane.

H18. The HT-independent export of parasite protein reporters to the host can occur.

H19. Non-HT-dependent export to the erythrocyte is based on charge.

Q2. How does the erythrocyte recognize putative protein cargo?

H20. PI3P and plamepsin V recycle back to ER once HT signal sorting and cleavage are completed.

Q3. Is phosphatidylinositol 3-kinase (PI3K) recruited in the secretory pathway?

H21. (Tawk, 2010). Blood cell infection increases PI3P levels.

Q4. What is the overall ratio of PI3P in ER to total cellular PI3P?

H22. The PI3P also functions in *Plasmodium infestans* ER.

H23. PI3P–ER binding is a generalized mechanism for pathogenic secretion in eukaryotic pathogens.

H24. Such mechanism aspects are targeted to disrupt pathogen–host interactions underlying disease.

Haldar group raised questions on a molecular mechanism of ART resistance in malaria.[33]

Q5. Do ART biochemical targets interact with genes identified by genome-wide association studies (GWAS)?

Q6. How do such targets interact with genes identified by GWAS?

H25. Molecular dynamics (MD) indicates DHA polar contacts with D1889 OH/Y1915 lactol O of *P. falciparum* PI3K (PfPI3K) at binding site.

Q7. Was ARS PI3K–AKT pathway inhibition due to PI3K/AKT direct inhibition, DHA metabolite, etc.?

H26. The DHA specifically targets PfPI3K.

H27. (Miotto, 2013). More than 1000 genes (e.g., PfPI3K) show clinical resistance to ARTs.

H28. Mutations decrease affinity for a protein substrate increasing its steady-state levels reducing ubiquitination/degradation.

Q8. Were PfPI3K level changes associated with PfKelch13 mutations in clinical/engineered laboratory strains?

H29. Analysis of clinical isolates from Cambodia indicated 1.5–2-fold rise in PfPI3K levels in resistant versus sensitive strains.

H30. Strain ANL-7 contained resistance mutation R539T: genome sequencing indicated contamination with resistant parasite strain.

H31. Ubiquitination drives degradation rather than a change in cellular localization.

H32. Elevation of kinase and/or its products provides a mechanism of ART resistance at this stage.

H33. The PfKelch13 mutation linked to resistance increases levels of PfPI3K.

H34. Equal PfPI3K amounts across two distinct nonisogenic strains do not show equivalent activity.

Q9. Did PfPI3K activity measured by PI3P production estimate resistance across nonisogenic strains?

H35. The VPS34 (a class III kinase) transgenic lines remained responsive to PfKelch13.

H36. As PI3P is signaling lipid, small changes in levels induce downstream activation of large resistance amplitudes.

Q10. Could additional cellular components influence PI3P/ring-stage survival assay (RSA) levels?

H37. In parasites, DHA blocks *P. falciparum* AKT (PfAKT) via inhibition of PfPI3K.

H38. (Yano, 1998). The PfAKT functions via a calmodulin-binding PH domain protein.

H39. PfAKT transgenic elevation induced PI3P 1.8-fold rise stimulated by feedback mechanisms.

H40. (Miotto, 2013). Secondary genes influence PfPI3K/PI3P levels explaining why AKT is under ART-selective pressure.

H41. PfPI3K absolute levels across nonisogenic strains do not measure PfPI3K activity/PI3P product and ART resistance.

H42. Fact that serum lipids be essential for intracellular blood stage parasite growth explains wide RSA range displayed by resistant, clinical strains.

Q11. Why do two resistant strains, for example, ANL-2/4, bear the same mutation C580Y and equivalent PfPI3K levels, but different PI3P-product amounts and resistance levels?

H43. PfKelch13 resistance mutations regulate PfPI3K activity to contribute to differences in RSA.

H44. (Painter, 2011; Cheeseman, 2012; Miotto, 2013; Takala-Harrison, 2013). PI3P influences host remodeling and apicoplast/food-vacuole functions, etc. implicated in ART resistance.

Burrows proposed Qs, hypotheses, and answers on malaria-running rings round ART.[34]

Q12. How is resistance conferred?

H45. A rise in PfPI3K levels confers resistance.

Q13. How is such kinase linked to *P. falciparum* Kelch13 (PfKelch13)?

H46. (Haldar, 2015). Mutations in PfKelch13 inhibit its association with PfPI3K.

H47. A drug that kills parasites in different life-cycle stages is driven by a single biotarget rather than different targets affecting different stages.

H48. DHA weak O–O bond is necessary for reversible PfPI3K inhibition.

H49. (Das, 2013). Resistance is overcome rising ART dose in ACT to inhibit PfPI3K more potently.

Q14. Why do parasites survive when levels of PfPI3K and PI3P rise?

Q15. Exactly, how does DHA inhibit PfPI3K?

Q16. How do PfPI3K and PfKelch13 bind together?

Q17. Why do mutations hinder such binding?

Q18. Why does polyubiquitination rise following PfPI3K–PfKelch13 binding?

Q19. Does tagging lead to intracellular signaling events or solely to complex destruction by proteasome?

Q20. Will this knowledge of the ring-stage target of DHA help people to design better drugs?

A20. As resistance is due to factors other than PfPI3K mutation, new PfPI3K inhibitors will not only need to be potent but also versus key ring-stage-resistant parasites.

Q21. Might parasite overcome PfPI3K inhibitors via mutations reducing PfKelch13–PI3K binding?

Q22. (Witkowski, 2013). Do they show PfPI3K inhibition/ring-stage resistance similar to DHA?

8.8 REDUCED ART *P. FALCIPARUM*-RING-STAGES SUSCEPTIBILITY IN CAMBODIA

Menard group proposed Hs/Q on reduced ART *P. falciparum*-ring stages susceptibility.[35]

H1. (Tucker, 2012). Resistant parasites tolerate more drug by exiting dormancy/resuming growth at greater rate than susceptible parental strains.

H2. (Tucker, 2012). Dormant resistant parasites present a higher survival rate after dormancy.

Q1. How do different findings relate to the reduced susceptibility of field isolates in Cambodia?

H3. ART resistance in Pailin (Cambodia) is associated with risen number of rings capable of entering transient developmental arrest and resuming growth after drug removal.

H4. Pailin parasites withstand DHA toxicity thanks to a risen number of rings able to enter quiescence/remain alive at high DHA concentrations.

H5. Such risen number of quiescent forms accounts for ability to recover earlier than susceptible strains and recrudesce at higher numbers than susceptible parasites.

Winzeler and Manary raised Qs on drug resistance genomics of anti-malarial drug ART.[36]

Q2. (Ferreira, 2013). What is ART resistance in *P. falciparum* really?

Q3. Should ARS treatment first failed/ultimately succeed be best described as *ART tolerance*?

Q4. Are there other ways to create resistance besides mutations in gene *P. falciparum* Kelch13 (*pfkelch13*)?

Q5. Is any additional mutation in other chromosomal region identified as under selection in population studies?

Q6. Which proteins does PfKelch13p interact with?

Q7. In addition, would these proteins interacting with PfKelch13p also be resistant determinants?

Q8. Will *pfkelch13* mutations make parasites resistant versus synthetic endoperoxides, for example, OZ439?

Q9. As ART has greatest effect on trophozoite/schizont-stage parasites, are there other genes, for example, *pffalcipain-2*, mutated in field samples?

Cravo et al. raised Qs on genomics contributing to fight ART-resistant malaria parasites.[37]

Q10. How is genomics contributing to the fight against ART-resistant malaria parasites?

Q11. How does the parasite develop resistance to ART and ARTDs?

Q12. How may resistant parasites be monitored/tracked in real time (RT) by molecular approaches?

Q13. How does the parasite evolve resistance to these drugs?

Q14. How are genomics approaches accelerating discovery of genetic mutations linked to ARTDs resistance?

Q15. What are the means by which they are used for RT molecular surveillance of resistance?
Wang et al. proposed Q/Pa on ART hem-activated promiscuous targeting in *P. falciparum*.[38]
Q16. What is the origin of Fe sources required for ART activation?
Q17. What are the exact targets of activated ART?
Q18. (O'Neill, Barton, and Ward, 2010). What is the molecular mechanism of action of ART?
Q19. How does ART kill the parasite?
Q20. How is ART activated?
Q21. Is either hem or free Fe^{2+} the prerequisite for drug activation?
Pa1. (Sigala and Goldberg, 2014). There are peculiarities and Pas in *Plasmodium* heme metabolism.
Q22. Is parasite's hem biosynthesis pathway involved in ART activation at the early ring stage?

8.9 MULTIDRUG-TOLERANCE INDUCTION IN *P. FALCIPARUM*: EXTENDED ART PRESSURE

Benoit-Vical group proposed Qs and Hs on induction of multidrug tolerance in *P. falciparum*.[39]

Q1. Does delayed parasite clearance, which exposes larger parasite numbers to ARTs for longer times, select higher grade resistance?
Q2. Does long-lasting ART pressure select a novel multidrug-tolerance profile?
Q3. Does ART resistance drive selection of higher grade resistance or resistance to partner drug?
H1. Reduced metabolic activity of quiescent older forms decreases toxicity of antimalarial drugs inhibiting parasite metabolic pathways (e.g., hemozoin formation, tetrahydrofolic acid synthesis).
Q4. Was tolerance to chloroquine associated with multidrug tolerance?
Q5. Is acquisition of multidrug tolerance a multistep process?
H2. An analysis of the cryopreserved intermediate time points of F32-ART lineage is needed.
Q6. Does finding that not only young but also older ring forms survived ART reflect different selection processes in field in Cambodia/during in vitro model?

H3. Sustained pressure on ART-resistant parasites drives resistance selection in older parasite stages and results in decayed efficacy of partner drug in field.

H4. The retained efficacy of atovaquone indicates alternative combination treatments.

8.10 ART/ARTDS AS A REPURPOSING ANTICANCER AGENT: WHAT DO PEOPLE NEED?

Yu group proposed Qs, Hs, and As on ART/ARTDs as a repurposing anticancer agent.[40]

Q1. What else do people need to do?

H1. Preclinical investigation/clinical experience provided evidence on ART/ARTDs anticancer effect.

H2. Major ARTs action mechanisms are due to toxic-free radicals generated by an endoperoxide moiety, cell-cycle arrest, apoptosis induction, and tumor-angiogenesis inhibition.

H3. ARTs are a new class of wide-spectrum antitumor drugs due to detailed efficacy/safety information.

H4. ARTs characteristics must be studied, for example, anticancer-effect pathways, efficient/specific drug delivery systems (especially crossing biobarriers), and data in clinical trials.

H5. The aim is to propose strategies to develop ARTs as a new class of cancer therapeutic agents.

H6. (Boichuk, 2014). Old approved drugs promise to have faster clinical trials than a brand-new drug.

H7. Beside anti-inflammatory and immunomodulatory functions, ARTs show anticancer effect (cf. Fig. 8.4).

Q2. How do ARTs kill tumor cells?

A2. ARTs play a role in cell swelling; once cytomembranes are destroyed, membrane permeability is altered, resulting in cells death.

H8. C-radical H: Toxic-free radicals generated by ART endoperoxide moiety via an Fe^{2+}-mediated reaction are essential to killing/attenuating tumor cells.

H9. (Efferth, 2006). Reactive oxygen species role is initiating factors in triggering cell damage.

H10. (Efferth, 2010). ARS interferes with some genes (e.g., Bub3, Mad3, Mad2), which are regulators of mitotic spindle checkpoint in G2/M phase.

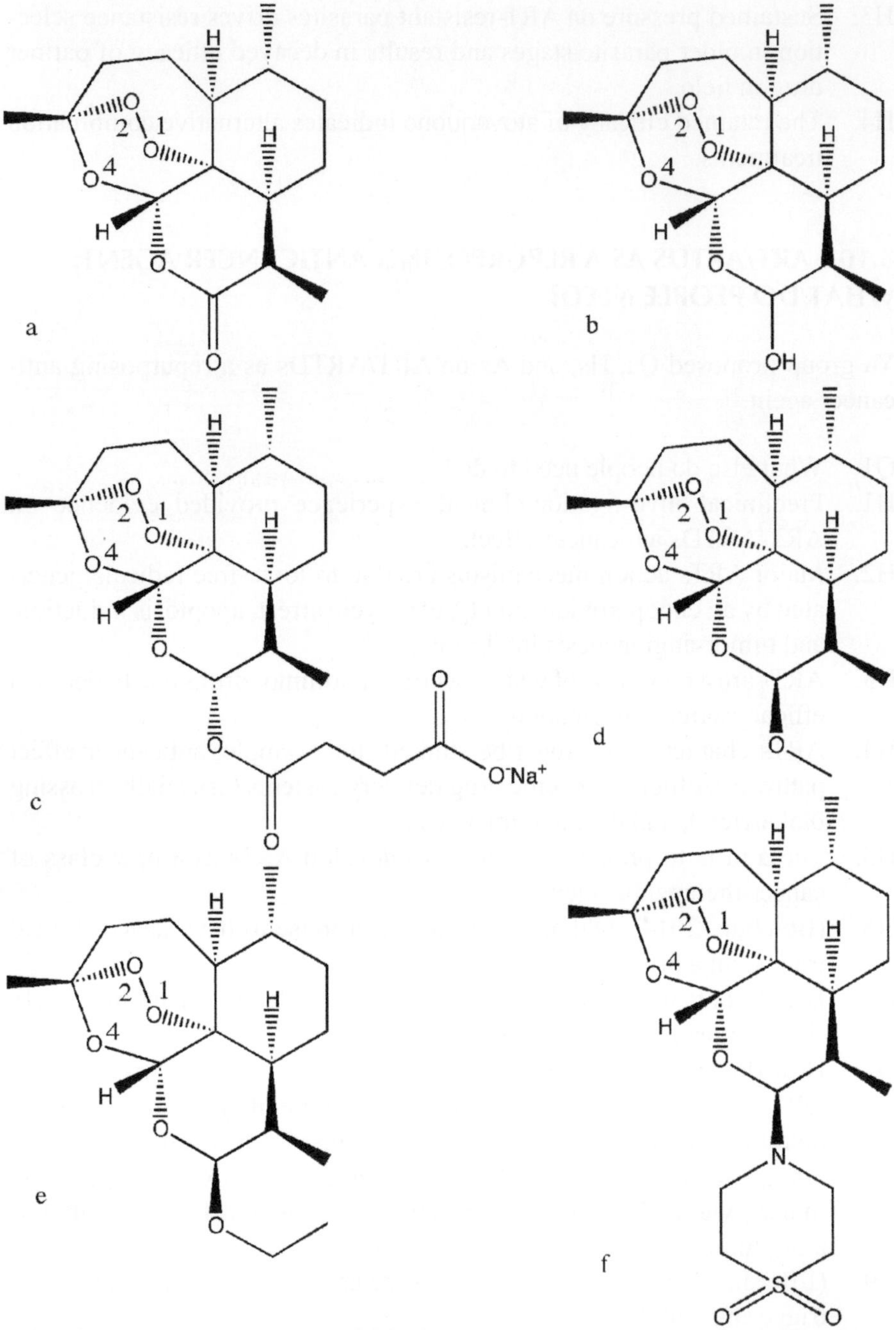

FIGURE 8.4 ART/ARTDs with anticancer activity: (a) ART, (b) DHA, (c) ARS, (d) ARM, (e) arteether, and (f) artemisone.

H11. (Mu, 2007, 2008; Efferth, 2015). ARTs that stimulated rises in intracellular Ca^{2+} levels and triggered oxidative damage are involved in apoptosis in tumor cells.

H12. DHA in inhibited angiogenesis largely relies on nuclear factor-κ gene binding (NF-κB) pathway.

H13. ARTDs can arouse embryotoxicity in experimental animals.

Q3. What needs to be emphasized?

A3. (Oliveira, 2013). ARTs toxicity is long-term effects and unrelated to short-term drug–plasma concentration.

H14. (Guo, 2016). Pathways that ARTs interfere with must be followed, for example, Fe-dependent H, anti-inflammatory effects on preventing tumor formation, chemo/radiotherapy synergistic/sensitizing mechanisms and ways to reverse multidrug resistance in tumor cells.

H15. Fe-dependent ART activation H was accepted due to fact that a high intracellular Fe concentration was essential for tumor cells continuous proliferation.

H16. ARTs anti-inflammatory effectiveness plays a role in malignant tumors occurrence/progression.

H17. Modified nanoparticles (NPs) are preferred because of ARTs' characteristic anticancer mechanism.

Q4. (O'Driscoll, 2015). What NPs do they pass via the biological barriers effectively?

Q5. (O'Driscoll, 2015). How NPs do they pass via the biological barriers effectively?

H18. For refractory patients with advanced cancer who had no response to existing treatments, single-arm trial design is considered via historical background data as contrast.

Q6. What kind of malignancy and chemotherapy agent can get the optimal results for the condition?

Q7. What drug combination with?

Q8. What is the optimal dose for the effect of enhanced sensitivity to radiotherapy?

8.11 FINAL REMARKS

From the present discussion, the following final remarks can be drawn.

1. Despite much research, ART remains the only known natural product to contain a 1,2,4-trioxane ring and *A. annua* continues to be the only known natural source. Phytochemical investigation of the species

revealed an abnormally wide range of other endoperoxides –O–O– and hydroperoxides –O–O–H, many of which were not tested for their antimalarial activity.[41]

2. Design, synthesis, and cytotoxicity of novel DHA-coumarin hybrids via click chemistry were reported.[42]

3. Further work is required to understand the molecular mechanism by which lactones inhibit c-Myb-dependent gene expression. Active lactones exist without an α-methylene lactone, for example, eremantholides, which allows the hypothesis that carbon-5 activation, is sufficient for the elicitation of antineoplastic activity, which recommends targeting the naturally occurring heliangolides and synthesizing simplified model compounds, containing the vinyl furanose moiety, to evaluate the hypothesis.

4. Work is in progress on quantitative structure–activity relationships of lactones as inhibitors of Myb-dependent gene expression and mechanisms of action.

ACKNOWLEDGMENTS

The authors acknowledge financial support from the Spanish Ministerio de Economía y Competitividad (Project No. BFU2013-41648-P), EU ERDF, Generalitat Valenciana (Project No. PROMETEO/2016/094), and Universidad Católica de Valencia *San Vicente Mártir* (Project No. PRUCV/2015/617).

KEYWORDS

- malaria
- hemoglobin
- ethereal extracts
- Chinese
- antimalarial substance

REFERENCES

1. Ghantous, A.; Gali-Muhtasib, H.; Vuorela, H.; Saliba, N. A.; Darwiche, N. What Made Sesquiterpene Lactones Reach Cancer Clinical Trials? *Drug Discov. Today* **2010**, *15*, 668–678.

2. Sanz, J. F.; Castellano, G.; Marco, J. A. Sesquiterpene Lactones from *Artemisia herba-alba*. *Phytochemistry* **1990**, *29*, 541–545.

3. Marco, J. A.; Sanz, J. F.; Falcó, E.; Jakupovic, J.; Lex, J. New Oxygenated Eudesmanolides from *Artemisia herba-alba*. *Tetrahedron* **1990**, *46*, 7941–7950.

4. Castellano, G.; Tena, J.; Torrens, F. Classification of Polyphenolic Compounds by Chemical Structural Indicators and its Relation to Antioxidant Properties of *Posidonia oceanica* (L.) Delile. *MATCH Commun. Math. Comput. Chem.* **2012**, *67*, 231–250.

5. Castellano, G.; González-Santander, J. L.; Lara, A.; Torrens, F. Classification of Flavonoid Compounds by Using Entropy of Information Theory. *Phytochemistry* **2013**, *93*, 182–191.

6. Castellano, G.; Lara, A.; Torrens, F. Classification of Stilbenoid Compounds by Entropy of Artificial Intelligence. *Phytochemistry* **2014**, *97*, 62–69.

7. Castellano, G.; Torrens, F. Information Entropy-based Classification of Triterpenoids and Steroids from *Ganoderma*. *Phytochemistry* **2015**, *116*, 305–313.

8. Torrens, F. A New Chemical Index Inspired by Biological Plastic Evolution. *Indian J. Chem., Sect. A* **2003**, *42*, 1258–1263.

9. Torrens, F. A Chemical Index Inspired by Biological Plastic Evolution: Valence-Isoelectronic Series of Aromatics. *J. Chem. Inf. Comput. Sci.* **2004**, *44*, 575–581.

10. Torrens, F.; Castellano, G. QSPR Prediction of Retention Times of Phenylurea Herbicides by Biological Plastic Evolution. *Curr. Drug Saf.* **2012**, *7*, 262–268.

11. Torrens, F.; Castellano, G. QSPR Prediction of Chromatographic Retention Times of Pesticides: Partition and Fractal Indices. *J. Environ. Sci. Health, Part B* **2014**, *49*, 400–407.

12. Torrens, F.; Castellano, G. Molecular Classification of Pesticides Including Persistent Organic Pollutants, Phenylurea and Sulphonylurea Herbicides. *Molecules* **2014**, *19*, 7388–7414.

13. Castellano, G.; Torrens, F. Quantitative Structure–Antioxidant Activity Models of Isoflavonoids: A Theoretical Study. *Int. J. Mol. Sci.* **2015**, *16*, 12891–12906.

14. Torrens, F.; Castellano, G. QSPR Prediction of Retention Times of Methylxanthines and Cotinine by Bioplastic Evolution. *Int. J. Quant. Struct. Prop. Relat.*, in press.

15. Torrens F.; Castellano G. QSPR Prediction of Retention Times of Chlorogenic Acids in Coffee by Bioplastic Evolution. In *Quantitative Structure–Activity Relationship*; Kandemirli, F., Ed.; InTechOpen: Vienna, Austria,2017; pp 45–61.

16. Castellano, G.; Redondo, L.; Torrens, F. QSAR of Natural Sesquiterpene Lactones as Inhibitors of Myb-dependent Gene Expression. *Curr. Top. Med. Chem.*, Submitted for Publication.

17. Torrens, F.; Redondo, L.; Castellano, G. Artemisinin: Tentative Mechanism of Action and Resistance. *Pharmaceuticals* **2017**, *10*, 20–24.

18. Collaboration Research Group for Qinghaosu. A Novel Kind of Sequiterpene Lactone–Artemisinin. *Chin. Sci. Bull.* **1977**, *22*, 142–142.

19. Anonymous. Antimalaria Studies on Qinghaosu. *Chin. Md. J.* **1979**, *92*, 811–816.

20. Tu, Y. Y.; Ni, M. Y.; Zhong, Y. R.; Li, L. N.; Cui, S. L.; Zhang, M. Q.; Wang, X. Z.; Liang, X. T. [Studies on the Constituents of *Artemisia annua* L. (authors' translation)] [Article in Chinese]. *Acta Pharm. Sin.* **1981**, *16*, 366–370.

21. Tu, Y. Y.; Ni, M. Y.; Zhong, Y. R.; Li, L. N.; Cui, S. L.; Zhang, M. Q.; Wang, X. Z.; Ji, Z.; Liang, X. T. Studies on the Constituents of *Artemisia annua* Part II. *Planta Med.* **1982,** *44,* 143–145.

22. Zhang, J., Ed. *A Detailed Chronological Record of Project 523 and the Discovery and Development of* Qinghaosu *(Artemisinin)*; Strategic Book: Houston (TX), 2013.

23. Klayman, D. L. *Qinghaosu* (Artemisinin): An Antimalarial Drug from China. *Science* **1985,** *228,* 1049–1055.

24. Brown, G. Artemisinin and a New Generation of Antimalarial Drugs. *Educ. Chem.* **2006,** *2006* (7), 1–7.

25. Krishna, S.; Bustamante, L.; Haynes, R. K.; Staines, H. M. Artemisinins: Their Growing Importance in Medicine. *Trends Pharmacol. Sci.* **2008,** *29,* 520–527.

26. Krishna, S.; Ganapathi, S.; Ster, I. C.; Saeed, M. E. M.; Cowan, M.; Finlayson, C.; Kovacsevics, H.; Jansen, H.; Kremsner, P. G.; Efferth; T.; Kumar, D. A Randomised, Double Blind, Placebo-Controlled Pilot Study of Oral Artesunate Therapy for Colorectal Cancer. *EBioMedicine* **2015,** *2,* 82–90.

27. Teuscher, F.; Gatton, M. L.; Chen, N.; Peters, J.; Kyle, D. E.; Cheng, Q. Artemisinin-Induced Dormancy in *Plasmodium falciparum*: Duration, Recovery Rates, and Implications in Treatment Failure. *J. Infect. Dis.* **2010,** *202,* 1362–1368.

28. Nosten, F. Waking the Sleeping Beauty. *J. Infect. Dis.* **2010,** *202,* 1300–1301.

29. Miller, L. H.; Su, X. Artemisinin: Discovery from the Chinese Herbal Garden. *Cell* **2011,** *146,* 855–858.

30. Klonis, N.; Crespo-Ortiz, M. P.; Bottova, I.; Abu-Bakar, N.; Kenny, S.; Rosenthal, P. J.; Tilley, L. Artemisinin Activity against *Plasmodium falciparum* Requires Hemoglobin Uptake and Digestion. *Proc. Natl. Acad. Sci. U.S.A.* **2011,** *108,* 11405–11410.

31. Klonis, N.; Xie, S. C.; McCaw, J. M.; Crespo-Ortiz, M. P.; Zaloumis, S. G.; Simpson, J. A.; Tilley, L. Altered Temporal Response of Malaria Parasites Determines Differential Sensitivity to Artemisinin. *Proc. Natl. Acad. Sci. U.S.A.* **2013,** *110,* 5157–5162.

32. Bhattacharjee, S; Strahelin, R. V.; Speicher, R. D..; Speicher, D. W.; Haldar, K. Endoplasmic Reticulum PI(3)P Lipid Binding Targets Malaria Proteins to the Host Cell. *Cell* **2012,** *148,* 201–212.

33. Mbengue, A.; Bhattacharjee, S.; Pandharkar, T.; Liu, H.; Estiu, G.; Strahelin, R. V.; Rizk, S. S.; Njimoh, D. L.; Ryan, Y.; Chotivanich, K.; Nguon, C.; Ghorbal, M.; Lopez-Rubio, J. J.; Pfrender, M.; Emrich, S.; Mohandas, N.; Dondorp, A. M.; Wiest, O.; Haldar, K. A Molecular Mechanism of Artemisinin Resistance in *Plasmodium falciparum* Malaria. *Nature (London)* **2015,** *520,* 683–687.

34. Burrows, J. Malaria Runs Rings Round Artemisinin. *Nature (London)* **2015,** *520,* 628–630.

35. Witkowski, B.; Khim, N.; Chim, P.; Kim, S.; Ke, S.; Kloeung, N.; Chy, S.; Duong, S.; Leang, R.; Ringwald, P.; Dondorp, A. M.; Tripura, R.; Benoit-Vical, F.; Berry, A.; Gorgette, O.; Ariey, F.; Barale, J. C.; Mercereau-Puijalon, O.; Menard, D. Reduced Artemisinin Susceptibility of *Plasmodium falciparum* Ring Stages in Western Cambodia. *Antimicrob. Agents Chemother.* **2013,** *57,* 914–923.

36. Winzeler, E. A.; Manary, M. J. Drug Resistance Genomics of the Antimalarial Drug Artemisinin. *Genome Biol.* **2014,** *15,* 544–1–12.

37. Cravo, P.; Napolitano, H.; Culleton, R. How Genomics is Contributing to the Fight against Artemisinin-Resistant Malaria Parasites. *Acta Trop.* **2015,** *148,* 1–7.

38. Wang, J.; Zhang, C. J.; Chia, W. N.; Loh, C. C. Y.; Li, Z.; Lee, Y. M.; He, Y.; Yuan, L. X.; Lim, T. K.; Liu, M.; Liew, C. X.; Lee, Y. Q.; Zhang, J.; Lu, N.; Lim, C. T.; Hua, Z.

C.; Liu, B.; Shen, H. M.; Tan, K. S. W.; Lin, Q. Haem-Activated Promiscuous Targeting of Artemisinin in *Plasmodium falciparum*. *Nat. Commun.* **2015**, *6*, 10111-1–11.

39. Ménard, S.; Haddou, T. B.; Ramadani, A. P.; Ariey, F.; Iriart, X.; Beghain, J.; Bouchier, C.; Witkowski, B.; Berry, A.; Mercereau-Puijalon; O.; Benoit-Vical, F. Induction of Multidrug Tolerance in *Plasmodium falciparum* by Extended Artemisinin Pressure. *Emerg. Infect. Dis.* **2015**, *21*, 1733–1741.

40. Li, Z.; Li, Q.; Wu, J.; Wang, M.; Yu, J. Artemisinin and its Derivatives as a Repurposing Anticancer Agent: What Else Do We Need to Do? *Molecules* **2016**, *21*, 1331-1–14.

41. Brown, G. D.; Liang, G. Y.; Sy, L. K. Terpenoids from the Seeds of *Artemisia annua*. *Phytochemistry* **2003**, *64*, 303–323.

42. Tian, Y.; Liang, Z.; Xu, H.; Mou, Y.; Guo, C. Design, Synthesis and Cytotoxicity of Novel Dihydroartemisinin–Coumarin Hybrids via Click Chemistry. *Molecules* **2016**, *21*, 758-1–15.

CHAPTER 9

CLASSIFICATION OF *CITRUS*: PRINCIPAL COMPONENTS, CLUSTER, AND META-ANALYSES

FRANCISCO TORRENS[1,*] and GLORIA CASTELLANO[2]

[1]Institut Universitari de Ciència Molecular, Edifici d'Instituts de Paterna, Universitat de València, P. O. Box 22085, E-46071 València, Spain

[2]Facultad de Veterinaria y Ciencias Experimentales, Departamento de Ciencias Experimentales y Matemáticas, Universidad Católica de Valencia San Vicente Mártir, Guillem de Castro-94, E-46001 València, Spain

*Corresponding author. E-mail: torrens@uv.es

CONTENTS

ABSTRACT

Citrus species belong to the agricultural biodiversity of the Valencia Region. However, the efficient mobilization of citrus biodiversity in innovative breeding schemes requires previous understanding of *Citrus* systematics, origin, genomic structure, and classification. Size criteria of seven *Citrus* species of the Botanic Garden Collection are classified by principal components analyses (PCAs) of size criteria and species cluster analyses (CAs), which agree. *Citrus* species group into four classes: genus *Citrus*, mandarins, papeda, and lime. Geometric PCA and species CA allow classifying them and concur. Different behaviour of species depends on fruit cross section and leaf surface area. Some criteria reduced the analysis to a manageable quantity from the enormous set of citrus features: fruit size in terms of cross section, leaf size in terms of surface area, etc. The meta-analysis was useful to rise the number of samples and variety of analyzed data. Different behaviour of citrus depends on fruit cross section and leaf surface area. Genus *Citrus*, mandarins, papeda, and lime are detected in separate classes. The classification of citrus shows maximum similarity between *Citrus clementina* and *Citrus depressa*. Maximum difference appears between *Citrus maxima*, *Citrus myrtifolia*, and *Citrus paradisi* because *Citrus myrtifolia* and *Citrus paradisi* are hybrids branching from ancestral species.

The Botanic Garden of the Universitat de València includes a collection of approximately 50 species of citrus, which range from the most well known and commonly grown (e.g., oranges, mandarins, lemons, and grapefruit) to others used as rootstock, widely in perfumery, for example, bergamot, or some others with important religious significance, for example, citron.[1] The collection contains others that are not cultivated for commercial purposes, illustrating the huge variety of citrus species. *Citrus* species cultivars contributed to the agricultural biodiversity of Valencia Region. The taxonomy of *Citrus* species is complex. Genetic analyses showed that only three species belong to genus *Citrus* [*Citrus maxima* (Burm.) Merr., *Citrus medica* L., and *Citrus reticulata* Blanco].[2] *Citrus* species are able to cross-breed, producing fruits with a wide range of morphological and organoleptic characteristics. However, the intervarietal genetic diversity is scarce, especially for sweet oranges, lemons, and grapefruits, which confers a substantial fragility versus emerging diseases, as shown by the ongoing major crisis in the Brazilian and Floridian citrus industries.[3–5] The efficient mobilization of citrus biodiversity in innovative breeding schemes will require prior knowledge of varietal group origins and genomic structures. Deciphering the phylogenomic structures of the secondary citrus species is essential before innovative conventional breeding strategies can be developed. Next-generation haplotyping was used to decipher nuclear genomic interspecific admixture in *Citrus* spp. via analysis of chromosome 2 (cf. Fig. 9.1).[6] Citrus fruits were clustered by genetic similarity [principal components analysis (PCA) of single-nucleotide polymorphism (SNP) diversity].

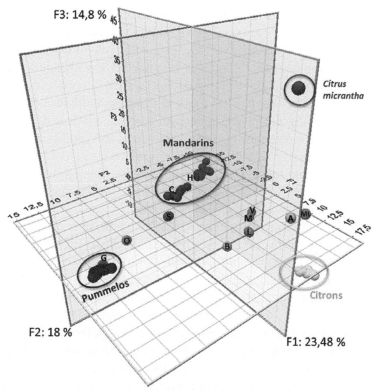

FIGURE 9.1 Organization of genotypic SNP diversity: All varieties and all SNP data were analyzed by PCA. ML: *Mexican* lime; A: *Alemow*; V: *Volkamer* lemon; M: *Meyer* lemon; L: regular and *sweet* lemons; B: bergamot; H: haploid clementine; C: clementines; S: sour oranges; O: Sweet oranges; G: Grapefruits. Source: Curk et al.[6] (Reprinted from Curk, F.; Ancillo, G.; Garcia-Lor, A.; Luro, F.; Perrier, X.; Jacquemoud-Collet, J. P.; Navarro, L.; Ollitrault, P. Next Generation Haplotyping to Decipher Nuclear Genomic Interspecific Admixture in Citrus Species: Analysis of Chromosome 2. *BMC Genet.* **2014,** *15,* 152. © Curk et al.; licensee BioMed Central. 2014. https://creativecommons.org/licenses/by/4.0/)

The citron (*C. medica* L., *etrog*) presents religious uses in both Judaism and Buddhism. It was one of the four fruits that the Jewish God allowed them to use in religious rites. Jews use it for a religious ritual during the Feast of Tabernacles; it is a Jewish symbol, which is found in various Hebrew antiques and archeological findings. Citrons used for ritual purposes cannot be grown by grafting branches. In Buddhism, a variety of citron native to China presents sections that separate into finger-like parts and is used as an offering in Buddhist temples. Antitumour promoting properties of edible plants from Thailand were reported.[7] Glyceroglycolipids from *C. hystrix* inhibited tumor-promoting activity.[8] Antibacterial activity of essential oils (EOs) was reported.[9] Citron cultivation, production, and uses in the Mediterranean Region were reviewed.[10] Anti-inflammatory activity of *Citrus bergamia* derivatives was revised.[11] The protective role of *Citrus* flavonoids neurodegenerative diseases (NDs) was discussed.[12] The chemical composition and in vitro antimicrobial, cytotoxic, and central nervous system (CNS) activities of EOs of *C. medica* cv. *liscia* and *rugosa* cultivated in Southern Italy were informed.[13] The antimicrobial activity confirmed their traditional uses as food-preserving agents and led them to hypothesize the use of EOs as antimicrobials. Alterations in adenylate cyclase (ADCY) 1 expression suggested a role for limonene in CNS effects. Traceability of satsuma mandarin (*Citrus unshiu* Marc.) honey was published via nectar/honey sac/honey pathways of headspace and semi-volatiles.[14]

In earlier publications, it was reported the molecular classification of yams,[15] soyabean, Spanish legumes, commercial *soyabean*,[16,17] fruits proximate and mineral content,[18] and food spices proximate content[19] by PCA, cluster analysis (CA) and meta-analyses. Polyphenols,[20] flavonoids,[21] stilbenoids,[22] triterpenoids, steroids,[23] isoflavonoids[24], natural sesquiterpene lactones,[25] artemisinin, and their derivatives[26,27] were analyzed. The main aim of the present report is to develop code-learning potentialities and, since citrus is more naturally described via varying size-structured representation, find approaches to structured information processing. In view of citrus nutritional and medical benefits, the objective was to categorize them with PCA/CA to differentiate vegetables groups and identify features of various plants. The next section presents the computational method. Following that, Sections 9.2 and 9.3 illustrate and discuss the calculation results. Finally, the last section summarizes our final remarks.

9.1 COMPUTATIONAL METHOD

PCA is a dimension reduction technique.[28–33] From original variables X_j, PCA builds orthogonal variables $\tilde{P}_j$ linear combinations of mean-centred ones $\tilde{X}_j = X_j - \bar{X}_j$ corresponding to eigenvectors of sample covariance matrix $S = 1/(n-1)\sum_{i=1}^{n}(x_i - \bar{x})(x_i - \bar{x})'$. For every loading vector $\tilde{P}_j$, matching eigenvalue $\tilde{l}_j$ of S tells how much data variability is explained: $\tilde{l}_j = \text{Var}(\tilde{P}_j)$. Loading vectors are sorted in decaying eigenvalues. First, k principal components (PCs) explain most variability. After selecting k, one projects p-dimensional data onto subspace spanned by k loading vectors and computes coordinates versus $\tilde{P}_j$, yielding scores:

$$\tilde{t}_i = \tilde{P}'(x_i - \bar{x}). \tag{9.1}$$

For every $i = 1, \ldots, n$ having trivially zero mean. With respect to original coordinate system, projected data point is computed fitting:

$$\hat{x}_i = \bar{x} + \tilde{P}\tilde{t}_i. \tag{9.2}$$

Loading matrix $\tilde{P}$ $(p \times k)$ contains loadings column-wise and diagonal one $\tilde{L} = (\tilde{l}_j)_j$ $(k \times k)$ eigenvalues. Loadings k explains variation:

$$\left(\sum_{j=1}^{k}\tilde{l}_j\right) \Big/ \left(\sum_{j=1}^{p}\tilde{l}_j\right) \geq 80\%. \tag{9.3}$$

CA encompasses different classification algorithms.[34,35] Starting point is $(n \times p)$ data matrix X containing p components measured in n samples. One assumes that the data were preprocessed to remove artefacts and the missing values were imputed. The CA organizes samples into a small number of clusters such that samples within cluster are similar. Distances l_q between samples $x, x' \in \mathfrak{R}^p$ are:

$$\|x - x'\|_q = \left(\sum_{i=1}^{p}|x_i - x'_i|^q\right)^{1/q}. \tag{9.4}$$

(e.g., Euclidean l_2, Manhattan l_1 distances). Comparing samples, Pearson's correlation coefficient (PCC) is advantageous:

$$r(x - x') = \frac{\sum_{i=1}^{p}(x_i - \bar{x})(x'_i - \bar{x}')}{\left[\sum_{i=1}^{p}(x_i - \bar{x})^2 \sum_{i=1}^{p}(x'_i - \bar{x}')^2\right]^{1/2}}. \tag{9.5}$$

where $\bar{x} = \left(\sum_{i=1}^{p}x_i\right)\Big/p$ is measure mean value for sample x.[36–42]

9.2 CALCULATION RESULTS

Seven species are studied from the Botanic Garden citrus collection genus (cf. Table 9.1), namely, three species of genus *Citrus* [pummelo *C. maxima* (Burm.) Merr., Chinotto orange *Citrus myrtifolia* Raf., grapefruit *Citrus paradisi* Macf.], two mandarins (clementine *Citrus clementina* Hort. ex Tan., shekwasha *Citrus depressa* Hay), one papeda (small-fruited papeda *Citrus micrantha* Wester), and one lime [Mexican lime *Citrus aurantifolia* (Christm.) Swing.]. Two size criteria are studied: fruit and leaf sizes. With regard to fruit size, *Citrus* species are very small (*C. micrantha*), small (*C. depressa*), medium (*C. clementina*), large (*C. paradisi*), and very large (*C. maxima*). With regard to leaf size, *Citrus* species are very small (*C. myrtifolia*), small (*C. aurantifolia*), medium (*C. clementina*), large (*C. paradisi*), and very large (*C. maxima*).

TABLE 9.1 *Citrus* Size Criteria: Fruit in Terms of Cross Section and Leaf Size in Terms of Surface Area.

No.	Species	Fruit cross section (cm)	Leaf surface area (cm²)
	Genus *Citrus*	–	–
1	Pummelo *C. maxima* (Burm.) Merr.	15.0	83.4
2	Chinotto orange *C. myrtifolia* Raf.	–	2.5
3	Grapefruit *C. paradisi* Macf.	10.0	45.0
	Mandarins	–	–
4	Clementine *C. clementina* Hort. ex Tan.	6.5	25.0
5	Shekwasha *C. depressa* Hay	3.5	–
	Papeda	–	–
6	Small-fruited papeda *C. micrantha* Wester	1.6	–
	Lime	–	
7	Mexican lime *C. aurantifolia* (Christm.) Swing	–	12.5

The dendrogram (binary tree) of citrus in terms of fruit cross section (cf. Fig. 9.2) shows three classes: (1,3)(4,5)(6)

Citrus maxima and *C. paradisi* present large/very large fruit cross section and are clustered into class 1; *C. clementina* and *C. depressa* show small/medium fruit size and are grouped into class 2; *C. micrantha* has very small fruit section and constitutes class 3. Genus *Citrus*, mandarins, and papeda are detected in separate classes. Maximum similarity results between *C. maxima* and *C. paradisi*, as well as between *C. clementina* and *C. depressa*.

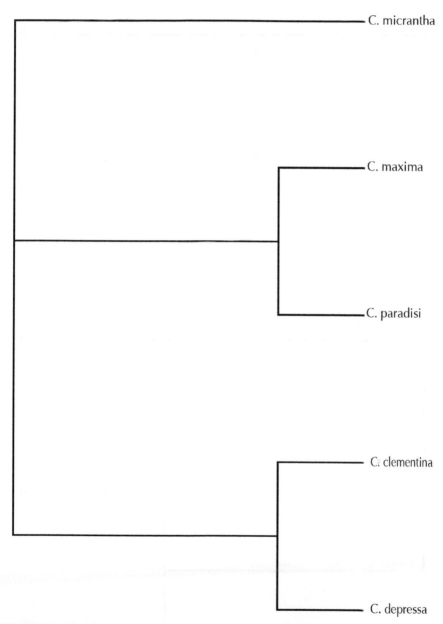

FIGURE 9.2 Dendrogram of citrus in terms of fruit cross section.

The dendrogram of citrus in terms of leaf surface area (cf. Fig. 9.3) shows four classes: (1,3)(2)(4)(7)

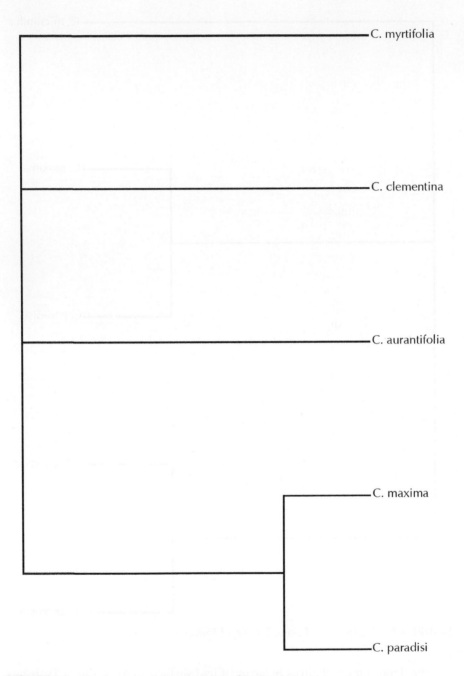

FIGURE 9.3 Dendrogram of citrus in terms of leaf surface area.

Citrus maxima and *C. paradisi* present large/very large leaf surface area and are clustered into class 1; *C. myrtifolia* shows very small leaf size and constitutes class 2; *C. clementina* has medium leaf area and forms class 3; *C. aurantifolia* presents small leaf size and makes class 4. Genus *Citrus*, mandarin, and lime are detected in separate classes. The classes are in partial agreement with the dendrogram of citrus in terms of fruit cross section (Fig. 9.2). Again, maximum similarity appears between *C. maxima* and *C. paradisi.*

The dendrogram of citrus in terms of fruit cross section and leaf surface area (cf. Fig. 9.4) shows four classes: (1,2,3)(4,5)(6)(7)

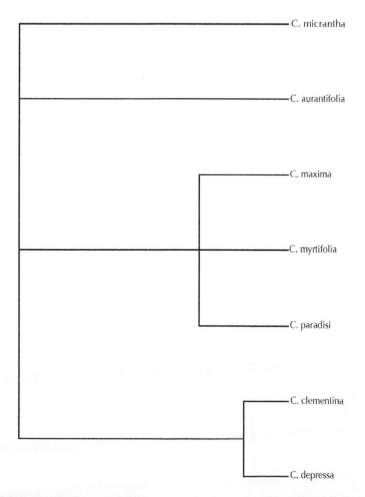

FIGURE 9.4   Dendrogram of citrus in terms of fruit cross section and leaf surface area.

Citrus maxima, C. myrtifolia, and *C. paradisi* present large/very large fruit cross section as well as very small/large/very large leaf surface area, and are clustered into class 1; *C. clementina* and *C. depressa* show small/ medium fruit size as well as medium leaf area, and are grouped into class 2; *C. micrantha* has very small fruit section and constitutes class 3; *C. aurantifolia* presents small leaf size and forms class 4. Genus *Citrus*, mandarins, papeda, and lime are detected in separate classes. The classes are in partial agreement with the dendrograms of citrus in terms of fruit cross section and leaf surface area (Figs. 9.2 and 9.3). One more time, maximum similarity appears between *C. clementina* and *C. depressa.*

The radial tree of citrus in terms of fruit cross section and leaf surface area (cf. Fig. 9.5) shows the same four classes above, in agreement with the dendrograms of citrus in terms of fruit cross section and leaf surface area (Figs. 9.2–9.4). Once more, maximum similarity results between *C. clementina* and *C. depressa.*

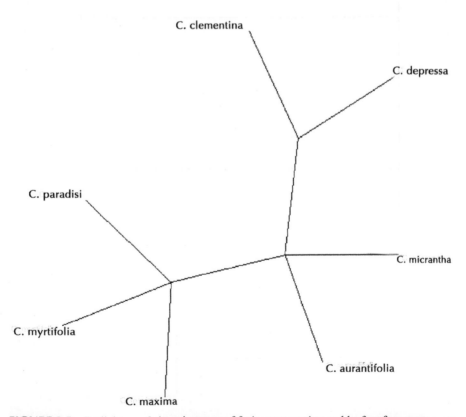

FIGURE 9.5   Radial tree of citrus in terms of fruit cross section and leaf surface area.

Splits graph (cf. Fig. 9.6) for seven citruses in Table 9.1 shows different behavior of citrus depending on fruit cross section and leaf surface area,[43] in qualitative agreement with the dendrogram of citrus in terms of fruit cross section and binary/radial trees in terms of fruit cross section/leaf surface area (Figs. 9.1–9.4). One more time, maximum similarity results between *C. clementina* and *C. depressa*. Maximum difference appears between *C. maxima, C. myrtifolia,* and *C. paradisi* because the last two are hybrids branching from ancestral species.

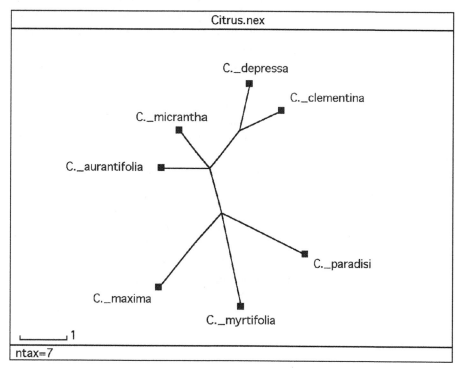

FIGURE 9.6   Splits graph of citrus in terms of fruit cross section and leaf surface area.

9.3   DISCUSSION

9.3.1   SYSTEMATICS

Citrus taxonomy is complex and botanists have not reached total agreement on it. They hybridize easily; even between different genera, spontaneous mutations frequently occur and, because of apomixis, most mutations are likely to be passed down, which gave rise to a huge number of varieties with varying degrees of commercial interest, which are usually difficult to identify. They and related genera belong to order *Geraniales*, suborder *Geraninas,* and family *Rutaceae*, which includes many genera and species, and 6 subfamilies, with *Citrus* and related genera belonging to monophyletic subfamily *Aurantioideae*, which is divided into two tribes: *Clauseneae* (5 genera) and *Citreae* (28 genera). The first comprises the most primitive genera (remote citroides) and the latter, the most recent (citroides, citrus). Subtribe *Citrinae* belongs to the latter tribe, which is subdivided into three groups: plants with primitive citrus fruit (6 genera), plants with citrus-like fruits (2 genera), and plants with true citrus fruit (6 genera among which is *Citrus*). Because of its complexity, genus *Citrus* gave rise to widely varied views among botanists. The two most important classification systems are Swingle and Reece (16 species)[44] and Tanaka (162 species)[45,46] but none is universally accepted. The species of most interest from the agricultural viewpoint are: Mexican/key lime (*C. aurantifolia*), Persian/Tahiti lime (*Citrus latifolia* L.), sour orange (*Citrus aurantium* L.), pummelo [*C. maxima* (L.) Osb.], lemon [*Citrus limon* (L.) Burm.], grapefruit (*C. paradisi*), mandarin (*C. reticulata*), sweet orange [*Citrus sinensis* (L.) Osb.], clementine (*C. clementina* Hort.), and satsuma [*C. unshiu* (Mak.) Marc]. The common name mandarin refers to a heterogeneous citrus group and one with the greatest genetic diversity. It is a controversial group and its classification is not clear; for example, Swingle and Reece considered mandarins as a single species (*C. reticulata*) while Tanaka established 36 species (*C. reticulata, Citrus deliciosa, Citrus tangerina, Citrus restini, C. unshiu, Citrus nobilis*, etc.). *Citrus reticulata* is not grown commercially and what one buys as mandarins on the market are hybrids of mandarin crossed with other species.

9.3.2 ORIGIN

Citrus species belong to the agricultural biodiversity of Valencia Region. However, the efficient mobilization of citrus biodiversity in innovative breeding schemes requires previous understanding of *Citrus* origins and genomic structures. The most widely accepted hypothesis, regarding the phylogenetic origin of citrus, is that most species of genus *Citrus* are direct or successive hybrids branching from four ancestral species (cf. Fig. 9.7): *C. medica* L. (citron), *C. reticulata* (mandarin), *C. maxima* (Burm.) Merr. (pummelo). and *C. micrantha* (papeda). As for secondary species, both sour and sweet oranges are originated from hybridization between pummelo and mandarin. Sour orange corresponds to direct hybridization, while sweet one comes from successive hybridizations, that is, crosses of mandarin with hybrids of mandarin and pummelo. Grapefruit comes from pummelo and sweet orange while clementines are a cross between mandarin and sweet orange. Lemons are hybrids of sour orange and citron while limes come from citron and papeda (*C. micrantha*).

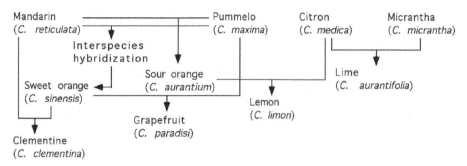

FIGURE 9.7   Genetic origin of the main commercial *Citrus* species.

9.3.3   BIOACTIVITIES OF CITRUS SPECIES EOS

Revista de Fitoterapia reported a special issue on cultural interbreeding in ethnopharmacology.[47] Zhang edited a book on Project 523.[48] Antimicrobial activity, phenolic content, and cytotoxicity of medicinal plant extracts were informed.[49] Cancer preventive agents, especially from food, are desirable for population, because of relatively low toxicities. Antibacterial activity of EOs versus respiratory tract pathogens showed that rapid evaporation was more effective than slow one. It was most efficient when exposed at high vapor concentration for a short time. In *C. medica* antimicrobial activity versus pathogens was because of a synergy among different components, more than that of a specific one; strong camphene effect on antimicrobial was hypothesized; EOs use as antimicrobials was hypothesized and their traditional utilizations as food-preserving agents were confirmed; ADCY1-expression alterations suggested a major role for limonene in effects on CNS. In order to limit the side effects of synthetic and biologic anti-inflammatory drugs, herbal medicines, nutraceuticals, and food supplements use rose as an alternative and/or complementary medicine to treat pathologies, for example, inflammation. Vegetables and fruits dietary intake prevents or delays NDs onset, properties which are mainly because of the presence of polyphenols, an important group of phytochemicals that are abundantly present in fruits, vegetables, cereals, and beverages. The main class of polyphenols is flavonoids, abundant in *Citrus* fruits.

9.4 FINAL REMARKS

From the present results and discussion, the following final remarks can be drawn.

1. Never blindly trust what you get (do your results have chemical sense?). Mathematical and statistical models are not the panacea (always ask yourself three questions: Why am I doing it? What the results might be? Will I be successful?). Use this knowledge in data analysis to guide your investigation or experimentation, not as an end in itself.
2. The hypothesis that variance in classification learning can be expected to decay as training set size rises was examined and confirmed.
3. *Citrus* species belong to the agricultural biodiversity of the Valencia Region. However, the efficient mobilization of citrus biodiversity in innovative breeding schemes requires previous understanding of *Citrus* origins and genomic structures.
4. Some criteria reduced the analysis to a manageable quantity from the enormous set of citrus features: fruit size in terms of cross section, leaf size in terms of surface area, etc. The principal components, cluster, and meta-analyses were useful to rise the number of samples and variety of analyzed data. Different behavior of citrus depends on fruit cross section and leaf surface area. Genus *Citrus*, mandarins, papeda, and lime are detected in separate classes.
5. *Citrus* species group into four classes: genus *Citrus*, mandarins, papeda, and lime. The classification of citrus shows maximum similarity between *C. clementina* and *C. depressa*. Maximum difference appears between *Citrus maxima*, *C. myrtifolia,* and *C. paradisi* because the last two are hybrids branching from ancestral species.
6. Interest in natural product led researchers and the wellness industry to explore natural sources for alternative and complementary medicine to prevent and/or treat inflammatory diseases.

ACKNOWLEDGMENTS

Francisco Torrens belongs to the Institut Universitari de Ciència Molecular, Universitat de València. Gloria Castellano belongs to the Departamento de Ciencias Experimentales y Matemáticas, Facultad de Veterinaria y Ciencias Experimentales, Universidad Católica de Valencia *San Vicente Mártir*. The authors thank support from the Spanish Ministerio de Economía y Competitividad (Project No. BFU2013-41648-P), EU ERDF, and Universidad Católica de Valencia *San Vicente Mártir* (Project No. PRUCV/2015/617).

KEYWORDS

- *Citrus*
- mandarins
- orange
- essential oils
- flavonoids

REFERENCES

1. Ancillo, G.; Medina, A. *Citrus: Botanical Monographs*; The Botanic Garden of the Universitat de València No. 2, Universitat de València: València, 2015; p 1.
2. Uzun, A.; Yesiloglu, T. Genetic Diversity in Citrus. In *Genetic Diversity in Plants*; Caliskan, M., Ed.; InTech: Rijeka (Croatia), 2012; pp 213–230.
3. Wang, N.; Trivedi, P. Citrus Huanglongbing: A Newly Relevant Disease Presents Unprecedented Challenges. *Phytopathology* **2013,** *103*, 652–665.
4. Grosser, J. W.; Dutt, M.; Omar, A.; Orbovic, V.; Barthe, G. Progress Towards the Development of Transgenic Disease Resistance in Citrus. *Acta Hortic.* **2011,** *892*, 101–107.
5. Texeira, D. C.; Ayres, J.; Kitajima, E. W.; Danet, L.; Jagoueix-Eveillard, S.; Saillard, C.; Bové, J. M. First Report of a Huanglongbing-like Disease of Citrus in Sao Paulo State, Brazil and Association of a New Liberibacter Species, *Candidatus* Liberibacter americanus, with the Disease. *Plant Dis.* **2005,** *89*, 107–107.
6. Curk, F.; Ancillo, G.; Garcia-Lor, A.; Luro, F.; Perrier, X.; Jacquemoud-Collet, J. P.; Navarro, L.; Ollitrault, P. Next Generation Haplotyping to Decipher Nuclear Genomic Interspecific Admixture in *Citrus* Species: Analysis of Chromosome 2. *BMC Genet.* **2014,** *15*, 152.
7. Murakami, A.; Kondo, A.; Nakamura, Y.; Ohigashi; H.; Koshimizu, K. Possible Antitumor Promoting Properties of Edible Plants From Thailand, and Identification of an Active Constituent, Cardamonin, of *Boesenbergia pandurata*. *Biosci. Biotechnol. Biochem.* **1993,** *57*, 1971–1973.
8. Murakami, A.; Nakamura, Y.; Koshimizu, K.; Ohigashi, H. Glyceroglycolipids from *Citrus hystrix*, a Traditional Herb in Thailand, Potently Inhibit the Tumor-promoting Activity of 12-O-tetradecanoylphorbol 13-acetate in Mouse Skin. *J. Agric. Food Chem.* **1995,** *43*, 2779–2783.

9. Inouye, S.; Takizawa, T.; Yamaguchi, H. Antibacterial Activity of Essential Oils and Their Major Constituents Against Respiratory Tract Pathogens by Gaseous Contact. *J. Antimicrob. Chemother.* **2001**, *47*, 565–573.

10. Klein, J. D. Citron Cultivation, Production and Uses in the Mediterranean Region. In *Medicinal and Aromatic Plants of the Middle East, Medicinal and Aromatic Plants of the World No. 2*; Yaniv, Z., Dudai, N., Eds.; Springer: Dordrecht, 2014; pp 199–214.

11. Ferlazzo, N.; Cirmi, S.; Calapai, G.; Ventura-Spagnolo, E.; Gangemi, S.; Navarra, M. Anti-inflammatory Activity of *Citrus bergamia* Derivatives: Where Do We Stand? *Molecules* **2016**, *21*, 1273.

12. Cirmi, S.; Ferlazzo, N.; Lombardo, G. E.; Ventura-Spagnolo, E.; Gangemi, S.; Calapai, G.; Navarra, M. Neurodegenerative Diseases: Might *Citrus* Flavonoids Play a Protective Role? *Molecules* **2016**, *21*, 1312.

13. Aliberti, L.; Caputo, L.; De Feo, V.; De Martino, L.; Nazzaro, F.; Souza, L. F. Chemical Composition and In Vitro Antimicrobial, Cytotoxic, and Central Nervous System Activities of the Essential Oils of *Citrus medica* L. cv. 'Liscia' and *C. medica* cv. 'Rugosa' Cultivated in Southern Italy. *Molecules* **2016**, *21*, 1244.

14. Jerkovic, I.; Prdun, S.; Marijanovic, Z.; Zekic, M.; Bubalo, D.; Sveçnjak, L.; Tuberoso, C. I. G. Traceability of Satsuma Mandarin (*Citrus unshiu* Marc.) Honey Through Nectar/Honey-sac/Honey Pathways of the Headspace, Volatiles, and Semi-volatiles: Chemical Markers. *Molecules* **2016**, *21*, 1302.

15. Torrens-Zaragozá, F. Molecular Categorization of Yams by Principal Component and Cluster Analyses. *Nereis* **2013**, *2013*(5), 41–51.

16. Torrens, F.; Castellano, G. From Asia to Mediterranean: Soya Bean, Spanish Legumes and Commercial Soya Bean Principal Component, Cluster and Meta-analyses. *J. Nutr. Food Sci.* **2014**, *4*(5), 98.

17. Torrens, F.; Castellano, G. Principal Component, Cluster and Meta-analyses of Soya Bean, Spanish Legumes and Commercial *Soya Bean*. In *Applied Chemistry and Chemical Engineering*; Haghi, A. K.; Ribeiro, A. C. F.; Pogliani, L.; Torrens, F.; Balkose, D.; Mukbaniani, O. V.; Pasbakhsh, P., Eds.; Apple Academic–CRC: Waretown (NJ) Vol. 5, in press.

18. Torrens-Zaragozá, F. Classification of Fruits Proximate and Mineral Content: Principal Component, Cluster, Meta-analyses. *Nereis* **2015**, *2015*(7), 39–50.

19. Torrens-Zaragozá, F. Classification of Food Spices by Proximate Content: Principal Component, Cluster, Meta-analyses. *Nereis* **2016**, *2016*(8), 23–33.

20. Castellano, G.; Tena, J.; Torrens, F. Classification of Polyphenolic Compounds by Chemical Structural Indicators and Its Relation to Antioxidant Properties of *Posidonia Oceanica* (L.) Delile. *MATCH Commun. Math. Comput. Chem.* **2012**, *67*, 231–250.

21. Castellano, G.; González-Santander, J. L.; Lara, A.; Torrens, F. Classification of Flavonoid Compounds by Using Entropy oF Information Theory. *Phytochemistry* **2013**, *93*, 182–191.

22. Castellano, G.; Lara, A.; Torrens, F. Classification of Stilbenoid Compounds by Entropy of Artificial Intelligence. *Phytochemistry* **2014**, *97*, 62–69.

23. Castellano, G.; Torrens, F. Information Entropy-based Classification of Triterpenoids and Steroids from *Ganoderma*. *Phytochemistry* **2015**, *116*, 305–313.

24. Castellano, G.; Torrens, F. Quantitative Structure-antioxidant Activity Models of Isoflavonoids: A Theoretical Study. *Int. J. Mol. Sci.* **2015**, *16*, 12891–12906.

25. Castellano, G.; Redondo, L.; Torrens, F. QSAR of Natural Sesquiterpene Lactones as Inhibitors of Myb-dependent Gene Expression. *Curr. Top. Med. Chem.* submitted for publication.

26. Torrens, F.; Redondo, L.; Castellano, G. Artemisinin: Tentative Mechanism of Action and Resistance. *Pharmaceuticals*, **2017**, *10,* 20–4-4.
27. Torrens, F., Redondo, L.; Castellano, G. Reflections on Artemisinin, Proposed Molecular Mechanism of Bioactivity and Resistance. In *Innovations in Physical Chemistry*; Haghi, A. K, Ed.; Apple Academic–CRC: Waretown (NJ), 2017; Vol. 2, in press.
28. Hotelling, H. Analysis of a Complex of Statistical Variables into Principal Components. *J. Educ. Psychol.* **1933**, *24,* 417–441.
29. Kramer, R. *Chemometric Techniques for Quantitative Analysis;* Marcel Dekker: New York, 1998; p 1.
30. Patra, S. K.; Mandal, A. K.; Pal, M. K. J. State of Aggregation of Bilirubin in Aqueous Solution: Principal Component Analysis Approach. *Photochem. Photobiol. Sect. A* **1999**, *122,* 23–31.
31. Jolliffe, I. T. *Principal Component Analysis*; Springer: New York, 2002; p 1.
32. Xu, J.; Hagler, A. Chemoinformatics and Drug Discovery. *Molecules* **2002**, *7,* 566–600.
33. Shaw, P. J. A. Multivariate Statistics for the Environmental Sciences; Hodder-Arnold: New York, 2003; p 1.
34. IMSL. *Integrated Mathematical Statistical Library (IMSL)*; IMSL: Houston, 1989.
35. Tryon, R. C. J. A Multivariate Analysis of the Risk of Coronary Heart Disease in Framingham. *Chronic Dis.* **1939**, *20,* 511–524.
36. Priness, I.; Maimon, O.; Ben-Gal, I. Evaluation of Gene-expression Clustering via Mutual Information Distance Measure. *BMC Bioinformatics* **2007**, *8,* 111.
37. Steuer, R.; Kurths, J.; Daub, C. O. Weise, J.; Selbig, J. The Mutual Information: Detecting and Evaluating Dependencies between Variables. *Bioinformatics* **2002**, *18*(Suppl. 2), S231–S240.
38. D'Haeseleer, P.; Liang, S.; Somogyi, R. Genetic Network Inference: From Co-expression Clustering to Reverse Engineering. *Bioinformatics* **2000**, *16,* 707–726.
39. Perou, C. M.; Sørlie, T.; van Eisen, M. B.; de Rijn, M.; Jeffrey, S. S.; Rees, C. A.; Pollack, J. R.; Ross, D. T.; Johnsen, H.; Akslen, L. A.; Fluge, O.; Pergamenschikov, A.; Williams, C.; Zhu, S. X. Lønning, P. E.; Børresen-Dale, A. L.; Brown, P. O.; Botstein, D. Molecular Portraits of Human Breast Tumours. *Nature* (London) **2000**, *406,* 747–752.
40. Jarvis, R. A.; Patrick, E. A. Clustering Using a Similarity Measure Based on Shared Nearest Neighbors. *IEEE Trans. Comput.* **1973**, *C22,* 1025–1034.
41. Page, R. D. M. *Program TreeView*; Universiy of Glasgow: Glasgow, UK, 2000; p 1.
42. Eisen, M. B.; Spellman, P. T.; Brown, P. O.; Botstein, D. Cluster Analysis and Display of Genome-wide Expression Patterns. *Proc. Natl. Acad. Sci. U. S. A.* **1998**, *95,* 14863–14868.
43. Huson, D. H. SplitsTree: Analyzing and Visualizing Evolutionary Data. *Bioinformatics* **1998**, *14,* 68–73.
44. Swingle, W.; Reece, P. The Botany of *Citrus* and Its Wild Relatives. In *The Citrus Industry: The Botany of Citrus and Its Wild Relatives*; Reuther, W.; Webber, H. J.; Batchelor, L. D., Eds.; University of California: Berkeley (CA), 1967; p 1.
45. Tanaka, T. *Species Problem in Citrus (Revisio Aurantiacearum IX)*; Japanese Society for Promotion of Science: Tokyo (Japan), 1954.
46. Tanaka, T. Fundamental Discussion of *Citrus* Classification. *Stud. Citrol.* **1977**, *14,* 1–6.
47. Fresquet, J. L., Aguirre, C., Eds., Special Issue: The Cultural Interbreeding in Ethnopharmacology: From Indigenous to Scientific Knowledges. *Rev. Fitoterapia* **2005**, *5*(S1), 1–271.

48. Zhang, J., Ed. *A Detailed Chronological Record of Project 523 and the Discovery and Development of Qinghaosu (Artemisinin)*; Strategic Book: Houston (TX), 2013; p 1.

49. Ghuman, S.; Ncube, B.; Finnie, J. F.; McGaw, L. J.; Coopoosamy, R. M.; Van Staden, J. Antimicrobial Activity, Phenolic Content, and Cytotoxicity of Medicinal Plant Extracts Used for Treating Dermatological Diseases and Wound Healing in Kwazulu-Natal, South Africa. *Front. Pharmacol.* **2016,** *7,* 320.

CHAPTER 10

BIODIVERSITY AND CONSERVATION

ANAMIKA SINGH[1] and RAJEEV SINGH[2,*]

[1]Department of Botany, Maitreyi College, University of Delhi, Delhi, India

[2]Department of Environment Studies, Satyawati College, University of Delhi, Delhi, India

*Corresponding author. E-mail: 10rsingh@gmail.com

CONTENTS

ABSTRACT

Biodiversity implies the variety and number of plants and animals in an ecosystem. It is a system within which the living organisms interact with each other along with microorganisms present. Diversity of ecosystem is mainly due to genetic variations among organisms and these variations are due to environmental factors. Species within any ecosystem are linked by food chains and loss of species or biodiversity indicates breaking of the link in the food chain. Biodiversity is actually a natural wealth of the planet and it fulfills all the requirements of any organism. Due to excess utilization of natural resources, the loss of biodiversity occurs, so there is a need of conservation and protection of environment.

10.1 INTRODUCTION

Biodiversity is a term refers to the variability of life on earth. The variability occurs between species and with species and it also reflects variations in ecosystem.[1] Biodiversity is measurer of different types of organisms present in an ecosystem. Biodiversity is actually variation of gene content of any ecosystem. These variations are due to adaptation of environment and it varies from equator to pole.[2,3] Biodiversity is just because of change in altitude and latitude of the earth. Biodiversity is actually distribution of flora and fauna on earth.[4,5] These flora and fauna can be of different varieties, number within a population. Strength of biodiversity or the variation in biodiversity depends upon the adaptation of environment. Drastic and rapid change in environment always leads to extinction of species which ultimately effects the environment.[6-8] In the term of biology, biodiversity is the total gene content of an ecosystem and species.[9] An advantage of this definition is that it seems to describe most circumstances and presents a unified view of the traditional types of biological variety previously identified.

10.2 MEASUREMENT OF BIODIVERSITY

Biodiversity is a multidimensional structure and it is really tough to measure.[10] An area can be defined as rich or poor in biodiversity.

Biodiversity can be measured as:

1. Number of genes, number of populations, and species or taxa in an area.
2. Evenness: It is the frequency of allele, number, and relative frequency, along with this genetic diversity as a measurement of heterozygosity of allele.
3. Difference: It is difference in alleles within a population. Generally, a population contains similar types of alleles but a bit diversion always creates difference level of genetic structure within a population. These differences cause generation of subspecies, It is having important evolutionary significant. Biodiversity can be measured by different methods but most important one is species richness.[10,11]

Species are the basic unit of biodiversity, thus, it makes biological sense to measure species richness rather than a higher taxonomic grouping. It is often easier to count the number of species compared to other measures of biodiversity.[11]

10.3   ELEMENTS OF BIODIVERSITY: BIODIVERSITY CAN BE VIEWED UNDER

1. *Genetic diversity* measures morphological diversity.[12]
2. *Species diversity* is actually taxonomic diversity.
3. *Community* or ecosystem diversity.

Each of this level is composed of three levels compositional diversity, structural diversity, and functional diversity.

a) *Compositional diversity*: Number of representative or species present in any level of diversity. Number of species present is a measure of compositional diversity.
b) *Structural diversity*: It represents sex ratio and age of the individuals of any species present.
c) *Functional diversity*: Variations in functions performed by the diversity. Functional diversity measures the number of functionally separate species such as feeding mechanism, predator, mortality predator–prey relationship, etc.

1. *Genetic diversity*: It is a variation in genetic makeup of individual or population within a species. Within a species, genetic diversity can be

observed by variation in growth, color, size, resistant to a disease, etc. Genetic diversity is an important factor which actually maintains the species diversity. It is total number of genetic characteristics in the genetic makeup of a species. It is a way for a population to adapt changes in concern with environment. Gene has alleles which contain specific characters and these characters adapts to a changing environments. Variation in a population is variation of alleles. With more variation, it is more likely that some individuals in a population will possess variations of alleles that are suited for the environment. Those individuals are more likely to survive to produce offspring bearing that allele. The population will continue for more generations because of the success of these individuals.[13] Genetic and phenetic diversity have been found in all species at the level of protein, DNA, and individual level. Nature is having a nonrandom diversity, heavily structured, and correlated with environmental variation and stress.[14] Genetic and species diversity is interdependent and there is delicate difference between it. Changes in species diversity cause changes in the environment, which leads to force, and adaptation to the remaining species. Changes in genetic diversity, mainly loss of species, always lead to a loss of biological diversity.[13] Loss of genetic diversity in species causes the extinction of species.[15,16] Genetic diversity is very important for the survival and adaptation of a species.[17] Change in the habitat of a population pressurizes the individual to adapt itself for the survival. It determines the ability to cope with an environmental challenge. Ultimately, it selects the species through natural selection.[14] Epidemic and disease cause loss of genetic diversity which effects biodiversity of any area.[18]

2. *Species diversity*: It is the number of different species by mean of abundance, distribution, function, or interaction. Species richness is number of species present in any area and it also determines the biodiversity of that area. Diversity among species is the measures of biodiversity but there are so many factors to be counted along with this, such as relative abundance and area where a particular species found. If an area is having 10 species and only 2 are dominating but for biodiversity, we cannot ignore rest 8 species as they also help to measure the biodiversity of that area.

Species diversity consists of two components:

1. Species richness: Number of species present in any area.
2. Species evenness: Abundance or frequency of the species is known as species evenness.[19-21]

3. *Community:* A community is the different groups of species staying together and interacts with each other. They share a common biotic (biological) and abiotic (physical) factors. A community differs by its habitat. A habitat is the set of resources used by a specific species. Community diversity is variation in type, structure, and functions. Ecosystem diversity is a higher level of diversity. Community diversity is an aggregation of different species and they are having separate demands of abiotic factors of any ecosystem. Community diversity largely depends upon abiotic factors such as fire, water, soil, topography, and geology.

Ecological diversity includes the variation in both terrestrial and aquatic ecosystems. Ecological diversity can also take into account the variation in the complexity of a biological community, including the number of different niches, the number of trophic levels, and other ecological processes. An example of ecological diversity on a global scale would be the variation in ecosystems, such as deserts, forests, grasslands, wetlands, and oceans. Ecological diversity is the largest scale of biodiversity, and within each ecosystem, there is a great deal of both species and genetic diversity.[22,23]

10.4 LOSS OF BIODIVERSITY: THE MEASURE OF LOSS OF SPECIES FROM ANY AREA OR ECOSYSTEM

10.4.1 CAUSES FOR LOSS OF BIODIVERSITY

a) *Habitat change*: Humans have the most effected habitat and they occupy and cover most of the areas of earth for their shelter, food, and development. As their demand of food, they cover most of the land for agriculture. Millions of areas have been cleared and utilized as agricultural fields which cause loss of habitat for number of species. Natural forest, wetlands, estuaries, and mangroves are continually utilized by humans and this causes serious loss of shelter for number of species. Wetlands are destroyed due to draining, filling, and pollution. Most of the species utilized the large areas such as beer and wild cats as they covers outer areas of forest for food search and interior of forest for breeding. It is known as habitat fragmentation. Loss of areas affects their natural habitat.

b) *Climate change*: Recent drastic changes in climate, especially in warmer regional with high temperatures, had significant impacts on biodiversity and ecosystems, which includes changes in species distributions, population sizes, the timing of reproduction or migration events, and an increase in the frequency of pest and disease outbreaks.

c) *Invasive species*: The spread of invasive alien species has increased because of increased trade and travel. While increasingly there are measures to control some of the pathways of invasive species, for example, through quarantine measures and new rules on the disposal of ballast water in shipping, several pathways are not adequately regulated, particularly with regard to introductions into freshwater systems.

d) *Overexploitation*: In a marine ecosystem, major loss of fish population is due to demand of fishes and overfishing. Demand for fish as food for people and as feed for aquaculture production is increasing, resulting in increased risk of major, long-lasting collapses of regional marine fisheries. 50% of the world's commercial marine fisheries are fully exploited while 25% are being overexploited. Frequent cutting down the trees in forest causes loss of shelter of small birds.\

e) *Pollution and global warming*: Due to high population and over demand of natural resources cause pollution which lead to loss of nature. Pollution can be of several types: air, water, noise, etc. and all affects the natural habitat, and it leads to loss of nature. Even nowadays, radiation is having high impact on plants and birds. Pollution causes global warning which effects natural life cycle of animals.\

f) *Poaching*: It is an illegal trade of wildlife products by killing endangered animals. Although, international ban on trade in products from endangered species smuggling of wildlife items such as furs, horns, tusks, specimens, and herbal products worth million of dollars per year continues, it also causes loss of species from natural habitat.

10.5 CONSERVATION STRATEGIES

These are the measures used to protect biodiversity:

1. *Maintain intact (viable) landscapes*: The intent of this strategy is to protect and improve the ecological integrity and long-term viability of the more intact (core) landscapes of the region. Within these areas, priority actions would be to: repair historic impacts, remove threats, and reinstate ecological processes.

2. *Reverse declines*: This strategy aims to reinstate ecosystems that have been differentially lost in locations, where this will meaningfully

contribute to stemming species declines and reinstating critical ecological processes (such as pollination). Within these areas, the priority actions are to reinstate open woodland systems and improve the habitat value of shrubby systems.

3. *Recover threatened species and ecological communities*: The intent of this strategy is to ensure the long-term persistence of species and ecosystems at immediate risk of extinction in the wild. The actions required to implement this work are specific to individual species and ecosystems, but typically focus on increasing distribution and abundance, and halting (or ideally reversing) declining trends. The nature of this work is guided by the current amount of knowledge.

4. *Control emerging threats*: This strategy aims to address threats to biodiversity before their impacts are fully realized. A couple of the more pervasive threats to the region include climate change and new invasive species.

5. *Protected areas*: The protected areas are biogeographical areas, where biological diversity along with natural and cultural resources are protected, maintained, and managed through legal and administrative measures. The demarcation of biodiversity in each area is determined on the basis of climatic and physiological conditions.
   In these areas, hunting, firewood collection, timber harvesting, etc. are prohibited so that the wild plants and animals can grow and multiply freely without any hindrance. Some protected areas are: cold desert (Ladakh and Spiti), hot desert (Thar), saline swampy area (Sunderban and Rann of Kutch), tropical moist deciduous forest (Western Ghats and northeast), etc. Protected areas include national parks, sanctuaries, and biosphere reserves. There are 37,000 protected areas throughout the world. As per World Conservation Monitoring Centre, India has 581 protected areas, national parks, and sanctuaries.

6. *National parks*: These are the small reserves meant for the protection of wildlife and their natural habitats. These are maintained by government. The area of national parks ranges between 0.04 and 3162 km. The boundaries are well-demarcated and circumscribed. The activities such as grazing forestry, cultivation, and habitat manipulation are not permitted in these areas.

7. *Sanctuaries*: These are the areas where only wild animals (fauna) are present. The activities such as harvesting of timbers, collection of forest products, cultivation of lands, etc. are permitted as long as these do not interfere with the project. That is, controlled biotic

interference is permitted in sanctuaries, which allows visiting of tourists for recreation. The area under a sanctuary remains in between 0.61 and 7818 km.

8. *Biosphere reserves*: Biosphere reserves or natural reserves are multipurpose protected areas with boundaries circumscribed by legislation. The main aim of the biosphere reserve is to preserve genetic diversity in representative ecosystems by protecting wild animals, traditional lifestyle of inhabitant, and domesticated plant/animal genetic resources. These are scientifically managed allowing only the tourists to visit.

Some important aspects of biosphere reserves are as follows:

1. These help in the restoration of degraded ecosystem.
2. The main role of these reserves is to preserve genetic resources, species, ecosystems, and habitats without disturbing the habitants.
3. These maintain cultural, social, and ecologically sustainable economic developments.
4. These support education and research in various ecological aspects.

10.6   CONCLUSION

Biodiversity is a wide term and it is specific for any particular area, and this diversity is due to effect of environment. The development of flora and fauna largely depends upon survival, and finally, the survival dominates and grows in next generation. Further, these species (plant and animals) modify their surroundings as per their needs.

KEYWORDS

- **biodiversity**
- **conservation**
- **genetic diversity**
- **natural selection**
- **community**

REFERENCES

1. Clark, M. R.; Schlacher, T. A.; Rowden, A. A.; Stocks, K. I.; Consalvey, M. *Science Priorities for Seamounts: Research Links to Conservation and Management. PLoS One* **2012,** *7,* e29232.

2. Tittensor, D. P.; Mora, C.; Jetz, W.; Lotze, H. K.; Ricard, D.; Berghe, E. V.; Worm, B. Global Patterns and Predictors of Marine Biodiversity Across Taxa. *Nature* **2010,** *466*(7310), 1098–1101.

3. Myers, N.; Mittermeier, R. A.; Mittermeier, C. G.; da Fonseca, G. A.; Kent, J. Biodiversity Hotspots for Conservation Priorities. *Nature* **2000,** *403*(6772), 853–858.

4. McPeek, M. A.; Brown, J. M. Clade Age and Not Diversification Rate Explains Species Richness Among Animal Taxa. *Am. Nat.* **2007,** *169*(4), E97–E106.

5. Rabosky, D. L. Ecological Limits and Diversification Rate: Alternative Paradigms to Explain the Variation in Species Richness Among Clades and Regions. *Ecol. Lett.* **2009,** *12*(8), 735–743.

6. Cockell, C. *Biological Processes Associated with Impact Events*, 1st ed.; Springer: Berlin, 2006; pp 197–219 (ESF IMPACT).

7. Algeo, T. J.; Scheckler, S. E. Terrestrial–Marine Teleconnections in the Devonian: Links Between the Evolution of Land Plants, Weathering Processes, and Marine Anoxic Events. *Philos. Trans. R. Soc. Lond. B Biol. Sci.* **1998,** *353*(1365), 113–130.

8. Bond, D. P. G.; Wignall, P. B. The Role of Sea-level Change and Marine Anoxia in the Frasnian–Famennian (Late Devonian) Mass Extinction. *Palaeogeogr. Palaeoclimatol. Palaeoecol.* **2008,** *263*(3–4), 107–118.

9. Davis, M. L. *Introduction to Environmental Engineering (Sie),* 4th ed.; McGraw-Hill Education (India) Pvt. Ltd., p 4 (accessed June 28, 2011).

10. Purvis, A.; Hector, A. Getting the Measure of Biodiversity. *Nature* **2000,** *405,* 212–219.

11. Gaston, K. J; Spicer, J. I. *Biodiversity: An Introduction,* 2nd ed.; Blackwell: UK, 2004.

12. Baco, A. R.; Cairns, S. D. Comparing Molecular Variation to Morphological Species Designations in the Deep-sea Coral *Narella* Reveals New Insights into Seamount Coral Ranges. *PLoS One* **2012,** *7,* e45555.

13. https://web.archive.org/web

14. *Nevo, E.* Evolution of Genome–Phenome Diversity Under Environmental Stress. *Proc. Natl. Acad. Sci. U. S. A.* **2001,** *98*(11), 6233–6240.

15. Groom, M. J.; Meffe, G. K.; Carroll, C. R. *Principles of Conservation Biology*, 3rd ed.; Sunderland, 2006 (http://www.sinauer.com/groom/).

16. Tisdell, C. Socioeconomic Causes of Loss of Animal Genetic Diversity: Analysis and Assessment. *Ecol. Econ.* **2003,** *45*(3), 365–376.

17. Frankham, *R.* Genetics and Extinction. *Biol. Conserv.* **2005,** *126*(2), 131–140.

18. www.cheetah.org

19. Hill, M. O. Diversity and Evenness: A Unifying Notation and Its Consequences. *Ecology* **1973,** *54,* 427–432.

20. Tuomisto, H. A Diversity of Beta Diversities: Straightening up a Concept Gone Awry. Part 1. Defining Beta Diversity as a Function of Alpha and Gamma Diversity. *Ecography* **2010,** *33,* 2–22.

21. Tuomisto, H. A Consistent Terminology for Quantifying Species Diversity? Yes, It Does Exist. *Oecologia* **2010,** *4,* 853–860.

22. McClain, C. R., Lundsten, L., Barry, J.; DeVogelaere, A. Assemblage Structure, But Not Diversity or Density, Change with Depth on a Northeast Pacific Seamount. *Mar. Ecol.* **2010,** *31*(Suppl. S1), 14–25.
23. Williams et al. Scales of Habitat Heterogeneity and Megabenthos Biodiversity on an Extensive Australian Continental Margin (100–1000 m depths). *Mar. Ecol.* **2010,** *31*, 222–236.

CHAPTER 11

ADVANCED OXIDATION PROCESSES, INDUSTRIAL WASTEWATER TREATMENT, AND INNOVATIONS OF ENVIRONMENTAL CHEMISTRY: A VISION FOR THE FUTURE

SUKANCHAN PALIT

Department of Chemical Engineering, University of Petroleum and Energy Studies, Post Office Bidholi via Premnagar, Dehradun 248007, Uttarakhand, India

E-mail: sukanchan68@gmail.com; sukanchan92@gmail.com

CONTENTS

ABSTRACT

The domain of environmental engineering science is moving from one paradigmatic shift over another. Environmental regulations, environmental restrictions, and industrial wastewater treatment are changing the scientific landscape of traditional and nontraditional environmental engineering techniques. The challenge and vision of science today are enigmatic and inspiring. Drinking water treatment and water purification are the areas of immense scientific endeavor which need to be reenvisioned and reenshrined with the passage of scientific history and time. Advanced oxidation processes (AOPs) are the nontraditional avenues of scientific endeavor. Ozone oxidation is also one visionary avenue of AOPs. The author pointedly focuses on the immense scientific potential and the wide success of AOPs particularly ozone oxidation. This treatise also stresses on the immense scientific vision of integrated advanced oxidation techniques. The world of scientific research pursuit in environmental engineering is moving toward zero-discharge norms. Technological vision and scientific objectives are the torchbearers toward a greater emancipation of AOPs today.

11.1 INTRODUCTION

The world of environmental engineering science and chemical process engineering is moving toward a newer paradigm of science and engineering. The domain of traditional and nontraditional techniques in environmental engineering science is crossing visionary frontiers. Environmental regulations and loss of norms of industrial pollution control have urged human society and scientific research pursuit to innovate and redefine environmental engineering. In such a crucial juncture of scientific history and time, advanced oxidation processes (AOPs), primarily ozonation, are changing the scientific landscape. Technological grandeur and versatile scientific objectives are the pillars and cornerstones of scientific innovation today. Zero-discharge norms are the coinwords of environmental engineering today. The wide vision and the effectivity of ozonation and other integrated AOPs are the prime focus of this treatise. Integrated AOPs and integrated environmental engineering techniques are the torchbearers toward a greater vision of environmental engineering and greater environmental sustainability today. Also, sustainable development is the other side of the visionary and far-reaching coin of environmental engineering today. In this treatise, the author repeatedly stresses on the different techniques, the immense success, and the wide scientific potential of the environmental engineering science.

11.2   THE AIM AND OBJECTIVE OF THE STUDY

Science and engineering in today's human civilization are moving at a rapid pace and surpassing visionary boundaries. Technological vision and scientific grandeur are ushering in a new era of scientific regeneration and scientific vision. Environmental engineering science needs to be reenvisioned and reenshrined as human mankind passes through the vicious and difficult phase. Loss of ecological and environmental biodiversity, the frequent environmental disasters, and the stringent environmental regulations will all lead a long and visionary way in the true emancipation of environmental engineering science and environmental sustainability today. The true challenge and the true success of scientific endeavor and research and development initiatives are ushering in a new eon of scientific vision and scientific forbearance. The author pointedly focuses on the intricate areas of environmental engineering techniques such as the chemistry of AOPs. The wide environmental engineering perspective is gleaned and brought forward to the scientific horizon with much scientific vision and scientific fortitude. The sole aim and objective of the study is to present AOP and ozonation in a wider scientific framework.

11.2.1   THE NEED AND THE RATIONALE OF THE STUDY

The scientific need and the scientific rationale of this study are immense, widespread, and far-reaching. Environmental engineering and the holistic world of sustainable development are gaining immense heights and surpassing visionary frontiers. Technology and science should go beyond sustainability whether it is energy or environment. The challenge of environmental engineering today is the greater realization of environmental sustainability. The author in this treatise widely observes the immense potential of AOPs, ozone oxidation, and also the wide domain of novel separation processes such as membrane science. Global water crisis today is in a state of immense catastrophe. Drinking water purification, industrial wastewater treatment, and water pollution control are the highest priorities of human civilization and scientific endeavor today. Technology needs to be reenvisioned and reenvisaged with every step of scientific endeavor. The world of drinking water treatment and water purification in the similar fashion needs to be rebuilt and revamped as human civilization moves toward a newer scientific eon.

11.3   WHAT IS ADVANCED OXIDATION PROCESS?

Industrial wastewater treatment and drinking water treatment today stand in the midst of deep scientific comprehension and scientific imagination. AOPs are revolutionizing the scientific landscape. Today, science is a huge colossus with a definite and unimaginable vision of its own. Wastewater treatment and AOPs/integrated AOPs today are the two opposite sides of the visionary coin. AOPs are the visionary scientific endeavor which are replete with scientific vision and effectiveness. Hazardous organic wastes represent the greatest challenges to environmental engineers and environmental scientists. AOPs are alternatives to the incineration of wastes, which has many disadvantages. Conventional incineration is widely thought to be a feasible alternative to landfill but as presently practiced, incineration may bring about serious problems due to release of toxic compounds such as polychlorinated dibenzodioxins (PCDDs) and polychlorinated dibenzofurans (PCDFs), into the environment via incinerator off-gas emissions and/or fly ash. [37]

The AOPs have proceeded along one of the two routes:

- Oxidation with O_2 in temperature ranges intermediate between ambient conditions and those found in incinerators, wet air oxidation (WAO) processes in the region of 1–20 MPa and 200–300°C.
- And the use of high-energy oxidants such as ozone and H_2O_2 and/or photons that are able to generate highly reactive intermediates—OH' radicals. [37]

In 1987, researchers defined AOPs as "near-ambient temperature and pressure water treatment processes which involve the generation of hydroxyl radicals in sufficient quantity to effect water purification." The hydroxyl radical (OH) is a powerful, nonselective chemical oxidant (Table 11.1), which acts very rapidly with most organic compounds. Once generated, the hydroxyl radicals aggressively attack virtually all organic compounds. Depending upon the nature of the organic species, two types of initial attacks are possible: The hydroxyl radical can abstract a hydrogen atom from water, as with alkanes or alcohols, or it can add itself to the contaminant, as in the case of olefins or aromatic compounds. [37]

TABLE 11.1 Relative Oxidation Power of Some Oxidizing Species.

Oxidizing species	Relative oxidation power
Chlorine	1.00
Hypochlorous acid	1.10
Permangate	1.24
Hydrogen peroxide	1.31
Ozone	1.52
Atomic oxygen	1.78
Hydroxyl radical	2.05
Positively charged hole on titanium dioxide, TiO_2^+	2.35

The attack by the OH' radical, in the presence of oxygen, initiates a complex cascade of oxidative reactions leading to mineralization of the organic compound. The exact routes are still not clear. For example, chlorinated organic compounds are oxidized first to intermediates, such as aldehydes and carboxylic acids, and finally to CO_2, H_2O, and the chloride ion. Nitrogen in organic compounds is usually oxidized to nitrate or to free N_2 and sulfur is oxidized to the sulfate. Cyanide is oxidized to cyanate, which is then further oxidized to CO_2 and NO_3^- (or, perhaps, N_2). [37]

As a general rule, the rate of destruction of a contaminant is approximately proportional to the rate constant for the contaminant with OH radical. Chlorinated alkenes are treated most efficiently because the double bond is very susceptible to hydroxyl attack. Saturated molecules (i.e., alkanes) react at a much slower rate and therefore, are more difficult to oxidize. AOPs are primarily helpful in oxidative destruction of compounds refractory to conventional ozonation or H_2O_2 oxidation. AOPs are extremely suited for destroying dissolved organic contaminants such as halogenated hydrocarbons (trichloroethane and trichlotroethylene), aromatic compounds (benzene, toluene, ethylbenzene, and xylene), pentachlorophenol (PCP), nitrophenols, detergents, pesticides, etc. [37]

11.4 THE SCIENTIFIC DOCTRINE OF OZONE OXIDATION

The world of ozone oxidation and AOPs is witnessing drastic and dramatic challenges as science and engineering are moving toward a newer scientific regeneration and a visionary scientific understanding. Global water crisis is in a state of immense catastrophe. Heavy metal groundwater contamination and arsenic groundwater contamination are today's challenging areas of scientific research pursuit. Ozone oxidation is one of the promising as well as emerging areas of environmental engineering science today.

11.5   THE CHEMISTRY AND MECHANISM OF AOP AND OZONE OXIDATION

AOPs, in a broad definition, are a set of chemical treatment procedures designed to remove organic (and sometimes inorganic) materials in water and wastewater by oxidation through reactions with hydroxyl radicals (OH'). In real-world applications of industrial wastewater treatment, this term usually refers more specifically to a subset of chemical processes that use ozone (O_3), hydrogen peroxide (H_2O_2), and/or UV light. One such type of process is termed as in situ chemical oxidation. [37]

Generally speaking, chemistry in AOPs could be essentially divided into three parts:

- formation of OH';
- initial attacks on target molecules by OH' and their breakdown to fragments;
- subsequent attacks by OH' until ultimate mineralization. [37]

The main reaction mechanism for oxidation of inorganic compounds is determined by the transfer of extra oxygen atom of ozone to the inorganic compounds. Summarizing, ozone oxidizes organic compounds selectively and partially. A large number of inorganic compounds are oxidized fast and completely. [37]

11.6   RECENT SCIENTIFIC ENDEAVOR IN THE FIELD OF AOPS

Al-Kdasi et al.[1] delineated with deep and cogent insight treatment of textile wastewater by AOPs. It is a comprehensive review targeting the furtherance of science and engineering of ozonation.[1] The use of conventional textile wastewater treatment processes becomes drastically and dramatically challenged to environmental engineers and scientists with evergrowing more and more restrictive effluent quality by water authorities. Traditional treatment such as biological treatment discharges will no longer be tolerated as 53% of 87 colors are identified as nonbiodegradable.[1] AOPs hold great and immense promises to provide alternative treatment opportunities for environmental protection and so the utmost need of this visionary technique. The domain of environmental engineering science today is treading a weary path toward scientific destiny and scientific profundity. The challenge of human scientific endeavor needs to be reenvisioned and rebuilt as environmental protection gains new heights. The authors of this treatise relentlessly delineate treatment efficiency for different AOPs and present their immense potential for the advancement of science and engineering. Textile wastewater includes a large variety of dyes and chemicals which cannot be degraded by conventional treatment procedures. Thus, there is an immense need and a scientific foray into the science of AOP.[1]

Goi[2] in her doctoral thesis enlarged and envisioned the concept of AOPs targeting its application for water and soil decontamination. AOPs, which involve the in situ generation of highly potent chemical oxidants such as hydroxyl radicals, have emerged as an important class of technology for accelerating the oxidation and destruction of a wide range of organic contaminants in polluted water and soil.[2] Technology needs to be rebuilt and revamped as the author widely observes the success and efficiency of AOPs/integrated AOPs in the destruction recalcitrant contaminants. The reaction rate constants of nitrophenols' oxidation in different AOPs systems (the Fenton, photo-Fenton, hydrogen peroxide/ultraviolet (UV), ozone, ozone/hydrogen peroxide, ozone/UV, ozone/hydrogen peroxide/UV) were effectively determined for oxidation processes optimization.[2] This treatise gives a wide glimpse of the immense potential and success of integrated AOPs with a definite target and vision toward furtherance of environmental engineering science.[2]

Technological ingenuity and the world of challenges in environmental engineering science are the forerunners of scientific validation and scientific vision today. The boundaries of environmental engineering are vast and versatile today. The author of this treatise deeply comprehends the success, the vision, and the wide scientific potential of AOPs and other nonconventional techniques. Environmental engineering paradigm is witnessing visionary challenges as human civilization and human scientific endeavor approach a new world of scientific regeneration. Integrated AOPs are the new areas of scientific research pursuit. Global water issues and global water technology initiatives are veritably changing the scientific scenario. Heavy metal groundwater contamination and more specifically, arsenic groundwater contamination are challenging the scientific domain and the scientific research pursuit.

Mofidi et al.[3] delineated with deep insight AOPs and UV photolysis for the treatment of drinking water.[3] This project investigated the use of UV light and ozone for micropollutant control in drinking water. The micro-pollutants' study included methyl tert-butyl ether (MTBE), N-nitrosodi-methylamine (NDMA), perchlorate, and bromate, and taste and odor (T and O) compounds 2-methylisoborneol (MIB) and geosmin. The work was investigated at the bench and pilot scale involving a pulsed UV lamp (43 L batch reactor) or an over–under ozone contact system.[3] UV disinfection is receiving increasing attention in the drinking water industry as a definitive method for utilities to comply with evergrowing disinfection regulations. Because most disinfectants provide wide and multiple benefits (e.g., both disinfection and oxidation), this study was conducted to gain larger scientific understanding whether UV light would provide the same benefits as ozone. Technological vision and scientific insight are immensely enhanced as this scientific investigation targets the fundamentals and basics of AOPs. Science and engineering are the forerunners toward a newer environmental order. This treatise gives wider glimpse to the immense potential and success of AOPs, especially ozonation technique.[3]

Mota et al.[4] reviewed AOPs and their application in petroleum industry. AOPs are far-reaching technologies based on the generation of highly reactive species, the hydroxyl radicals, used in oxidative degradation procedures for organic compounds dissolved or dispersed in aquatic media.[4] These processes are promising as well as evergrowing alternatives for decontamination of media containing dissolved recalcitrant organic compounds, which cannot be removed by conventional methods. Technological and academic rigor are at its visionary best as the authors delve deep into the world of application domain of AOPs. The present chapter widely observes a series of AOPs, analyzing the aspects related to each type of process, such as the interference of external agents and the ideal operating conditions, based on the analysis and the comparison of different AOPs.[4] The path of science in AOPs is challenged today. The authors repeatedly and rigorously delve deep into the success of advanced oxidation techniques targeting the furtherance of science and engineering.[4]

Grote[5] in a visionary report delineated application of AOPs in water treatment.[5] Conventional oxidation processes are used in water treatment to disinfect water, to reduce toxins, odor, and color, or to reduce manganese and iron levels in potable water. These processes may not destroy all the toxins and have a disastrous potential to create dangerous disinfection by-products (DBPs). AOPs utilize the strong oxidizing power of hydroxyl radicals that can reduce organic compounds to harmless end products. Science and engineering of AOPs are changing the scientific scenario today.[5]

Sharma et al.[6] delineated with deep details AOPs for wastewater treatment. AOPs constitute a promising technology for the treatment of industrial wastewaters containing noneasily removing organic compounds. The success and potential of science and engineering are being emboldened and enshrined as scientific research pursuit enters into a new eon. All AOPs are designed to produce hydroxyl radicals. It is the hydroxyl radical that acts with high efficiency to destroy the organic compounds. AOP combine ozone (O_3), UV, hydrogen peroxide (H_2O_2), and/or catalyst to provide a powerful water treatment solution for the reduction and/or removal of residual organic compounds as measured by chemical oxygen demand (COD), biological oxygen demand (BOD), or total organic carbon (TOC).[6] This treatise targets efficient AOPs developed to decolorize and/or degrade organic pollutants for environmental protection. Fundamentals and primary methods of applications such as Fenton, electroFenton, ozonation, and UV radiation are discussed with cogent insight.[6]

Jose[7] discussed in details the combination of AOPs with biological treatment for the remediation of water polluted with herbicides.[7] Water is an irreplaceable asset of human civilization. Water reuse technology is thus the utmost need of the hour. Technology of water science and engineering is today surpassing visionary frontiers and is gaining new heights in this century. Global water shortage and global water research and development initiatives need to be reenvisioned and reenvisaged as science moves forward. Water quality stands as a major imperative to the emancipation of environmental engineering science today.[7] The challenge of research and development initiatives needs to be redrawn and reenshrined. The nonbiodegradable anthropogenic pollution discharged in the water systems deteriorates the water quality, and so its removal and degradation are of imminent need. Sustainable access to clean drinking water is the main objective of human endeavor, scientific rigor, and progress of human society. The aim of this study is to bring new insights into the general problem of water quality improvement by finding new ways of treating polluted water with different herbicides. The introduction to the review targets AOPs used in this project, the biological treatment avenues, the natural interferences found in wastewater, and some different and versatile strategies used to assess the efficiency of the coupling of chemical and biological treatment avenues.[7]

Van der Helm[8] discussed in deep details in a widely observed thesis Integrated Modeling of Ozonation for Optimization of Drinking Water Treatment. The author gives a wide overview of the present water purification technologies in the Netherlands.[8] In the Netherlands, drinking water treatment plants are robust and designs are based on the performance of individual processes with preset boundary conditions. It is an excellent example of water treatment research and development initiatives in the developed world. In the operation of drinking water plants, the processes are usually optimized individually on the basis of "rule of thumb" and operator knowledge and experience. However, an optimal operation involves a number of water quality parameters, costs, and environmental impact.[8] Thus, a robust operation with optimization of different parameters is the hallmark of this study. An integrated approach to the optimization of drinking water plants is today ushering a new vision to environmental engineering science globally. The success of novel separation processes such as membrane science needs to be reenvisioned as optimization science gains new scientific heights. Applied mathematics and optimization science are the cornerstones of water treatment scientific strategy today.[8]

11.7 VISIONARY SCIENTIFIC RESEARCH PURSUIT IN THE FIELD OF OZONE OXIDATION

Scientific research pursuit in the field of ozone oxidation is versatile and far-reaching. The vision of AOPs and the wide applications of ozone oxidation will lead a long way in the true realization and the true emancipation of environmental science and environmental engineering today. Water technology and green engineering are at a state of immense catastrophe. Global water challenges and groundwater heavy metal remediation need to be readdressed and reenvisioned as science and engineering witness paradigmatic shift. Biodegradability and application of nontraditional environmental engineering techniques are the forerunners of greater realization of environmental sustainability today.

The United States Environmental Protection Agency Report[9] deeply comprehends the success and immense potential of ozone disinfection in a water technology fact sheet.[9] Disinfection science is considered to be the primary mechanism for the inactivation/disinfection of pathogenic organisms to prevent the spread of waterborne diseases to downstream users and the environment. It is vital that wastewater be adequately treated prior to disinfection in order for any disinfectant to be effective. Technology is so progressive in today's world.[9] Ozone is produced where oxygen (O_2) molecules are dissociated by an energy source into oxygen atoms and subsequently collide with an oxygen molecule to form an unstable gas, ozone (O_3) which is used to disinfect wastewater.[9] Science, engineering, and technology are in the path of immense scientific rejuvenation and scientific regeneration.

Most wastewater treatment plants generate ozone by imposing high-voltage alternating current (6–20 kV) across a dielectric discharge gap that contains an oxygen-bearing gas.[9]

Khadre et al.[10] discussed with deep and cogent insight microbiological aspects of ozone applications in food. Ozone is a powerful antimicrobial agent that is suitable for application in food in the gaseous and aqueous states.[10] Molecular ozone or its decomposition products (e.g., hydroxyl radical) inactivate microorganisms rapidly by reacting with intracellular enzymes, nucleic material and components of their self-envelope, spore coats, or viral capsids. Science of microbiology is today integrated with chemical process engineering and environmental engineering science. A scientist's vision is enhanced and technological objective redefined as science and engineering enters into a newer paradigm and a newer phase. Microbiological aspects of ozone applications are far-reaching and new. Science needs to gain high and visionary heights as human scientific research pursuit witnesses paradigmatic shift.[10] Combination of ozone with appropriate initiators (e.g., UV or H_2O_2) results in AOPs that are potentially effective against the most resistant microorganisms; however, applications in food need to be widely approved. Technology and science are moving toward a newer era of immense scientific and academic rigor. Food processing innovations need to be emboldened and enhanced as technology advances. In order to meet the consumers' demand of fresher and safer ready-to-eat products, technology needs to be veritably targeted. High-pressure processing, pulsed electric field, and high-intensity pulsed light are some of the new and emerging technologies. Today, attention is now focused on ozone as a powerful sanitizer that may meet expectations of the industry, approval of regulatory agencies, and wide acceptance of the consumer. Microbiology is in a state of immense scientific vision as interdisciplinary areas of science gain new heights. Regulatory agencies in the United States have been hesitant in the past to approve the use of ozone for treatment of drinking water and direct food applications. Now, the scenario has changed with new vision and newer innovation. Currently, there are more than 3000 ozone-based water treatment installations throughout the world and more than 300 potable water treatment plants in the United States. The vision and challenge of science are wide and clear in the scientific landscape.[10] Ozone can now be used as an effective technology in the water purification and drinking water treatment. Major advantages of ozone made it one of the most important and primary technologies attracting the attention of the food industry. Ozone is one of the potent sanitizers known. Excess ozone autodecomposes rapidly to produce oxygen and thus, it leaves no residue in the food. Food processers who are introducing ozone in their facility and researchers who are exploring the use of ozone in food processing are in need of relevant and concise information about this sanitizer. Thus, there is an utmost need of ozonation research.[10]

Cervi[11] widely observed formation and removal of aldehydes as ozonation by-products in a pilot-scale water treatment plant. This thesis is a visionary avenue of scientific endeavor. The vision and challenge of science are widely proven and enhanced with every step of scientific rigor. Ozonation or ozone oxidation is the relevant green technology today.[11] Today, science is a huge pillar with vision and objectives at its zenith. This study investigated the effect of water treatment plant configurations on the formation and removal of aldehydes in a pilot water treatment plant in Windsor, Canada. The plant was operated in nonozonation, precoagulation ozonation, and postsedimentation ozonation treatment configurations.[11] Some parameters such as ozone dose and flow rate through the anthracite/sand filters were studied to determine the effect of formation and removal of aldehydes in the pilot plant. Technology is vastly progressive surpassing difficult scientific frontiers.[11] This treatise gives a wide view of the immense scientific potential of ozonation technique. It was found that formaldehyde, acetaldehyde, glyoxal, and methylglyoxal were the main aldehyde species formed as a result of ozonation with formaldehyde. Aldehyde formation was found to increase as ozone dose was increased. The challenge of technology is opening the windows of innovation in ozone oxidation as science of ozonation moves forward. Ozone was first used in the drinking water treatment in 1893 in Oudshoom in the Netherlands.[11] By 1990s, there were close to 75 water treatment plants in Canada and the United States with ozonation facilities and many more plants exist worldwide, especially in the Western Europe. Technological challenges and scientific research motivation expanded within the course of time in this century as environmental protection widely observed dramatic and drastic changes.

The author in this widely observed treatise delineates the immense importance and the wide imperatives of industrial wastewater treatment, water purification, and drinking water treatment. Scientific and technological challenges are today in the path of immense scientific rejuvenation and wide scientific vision. Ozonation research and development initiatives are today changing the scientific landscape and ushering in a new eon in the field of AOPs.

McElroy et al.[12] delineated in a well-researched report the determination of ozone by UV analysis. An overview of the health and environmental aspects of ozone is summarized and the scope and availability are delineated with deep details. Ozone (O_3), a colorless gas, has both beneficial and detrimental effects on human health and the environment.[12] Ozone, naturally occurring in the upper atmosphere, protects human mankind against skin cancer caused by UV radiation from the sun. Ozone in smog is formed by sunlight reacting with the oxides of nitrogen (NO_x) and volatile organic compounds (VOCs) discharged into the air from gasoline vapors, solvents, fuel combustion products, and consumer products.[12] The wide scientific imagination and the avenues of scientific endeavor will lead a long way in the true emancipation of environmental engineering science today. The analytical principle is based on absorption of UV light by the ozone molecule and the subsequent use of photometry to measure the reduction of quanta of light reaching the detector at 254 nm^{-1}. The degree of reduction depends on the pathlength of the UV sample cell, ozone concentration introduced in the sample cell, and the wavelength of the UV light.[12] Science of ozonation is witnessing a new dawn of scientific endeavor and a new eon of scientific emancipation.[12]

Technological vision and scientific objectives are of vital importance in the progress of scientific and academic rigor today. Environmental engineering science and environmental engineering techniques are in the path of immense scientific vision. Technological and scientific validation need to be reenvisioned and restructured as the world of science and engineering moves toward zero-discharge norms. The author in this treatise widely observes the success and immense potential of ozonation of dyes as a target toward the furtherance of science of AOPs.

Joshi et al.[13] discussed lucidly in a review article the color removal in textile effluents. The vast environmental issues associated with residual color in textile effluents have posed a major and drastic challenge to environmental scientists as well as textile coloration processes. Science and technology of ozonation and AOPs are witnessing crucial challenges as human scientific endeavor in environmental engineering is reenvisioning and restructuring itself. Increasing environmental regulations and zero-discharge norms are changing the scientific landscape. Dyes are highly dispersible esthetic pollutants and are difficult to treat as most dyes are highly stable molecules made to resist degradation by light, chemical, and biological treatments. Technology and engineering science are rigorously challenged today.[13] Global water concerns and challenges are drastically changing the scientific landscape and are ushering in a new era in environmental engineering.

Technological advancements are at a state of immense catastrophe as environmental engineering science faces the crisis of our times. The need for environmental sustainability is evergrowing as the concept of zero-discharge norms gains new and visionary heights. In this treatise, the author widely observes the success of AOP and the immense scientific vision behind it. Technological objectives and scientific candor are the visionary pillars of scientific research pursuit today.

Johansson[14] delineated with lucid and cogent insight treatment of dye wastewater with ozone solution. The textile industry annually produces large amounts of wastewater which contains several different compounds from the production. This factor is of immense importance in the wide domain of application of nonconventional environmental engineering techniques such as AOPs. There are several ways to treat wastewater from textile factories in order to remove the remains of the dyes. The widely common ways are physical, chemical, and biological treatment. One visionary scientific endeavor is ozone oxidation of dyes. Ozonation does not contribute to secondary pollution. The vision, the challenge, and the immense application potential of ozonation need to be readdressed to the utmost in the present century. This project targets making a small-scale equipment and optimizing some parameters in the treatment procedures of industrial wastewater. The parameters to be investigated are the pH, oxygen flow rate, and the mixing ratio between wastewater and ozone solution.[14]

Khan et al.[15] delineated with lucid details the advanced oxidative decolorization of red CI-5B, its effects of dye concentration, process optimization, and reaction kinetics. Technology of ozonation is slowly evolving as the treatment of ozone in dye degradation reaches visionary heights. The success and challenge of ozone oxidation are immense as scientific and academic rigor surpass visionary frontiers. Process conditions (dye concentration, pH, and oxidant dose) were optimized for UV, O_3, H_2O_2/UV, O_3/UV, H_2O_2/O_3, and H_2O_2/O_3/UV to treat red CI-5B dye of various concentrations (100, 300, and 500 mg/L).[15] Ozonation resulted in the color removal of more than 90%, whereas H_2O_2/O_3 showed no advantage over the O_3 alone.[15] However, H_2O_2/UV was found to be very suitable as it gave almost 100% decolorization in a relatively short residence time. The challenge of science of ozonation is opening up new windows of innovation and scientific vision in decades to come. Decolorization rate for all the processes was reduced to half when the dye concentration was increased from 100 to 300 mg/L.[15] A deep investigation of comparison of rate constants revealed that H_2O_2/UV is four times faster than UV alone.[15] Integrated AOPs are the challenge and wide vision of science and engineering today. Textile engineering and chemical process engineering are embarking on a newer future scientific endeavor. The scientific rigor of textile chemistry and environmental engineering science is wide and versatile. This treatise pinpoints the immense challenges and the vast and versatile potential of integrated AOPs. Wet processes for textile production are one of the largest water-consuming and -polluting sources. The production of textiles requires a significant amount of water for preparation, dyeing, washing, and finishing stages from 5 to 40 times fiber weight. It highly depends on machinery type, treatment type, and operator care.

Grande[16] in his doctoral thesis widely observed treatment of textile wastewater by ozonation. Wet processes for textile production are one of the largest water-consuming and -polluting sources. Technological advancements have reached a pinnacle with the evolution of science of ozonation of azo as well as anthraquinone dye. The production of textiles requires a significant amount of water for preparation, dyeing, washing, and finishing stages from 5 to 40 times fiber weight. It solely depends on machinery type, treatment type, and operator care. Consequently, large amounts of wastewater must be treated before discharging in the environment. The wastewater produced can be widely classified as follows:

- process wastewater discharged from different process steps, such as desizing, bleaching, dyeing, printing, etc.;[16]
- wash off water with high flow rate and medium pollutant load;[16]
- utility cooling waters with temperature as sole pollution parameter.

Textile wastewater characteristics show a great variety depending on the types of inorganic and organic compounds used in each process. Generally, wastewater contains noticeable color, high concentration of COD, and heavy metals typically of some dye structures. The organic and inorganic contaminant of the textile wastewaters must be removed by appropriate treatments according to environmental rules and regulations. Today, technology of ozonation is highly advanced, visionary, and far-reaching. Ahead of designing a wastewater treatment system, it is important to take into account what the goals are:

- preliminary simplified treatment to reach the threshold values of the parameters and discharge water in a sewage collector upstream of a consortium final process;[16]
- complete treatment to reach the minimum requirements given by law to discharge treated water into surface aquatic system;[16]
- advanced treatment to generate process water and address clean water free from contaminants.[16]

The technologies adopted by the researcher are:

- activated sludge reactor,
- bioflotation,
- continuous backwash biological aerated filter,
- flow-jet aerated reactor.

Garcia-Morales et al.[17] discussed with cogent insight the integrated AOP (ozonation) and electrocoagulation (EC) treatments for dye removal in denim effluents.[17] The highlights of this research are to study the removal of indigo carmine dye used in industrial denim dyeing processes. For this purpose, integrated AOPs comprising ozone and EC techniques were used. Technological and scientific motivation reached its scientific pinnacle as industrial wastewater treatment domain reaches a newer paradigm. The challenge and vision of ozonation science are readdressed and reaffirmed in this widely observed researched work. In this research work, after ozone was applied, 64% color removal, 78% turbidity removal, and 3% COD reduction was observed. The synergy between the two treatments yielded increased removal of color, COD, and turbidity efficiencies as compared with the results obtained with the single treatments.[17]

Multani et al.[18] delineated with deep details the removal of color and COD from reactive green-19 dyeing wastewater using ozone. Ozone is primarily used in AOPs and it pointedly focuses on a set of chemical procedures designed to remove organic and inorganic materials in dye industry wastewater.[18] The avenues of scientific research pursuit in ozonation and AOPs are changing the scientific and engineering frontiers. Reactive-19 effluent is treated with the ozonation process. The experimental operating procedures are dye concentration (200, 300, and 400 ppm), time of ozone passing (20, 40, and 60 min), and pH (4, 7, and 10).[18]

Muhammad et al.[19] discussed in minute details the decolorization and removal of COD and BOD from raw and biotreated textile dyebath effluent through AOPs. In this paper, a comparative study of the treatment of raw and biotreated (upflow anaerobic sludge blanket, UASB) textile dyebath effluent using AOPs is presented. The AOPs applied on raw and biotreated textile dyebath effluent, after characterization in terms of COD, color, BOD, and pH, were ozone, UV, UV/hydrogen peroxide, and photo-Fenton. Science and engineering of nontraditional environmental engineering techniques are widely surpassing visionary boundaries. The authors found out that the decolorization of raw dyebath effluent was 58% in case of ozonation. Biodegradability was improved by applying AOPs after the biotreatment of dyebath effluent.[19]

Fahmi et al.[20] delineated with cogent insight the characteristic of color and COD removal of azo dye by AOP and biological treatment. In this detailed study, the characteristic of COD and color removal of azo dye by AOPs and biological treatment were evaluated for applying in azo dye industrial effluent treatment. Technology needs to be reenvisioned and reenvisaged at each step of scientific endeavor in the field of environmental engineering science. Reactive red 120 has been selected amongst azo dyes due to its high solubility in aquatic environment. Wastewater from dye industrial effluent is a complex synthetic organic that should be treated properly to reduce its COD, color, and toxicity.[20] Amongst the wastewaters, azo dye effluents from textile, plastic, leather, and paper industries are the largest and the vital components which need to be effectively treated. There are a variety of methods such as coagulation and adsorption but not effective. Thus, there is a need for chemical oxidation such as AOPs.[20]

Venkatesh et al.[21] discussed in deep details the treatment and degradation of synthetic azo dye solution containing acid red 14 by ozone. Treatment and degradation of synthetic azo dye solution of acid red 14 dye, as an azo dye, was studied in a batch reactor using ozonation. Ozonation as AOP is one of the most effective means of decolorizing dye wastewater due to its strong oxidizing property by breaking down the functional group that produces color on azo dye and has been shown to achieve high color removal. Acid red 14 has been selected amongst azo dyes due to its high solubility in the aquatic environment.[21]

Selcuk et al.[22] comprehended in deep details the ozone preoxidation of a textile industry wastewater for acute toxicity removal. Today, the technology is moving at a rapid pace and crossing visionary frontiers. Ozonation of dyes is a visionary case study of the entire ozonation process. The author in this entire treatise pointedly focuses on the wide and visionary world of chemical oxidation with immense stress on ozonation.[22] A scientist's immense vision, scientific and academic rigor, and the futuristic vision of science and technology will all lead a long way in the true realization and true emancipation of environmental engineering science. In this work, preozonation for degradation, decolorization, and detoxicifying of a raw textile wastewater collected in a textile finishing industry (Istanbul, Turkey) is widely investigated. Differing from the previous studies, a low ozone(O_3) flow rate (9.6 mg min^{-1}) was applied at the original pH of wastewater.[22] The effect of pH varying from 5 to 11 and the H_2O_2 dose of 600 mg L^{-1} on ozone oxidation was also investigated.[22]

Poznyak et al.[23] discussed in details the treatment of textile industrial dyes by simple ozonation with water recirculation. In this study, three textile dyes were destroyed by ozone in water solution. These dyes were selected because of their complex chemical structure and extended application in textile industry. Decomposition of dyes by simple ozonation was observed at the initial pH of the aqueous colorants solutions UV–vis analysis was used for preliminary control of the degree of decomposition.[23] High-performance liquid chromatography (HPLC) analysis was used to identify intermediates and final products formed during ozonation. Technological advancements and scientific motivation are in the path of immense scientific regeneration and scientific vision as nonconventional environmental engineering techniques gain new heights to alleviate global water crisis. Industrial wastewater treatment and AOPs are the two opposite sides of the visionary coin of environmental engineering. In this research study, the color disappears completely for all dyes after 1.5–2 min of ozonation.[23]

Lambert et al.[24] discussed in lucid details the ozone EC processes for treatment of dye in leather industry wastewater. The authors in this treatise did a comparative study. The aim of this study was to investigate the feasibility of two decontaminating processes, ozonation and EC, to decolorize wastewater generated by humid finishing leather production.[24] With the primary objective to determine if one or both the processes could remove color, bench-scale experiments were conducted with three different colorants, representative of the main dye groups: diazo (CI Direct Blue1), anthraquinone (CI Green G), and aniline (CI Fast Red B Base). As one of the AOPs, ozonation has an evergrowing and promising importance.[24] Technology of AOP is today surpassing visionary frontiers. In general, ozone oxidation pathways include direct oxidation by ozone or indirect oxidation by hydroxyl radicals (OH'). Direct oxidation involves degradation of organic compounds by ozone molecules under acidic conditions while indirect oxidation considers a degradation mechanism of organics throughout by hydroxyl radicals under basic conditions.[24] The EC process consists of electro-dissolve soluble anodes (Fe or Al), generating metallic hydroxide flocs. The technological success and scientific articulation are more pronounced as AOP gains newer visionary heights with the passage of scientific history and time.[24]

Tabrizi[25] did a comprehensive research work on decoloration of reactive dyes with ozone in a semibatch and a continuous stirred-tank reactor. The aim of this doctoral thesis was to model and interpret the behavior of ozone when utilized to decolorize reactive dye solutions in textile dyeing applications. The target is to eliminate the waste from industrial effluents and for the ultimate vision for the conservation of water.[25] Ozonation experiments were performed in a semibatch and continuous stirred-tank reactors in steady- and unsteady state operations. The author in his doctoral thesis devised new technologies and investigated the feasibility of AOPs in treating industrial effluents in South Africa. Global water research and development initiatives are in the path of newer scientific vision and deep scientific determination. The eye of green technology and the targets of green science are opening up new vistas of technological challenges in water initiatives in decades to come.[25]

11.8   SCIENCE AND TECHNOLOGY OF WATER PURIFICATION AND THE VISION FOR THE FUTURE

The world of water purification today veritably stands in the midst of deep scientific vision and scientific introspection. Global water crisis is in the verge of immense disaster. Technology and engineering have few answers to the immense global water challenges and the wide global water research and development initiatives. The author in this treatise repeatedly points out toward the success of environmental engineering techniques in alleviating water crisis. Arsenic groundwater remediation and provision of clean drinking water need to be readdressed and reenvisaged with the passage of scientific history and time. Science today is a huge colossus with a definitive vision of its own. Technological vision and immense scientific objectives are the torchbearers of today's scientific endeavor. Water catastrophes will surely open up new avenues of challenges in the scientific horizon in years to come.

11.9   CHALLENGES AND VISION IN THE FIELD OF WATER PURIFICATION

The challenges and the vision in the field of water purification are vast and versatile. Scientific research pursuit in the world of water science and water technology are today reaching its zenith of scientific forbearance. The world of science and technology today is crossing visionary boundaries. Chemical process engineering and environmental engineering science need to be reenvisioned and rebuilt with immediate effect.

One of the most pervasive problems afflicting people throughout the world is inadequate supply of clean water and proper sanitation. Problems with water are expected to worse in the coming decades with water scarcity occurring globally and growing to alarming heights. This problem stands in water-rich countries also.

11.10 SCIENTIFIC RESEARCH PURSUIT IN THE FIELD OF WATER PURIFICATION

The world of global water challenges and the science of water purification are ushering in a new eon in the furtherance of science. Provision of potable and clean drinking water is the visionary challenge of our times. The author repeatedly points out the need, the visionary targets, and the immense success of conventional and nonconventional water purification techniques and environmental techniques.

Shannon et al.[26] discussed with deep and cogent insight the science and technology for water purification in the coming decades. Water solutions need to be in the high agenda of civil society scientific endeavor. One of the pervasive and enigmatic problems afflicting people throughout the world is inadequate supply of clean water and sanitation. Technology is challenged and scientific frontiers overcome.[26] Problems with water are expected to grow worse in the coming decades with water scarcity occurring globally, even in regions currently considered water-rich. The authors in this treatise widely observe some of the science and technology being developed to improve the disinfection and decontamination of water as well as worldwide efforts to increase potable water supplies. The authors targeted disinfection, decontamination, reuse and reclamation, desalination, and effective water reuse.[26] The immensely wide and many problems worldwide associated with the lack of clean, freshwater are widely known: 1.2 billion people lack access to safe drinking water, 2.6 billion people have little or no sanitation, millions of people die annually—3900 children a day—from diseases transmitted through unsafe water or human excreta.[26] The effective control of waterborne pathogens in drinking water calls for the innovation of new disinfection strategies, including multiple-barrier approaches that provide reliable physicochemical removal (e.g., coagulation, flocculation, sedimentation, and media or membrane filtration). The overarching goal of scientific endeavor and the furtherance of science is to detect and remove toxic substances from water affordably and robustly. Technological advancements should go parallel with scientific validation. The challenge of science in decontamination is to grow immense confidence in nonconventional techniques such as AOPs. The evergrowing goal for the future of reclamation and reuse of water is to capture water directly from nontraditional sources such as industrial and municipal wastewaters and restore to potable quality. Desalination is another avenue of water research. The future of desalination is to increase the freshwater supply via desalination of seawater and saline aquifers.[26]

Ray et al.[27] discussed with deep details the novel photocatalytic reactor for water purification. A novel photocatalytic reactor design for water purification is characterized by the use of new extremely narrow diameter lamps, thus allowing for much higher surface area for catalyst coating per unit reactor volume and consequently for much higher specific reactor capacity. Experiments in a reactor containing 21 novel U-shaped lamps coated with catalyst showed a 695% increase in efficiency of the reactor performance in comparison with a classical annular reactor and 259% in comparison with a slurry reactor. The challenge and targets of science and technology of water purification need to be reenvisioned and reenvisaged as global water research and development initiatives ushers in a new futuristic vision.[27] Water treatment processes based on aqueous phase hydroxyl radical chemistry are becoming powerful oxidation methods that can completely destroy (mineralize) almost all toxic organic compounds in water. Heterogeneous photocatalysis is one of the AOPs that couples low-energy UV light with semiconductors acting as photocatalysts and has emerged as a viable and feasible alternative.[27] Photocatalytic reactions are promoted by solid photocatalyst particles, which are either dispersed in the liquid or immobilized on a surface. In the recent years, several review articles have appeared on photocatalytic water treatment with generally positive results. The major challenges in the development of photocatalytic reactor are the problems of scale-up of multiphase photocatalytic reactors than that of conventional chemical reactors or homogeneous photoreactors.[27] The overall rate of reaction of photocatalytic processes which usually follow Langmuir–Hinshelwood kinetics is slow compared to conventional chemical reaction rates due to low concentration levels of the pollutants.[27] In this research work, a new photocatalytic reactor design concept with a 100–150-fold increase in surface area per unit volume of reaction liquid inside the reactor relative to a classical annular reactor design and a 10–20-fold increase relative to immersion-type reactor with classical lamps has been proposed and designed. Technology of photocatalytic reactor is moving today toward newer dimension and a newer visionary age as global water crisis reenvisions itself toward innovations and challenges. This treatise will open up new vistas of scientific research pursuit in the years to come.[27]

Dijkstra et al.[28] discussed with lucid details an experimental comparison of three reactor designs for photocatalytic water purification.[28] The photocatalytic degradation of formic acid was compared for a suspended system, an immobilized system with coated wall, and an immobilized system packed with coated glass beads. Heterogeneous photocatalysis is a novel water purification method in the group of advanced oxidation technologies. In this technique, a semiconductor, mostly TiO_2, is illuminated with UV-A light and generates electron hole pairs, which in turn can generate hydroxyl radicals. Regarding the catalyst configuration, two operating modes for photoreactors can be identified with the catalyst suspended or with the catalyst immobilized on a carrier material. When these two are compared, the suspended system often has been found as more efficient which can be attributed to the absence of mass transfer limitation and the large specific surface area of the nanometer-scale particles. However, the need for a separation step to retrieve the catalyst after the purification is a great disadvantage of this system compared to the immobilized system.[28]

11.11 INNOVATIONS OF ENVIRONMENTAL CHEMISTRY AND THE VISION FOR THE FUTURE

Innovations in environmental chemistry are today vast and versatile. Environmental engineering science and chemical process engineering are today ushering in a newer visionary eon in the field of human scientific endeavor. Environmental chemistry can be termed as the study of sources, reactions, transport, effects, and fates of chemical species in the air, soil, and water environments; and the effect of human activity and biological activity on these. Today, environmental chemistry and green chemistry are two opposite sides of the visionary coin of science. AOPs, integrated AOPs, and other chemical oxidation processes are the challenging areas of green chemistry today. The success and the vision of chemical process engineering and environmental engineering science are today ushering in a newer visionary era of scientific vision and scientific forbearance. In this entire treatise, the author repeatedly focuses on green chemistry, the wide avenues of green engineering, and the futuristic vision of novel separation processes.

11.12 ENVIRONMENTAL SUSTAINABILITY AND ENVIRONMENTAL ENGINEERING SCIENCE

Sustainability issues such as environmental and energy sustainability are challenged today as scientific research pursuit surpasses visionary boundaries. The challenge and vision of environmental engineering are the utmost need of the hour. Sustainable development with respect to energy and environment is the immediate need and is imperative toward the furtherance of science. Environmental engineering is today in a state of immense catastrophe. Environmental science and nuclear science and engineering are the two opposite sides of the visionary coin. Nuclear disasters are destroying the scientific landscape. In such a crucial situation of scientific history and time, environmental engineering science assumes immense importance. Research and development initiatives in water technology are the forerunners toward a greater visionary eon of tomorrow. A scientist's immense vision, the widely observed fruits of technology, and the immense academic rigor will go a long way in the true realization of environmental engineering science and environmental sustainability. Environmental disasters and loss of ecological biodiversity are the forerunners toward a newer visionary era of environmental protection. The challenge and vision of science and engineering are groundbreaking and veritably surpassing wide visionary frontiers. Sustainable development today is linked by an unsevered umbilical cord with progress of human civilization and the wide world of scientific rigor.

11.13　THE CHALLENGE AND VISION OF SCIENCE IN INDUSTRIAL WASTEWATER TREATMENT

Industrial wastewater treatment today is in the path of newer scientific regeneration and deep scientific vision. The challenge of zero-discharge norms is of utmost importance in the progress of environmental engineering science. Global water challenges and groundwater remediation are changing the scientific landscape. Scientific forbearance and scientific fortitude are challenged today and need to be reenvisioned in the path of immense academic rigor. Wastewater engineering and green technology are the coinwords of today. The wide vision of scientific candor and scientific cognizance in wastewater engineering and environmental engineering is opening up new avenues of future thoughts and future trends in research in the years to come.

The challenge and vision of nonconventional environmental engineering techniques are surpassing visionary boundaries today. AOPs and other nonconventional environmental engineering techniques are challenged today and need to be revamped and rebuilt. Science today is a huge colossus with a wide vision of its own. Industrial wastewater treatment and water purification are the need of the hour today. Environmental protection, industrial water pollution control, and the success of environmental sustainability will all lead a long way in the true emancipation of environmental engineering science in today's scientific horizons.

11.14 WATER PURIFICATION, DRINKING WATER TREATMENT, AND AOPS

Water purification and drinking water treatment today stand in the midst of immense scientific vision and scientific introspection. Global water crisis is in a state of immense distress and unbelievable catastrophe. Heavy metal groundwater contamination and arsenic groundwater contamination are the vexing and enigmatic issues facing human civilization and human scientific endeavor today. Nonconventional environmental engineering techniques such as advanced oxidation techniques need to be rebuilt and readdressed with the furtherance of science and the scientific progress of human civilization. Progress of human scientific research pursuit, the challenges of sustainable development and, the immense academic rigor will all lead a long way in the true realization and deep introspection of environmental engineering science.

11.15 GROUNDWATER CONTAMINATION, ARSENIC AND HEAVY METAL GROUNDWATER REMEDIATION, AND THE VISION FOR THE FUTURE

Groundwater contamination is a widespread global water catastrophe. The challenge, the vision. and the targets of scientific research pursuit need to be rebuilt and reenvisaged for the furtherance of science. Arsenic and other heavy metal groundwater contamination are challenging the scientific domain. Research and development initiatives in global water challenges are the utmost need of the hour. Science today is a huge colossus with a definite vision of its own. Heavy metal is a general collective term, which applies to the group of metals and metalloids with atomic density greater than 4000 kg m^{-3} or five times more than water and they are natural components of the earth's crust. Although some of them act as essential micronutrients for living beings, at higher concentrations, they can lead to severe poisoning. In the environment around us, the heavy metals are generally more persistent than organic contaminants such as pesticides or petroleum by-products. They can become mobile in soils depending on the soil pH and its speciation. The vision for the future is enhanced as global water research and development initiatives gain newer visionary heights.

11.16 SCIENTIFIC ENDEAVOR IN THE FIELD OF ARSENIC REMEDIATION

Arsenic groundwater remediation needs to be readdressed and rebuilt with the passage of scientific research pursuit, scientific history, and time. Science and technology are in the path of newer scientific rejuvenation and scientific fortitude.

Welch et al.[29] delineated with deep details arsenic in groundwater in the United States and its occurrence and geochemistry. Concentrations of naturally occurring arsenic in groundwater vary regionally due to a combination of climate and geology. Although slightly less than half of 30,000 arsenic analyses of groundwater in the United States were less than 1 µg/L, about 10% exceeded 10 µg/L. This is a major challenge in the scientific scenario globally. At a broad regional scale, arsenic concentration exceeding 10 µg/L appear to be more frequently observed in the western United States than in the eastern half.[29] Arsenic in drinking water can veritably impact human health and is considered one of the prominent environmental causes of cancer mortality in the world. The authors delineated in details arsenic occurrence in relation to anthropogenic sources, arsenic occurrence in relation to natural sources, arsenic occurrence in atmospheric precipitation and surface water, arsenic occurrence in groundwater, geochemical controls on arsenic in ground water, arsenic reaction concepts, and a widely observed research endeavor into arsenic occurrence and geochemistry.[29]

Wang et al.[30] deeply comprehended in a review paper the natural attenuation processes for remediation of arsenic contaminated soils and groundwater.[30] Technological and deep scientific validation are gaining immense heights in water science and technology in this decade. This preview presents a wide preview on the hazardous arsenic contamination in many countries. Natural attenuation of arsenic contaminated soils and groundwater may be a cost-effective in situ remedial option. It definitely relies on the site intrinsic assimilative capacity and allows in-place cleanup. Natural attenuation processes are witnessing immense scientific cognizance as groundwater remediation stands as a major pillar to the furtherance of water technology. In another visionary avenue, microbial activity may catalyze the transformation of As species or mediate redox reactions thus influencing As mobility. This treatise widely observes the success of today's technology in groundwater remediation.[30]

Su et al.[31] delineated in deep details arsenate and arsenite removal by zerovalent iron and its kinetics, redox transformation, and implications for in situ groundwater remediation. Arsenic, a known carcinogen in humans, is often found in contaminated groundwater as a result of weathering of rocks, industrial waste discharges, agricultural use of arsenical herbicides, pesticides, etc.[31] Science of arsenic groundwater remediation is evolving into a new dimension of future scientific pursuit. High naturally occurring As (greater than the proposed EPA maximum contaminant level for As in drinking water of 0.01 mg/L) in well water supplies has been reported as a widespread health hazard in Bangladesh, Taiwan, and elsewhere in the world. The authors discussed in deep details arsenic removal kinetics and the futuristic vision of arsenic groundwater remediation.[31]

Mukherjee et al.[32] comprehended and delineated the arsenic groundwater crisis in various countries in Asia. The widespread incidence of high concentrations of arsenic in drinking water has emerged as a major public health issue.[32] With newer-affected sites discovered during the last decade, a significant change has been observed in the arsenic contamination scenario throughout the world, especially in the Asian countries. Along with the present situation in severely affected countries in Asia, such as Bangladesh, India, and China, recent instances from Pakistan, Myanmar, Afghanistan, Cambodia, etc., are presented with lucid details. Adverse health effects of arsenic depend strongly on the dose and duration of exposure.[32] Specific health and dermatological effects are highly characteristics of wide exposure to arsenic. Science and technology are challenged today at this largest environmental disaster throughout the world. Sustainable development and environmental sustainability are in a state of disaster in South Asia with the devastating issue of arsenic groundwater contamination. The authors specifically point out to the environmental disaster in Afghanistan, Argentina, Australia, Bangladesh, China, Finland, Germany, and different states of India and the Indian subcontinent.[32] This treatise pointedly targets the alarming situation of arsenic contamination and the success of science and environmental engineering techniques.[32]

Chakraborti et al.[33] presented the status of arsenic groundwater contamination in Bangladesh in a 14-year study report. Since 1996, 52,202 water samples from hand tube wells were analyzed for arsenic (As) by flow injection hydride generation atomic absorption spectrometry (FI-HG-AAS) from all 64 districts of Bangladesh; 27.2% and 42.1% of the tubewells had As above 50 and 10 µg/L, respectively; 7.5% contained As above 300 µg/L, the concentration predicting overt arsenical skin lesions. This paper is an eye-opener to the entire scientific world. The groundwater of 50 districts contained As above the Bangladesh standard for As in drinking water (50 µg/L) and 59 districts had As above the WHO guideline value (10 µg/L). Technology is widely challenged as human scientific endeavor witnesses devastating challenges with the growing concern for arsenic groundwater crisis.[33]

Hashim et al.[34] gave a detailed scientific review on remediation technologies for heavy metal contaminated water. The contamination of groundwater by heavy metal, originating either from natural soil sources or from anthropogenic sources, is a matter of immense global concern.[34] Thus, remediation of contaminated groundwater is of highest priority for the furtherance of science and progress of human civilization. In this treatise, 35 approaches for groundwater treatment have been reviewed and deeply classified under three categories: chemical, biochemical/biological/bioabsorption, and physicochemical treatment processes.[35–37]

11.17   NANOTECHNOLOGY AND WASTEWATER TREATMENT

Industrial wastewater treatment today stands in the crucial juncture of scientific introspection and scientific vision. Provision of clean drinking water to meet basic human needs stands as a major pillar of human civilization in the 21st century. The challenge and vision go beyond scientific imagination and deep scientific understanding. Enhanced population growth, global climate change, and water quality degradation stand as a major challenge in the progress of scientific rigor. Technological validation and scientific splendor are witnessing immense test of our times. Rakhi et al.[38] reviewed applications of nanotechnology in wastewater treatment. The review[38] covered candidate nanomaterials, properties, and mechanisms that enable the applications, barriers, and research needs of the domain of nanotechnology and wastewater treatment.[38] Nanomaterials are typically defined as materials smaller than 100 nm in at least one dimension. At this scale, materials often possess novel size-dependent properties of nanomaterials, such as fast dissolution, high reactivity, and strong absorption, which relate to the high-specific surface area. Wastewater applications in the field of nanotechnology are in adsorption, carbon-based nanoabsorbents, heavy metal removal, metal-based nanoabsorbents, polymeric nanoabsorbents, and zeolite domains.[35–37]

11.18   FUTURE RESEARCH TRENDS AND FUTURE FLOW OF THOUGHTS

Science and technology today have a wide vision. Environmental engineering science is veritably surpassing visionary scientific boundaries. Today, global water issues are changing the face of human scientific endeavor. Technology and engineering science need to be reenvisioned and rebuilt with each step of scientific life. This treatise gives wide glimpses into the world of nonconventional environmental engineering techniques such as AOPs. The other avenue of research pursuit needs to be stressed upon in the area of novel separation processes such as membrane science. Futuristic vision should be targeted toward scientific endeavor in the field of membrane separation processes. Challenges and barriers of scientific validation are immense and evergrowing. Areas which are latent in scientific research pursuit need to be more profound and replete with deep scientific vision. Future research trends and future flow of thoughts should be more replete with scientific profundity and deep scientific comprehension.[35-37]

11.19   FUTURE OF AOPS AND FUTURISTIC VISION OF INDUSTRIAL WASTEWATER TREATMENT

Industrial wastewater treatment and nonconventional environmental engineering techniques are two opposite sides of the visionary coin today. AOPs and other integrated techniques are ushering in a new visionary eon in the field of environmental engineering science. Global water crisis and heavy metal groundwater contamination need to be reenshrined and reenvisaged with each step of scientific research pursuit. Membrane separation processes are the other avenues of environmental engineering pursuit which need to be tackled with utmost importance. The challenge and vision of science and engineering in the field of environmental science are today evergrowing and far-reaching. Environmental regulations and the concept of zero-discharge norms have urged the scientific scenario to devise and innovate new technologies. One such technology is AOPs. The author repeatedly stresses upon the success of various water purification technologies with the immense vision of environmental sustainability and the march of science. The overarching goal of human civilization today is the provision of basic human needs such as water and sustainability. In such a critical juncture, environmental science and technology are ushering in a new dawn of scientific research pursuit. The future today lies in the hands of environmental engineering science as human civilization moves forward.

11.20   CONCLUSION

The world of environmental engineering science and chemical process engineering is crossing all the visionary frontiers. Technology and science are progressing forward with immense scientific vision and scientific forbearance. The authors pointedly focus on the immense potential and success of AOPs with a vision toward the furtherance of science and engineering. The wide scientific rigor, the immense scientific urge to excel, and the targets of science will lead a long way in the true realization and true emancipation of environmental engineering science today. Science of environment today stands in the midst of deep introspection and scientific vision. Future trends in research stand in the midst of innovation and deep comprehension. The futuristic trends in research endeavor should focus on the novel separation processes such as membrane science. This area needs to be explored with utmost need as global water challenges cross visionary frontiers. Scientific vision, scientific profundity, and scientific fortitude are witnessing newer challenges as environmental engineering moves forward. This treatise gives a wide and definite glimpse on the success of environmental engineering techniques in treating industrial wastewater treatment and the furtherance of scientific vision and scientific innovation.

ACKNOWLEDGMENT

The author wishes to acknowledge the contributions of chancellor, vice-chancellor, faculty, staff, and students of University of Petroleum and Energy Studies, Dehradun, India without whom this writing project would not have been complete. The author also wishes to acknowledge the support and contribution of Shri Subimal Palit, late father of the author and an eminent textile engineer from India who taught the author the rudiments of chemical engineering.

KEYWORDS

- advanced
- oxidation
- oxygen
- hydroxyl
- wastewater
- pollution

REFERENCES

1. Al-Kdasi, A.; Idris, A.; Saed, K.; Guan, C. T. Treatment of Textile Wastewater by Advanced Oxidation Processes—A Review. *Glob. Nest Int. J.* **2004,** 6(3), 222–230.
2. Goi, A. Advanced Oxidation Processes for Water Purification and Soil Remediation. Doctoral Thesis, Department of Chemical Engineering, Tallinn University of Technology, Estonia, 2005.
3. Mofidi, A. A.; Min, J. H.; Palencia, L. S.; Coffey, B. M.; Liang, S.; Green, J. F. *Advanced Oxidation Processes and UV Photolysis for Treatment of Drinking Water*; Report of California Energy Commission, Sacramento, California, USA, 2002.
4. Mota, A. L. N.; Albuquerque, L. F.; Beltrame, L. T. C.; Chiavone-Filho, O.; Machulek, A., Jr.; Nascimento, C. A. O. Advanced Oxidation Processes and Their Application in the Petroleum Industry: A Review. *Braz. J. Oil Gas* **2008,** 2(3), 122–142.
5. Grote, B. In *Application of Advanced Oxidation Processes (AOPs) in Water Treatment*, 37th Annual Queensland Water Industry Operations Workshop, Parklands, Gold Coast, Australia, 2012.
6. Sharma, S.; Ruparelia, J. P.; Patel, M. L. In *A General Review on Advanced Oxidation Processes for Wastewater Treatment*, International Conference on Current Trends in Technology, Gujarat, 2011.
7. Jose, M. Combination of Advanced Oxidation Processes with Biological Treatment for the Remediation of Water Polluted with Herbicides. Doctoral Thesis, Universitat Autonoma de Barcelona, June 2007.
8. Van der Helm, A. W. C. Integrated Modeling of Ozonation for Optimization of Drinking Water Treatment. Doctoral Thesis, Technische Universiteit, Delft, Netherlands, 2007.
9. EPA. United States Environmental Protection Agency Report, Wastewater Technology Fact Sheet, Ozone Disinfection, Office of Water: Washington, DC, USA, 1999.
10. Khadre, M. A.; Yousef, A. E.; Kim, J.-M. Microbiological Aspects of Ozone Applications in Food: A Review. *J. Food Sci.* **2001,** 66(9), 1242–1252 (Concise Reviews in Food Science).
11. Cervi, S. Formation and Removal of Aldehydes as Ozonation By-products in a Pilot Scale Water Treatment Plant. Master of Applied Science Thesis, University of Windsor, Canada, 1996.
12. McElroy, F.; Mikel, D.; Nees, M. Determination of Ozone by Ultraviolet Analysis, A New Method for Volume II, Ambient Air Specific Methods. In *Quality Assurance Handbook for Air Pollution Measurement Systems*; Ventura County APCD, 1997.
13. Joshi, M.; Bansal, R.; Purwar, R. Color Removal from Textile Effluents. *Indian J. Fibre Text. Res.* **2004,** 29, 239–259.
14. Johansson, H. P. *Treatment of Dye Wastewater with Ozone Solution*; International Summer Water Resources Research School, Lund University, Sweden and Xiamen University: China, 2013.
15. Khan, H.; Ahmad, N.; Yasar, A.; Shahid, S. Advanced Oxidative Decolorization of Red CI-5B: Effects of Dye Concentration, Process Optimization and Reaction Kinetics. *Pol. J. Environ. Stud.* **2010,** 19(1), 83–92.
16. Grande, G. A. Treatment of Wastewater from Textile Dyeing by Ozonization. Doctoral Thesis, Politecnico di Torino, 2015.
17. Garcia-Morales, M. A.; Roa-Morales, G.; Barrera-Diaz, C.; Miranda, V. M.; Hernandez, P. B.; Silva, T. B. P. Integrated Advanced Oxidation Process (Ozonation) and

Electrocoagulation Treatments for Dye Removal in Denim Effluents. *Int. J. Electrochem. Sci.* **2013,** *8,* 8752–8763.

18. Multani, M. Y.; Shah, M. J. Removal of Color and COD from Reactive Green-19 Dyeing Wastewater Using Ozone. *Int. J. Eng. Sci. Res. Technol.* **2014,** *3*(2), 699–704.

19. Muhammad, A.; Shafeeq, A.; Butt, M. A.; Rizvi, Z. H.; Chughtai, M. A.; Rehman, S. Decolorization and Removal of COD and BOD from Raw and Biotreated Textile Bath Effluent Through Advanced Oxidation Processes (AOPs). *Braz. J. Chem. Eng.* **2008,** *25*(3), 453–459.

20. Fahmi, M. R.; Abidin, C. Z. A.; Rahmat, N. R. *In Characteristic of Color and COD Removal of Azo Dye by Advanced Oxidation Process and Biological Treatment,* International Conference on Biotechnology and Environment Management, IPCBEE, 2011, Vol. 18.

21. Venkatesh, S.; Quoff, A. R.; Pandey, N. D. Treatment and Degradation of Synthetic Azo Dye Solution Containing Acid Red 14 by Ozone. *J. Innov. Res. Solut.* **2014,** *1*(1).

22. Selcuk, H.; Meric, S. Ozone Pre-oxidation of a Textile Industry Wastewater for Acute Toxicity Removal. *Glob. Nest J.* **2006,** *8*(2), 95–102.

23. Poznyak, T.; Colindres, P.; Chairez, I. Treatment of Textile Industrial Dyes by Simple Ozonation with Water Circulation. *J. Mex. Chem. Soc.* **2007,** *51*(2), 81–86.

24. Lambert, J.; Vega, M. M.; Isarein-Chavez, E.; Peralta-Hernandez, J. M. Ozone and Electrocoagulation Processes for Treatment of Dye in Leather Industry Wastewater: A Comparative Study. *Int. J. Emerg. Technol. Adv. Eng.* **2013,** *3*(7), 1–9.

25. Tabrizi, M. T. F. Decolouration of Reactive Dyes with Ozone in a Semi-batch and a Continuous Stirred Reactor. Doctoral Thesis, University of the Witwatersrand, 2013.

26. Shannon, M. A.; Bohn, P. W.; Elimelech, M.; Georgadis, J. G.; Marinas, B. J.; Mayes, A. M. Science and Technology for Water Purification in the Coming Decades. *Nature* **2008,** *452,* 301–310.

27. Ray, A. K.; Beenackers, A. A. C. M. Novel Photocatalytic Reactor for Water Purification. *Am. Inst. Chem. Eng. J.* **1998,** *44*(2), 477–483.

28. Dijkstra, M. F. J.; Buwalda, H.; De Jong, A. W. F.; Michorius, A.; Winkelman, J. G. M.; Beenackers, A. A. C. M. Experimental Comparison of Three Reactor Designs for Photocatalytic Water Purification. *Chem. Eng. Sci.* **2001,** *56,* 547–555.

29. Welch, A. H.; Westjohn, D. B.; Helsel, D. R.; Wanty, R. B. Arsenic in Groundwater of the United States: Occurrence and Geochemistry. *Groundwater* **2000,** *38*(4), 589–604.

30. Wang, S.; Mulligan, C. N. Natural Attenuation Processes for Remediation of Arsenic Contaminated Soils and Groundwater. *J. Hazard. Mater.* **2006,** *B138,* 459–470.

31. Su, C.; Puls, R. W. Arsenate and Arsenite Removal by Zerovalent Iron: Kinetics, Redox Transformation, and Implications for In Situ Groundwater Remediation. *Environ. Sci. Technol.* **2001,** *35,* 1487–1492.

32. Mukherjee, A.; Sengupta, M. K.; Amir Hossain, M.; Ahamed, S.; Das, B.; Nayak, B.; Lodh, D.; Rahman, M. M.; Chakraborti, D. Arsenic Contamination in Groundwater: A Global Perspective with Emphasis on the Asian Scenario. *J. Health Popul. Nutr.* **2006,** *24*(2), 142–163.

33. Chakraborti, D.; Rahman, M. M.; Das, B.; Murrill, M.; Dey, S.; Mukherjee, S. C.; Dhar, R. K.; Biswas, B. K.; Chowdhury, U. K.; Roy, S.; Sorif, S.; Selim, M.; Rahman, M.; Quamruzzaman, Q. Status of Groundwater Arsenic Contamination in Bangladesh: A 14 Year Old Study Report. *Water Res.* **2010,** *44,* 5789–5802.

34. Hashim, M. A.; Mukhopadhyay, S.; Sahu, J. N.; Sengupta, B. Remediation Technologies for Heavy Metal Contaminated Groundwater. *J. Environ. Manag.* **2011,** *92,* 2355–2388.

35. Cheryan, M. *Ultrafiltration and Microfiltration Handbook*; Technomic Publishing Company, Inc.: USA, 1998; pp 1–28.
36. Palit, S. Filtration: Frontiers of the Engineering and Science of Nanofiltration—A Far-reaching Review. In *CRC Concise Encyclopedia of Nanotechnology*; Ortiz-Mendez, U., Kharissova, O. V., Kharisov, B. I., Eds.; Taylor and Francis: USA, 2016; pp 205–214.
37. Palit, S. Advanced Oxidation Processes, Nanofiltration, and Application of Bubble Column Reactor. In *Nanomaterials for Environmental Protection*; Kharisov, B. I., Kharissova, O. V., Dias, H. V. R., Eds.; Wiley: USA, 2015; pp 207–215.
38. Rakhi, M. S.; Suresh Babu, G.; Premalatha, M. Applications of Nanotechnology in Wastewater Treatment: A Review. *Imp. J. Interdiscip. Res.* **2016,** *2*(11), 1500–1511.

PART III
Innovations in Applied Sciences

CHAPTER 12

NICKEL STEARATE HYDROSOLS FOR HYDROPHOBIC CELLULOSE

CEMAL İHSAN SOFUOĞLU[1], FATMA ÜSTÜN[2], BURCU ALP[2], and DEVRIM BALKÖSE[2,*]

[1]Department of Computer Engineering, İzmir Institute of Technology, Gulbahce Urla, İzmir, Turkey

[2]Department of Chemical Engineering, Izmir Institute of Technology, Gulbahce, Urla, İzmir, Turkey

*Corresponding author. E-mail: devrimbalkose@gmail.com

CONTENTS

ABSTRACT

The nickel stearate hydrosols were prepared by mixing aqueous sodium stearate and nickel chloride solutions in the present study. Nickel stearate powder with 50 nm crystal size and 11 μm particle size was obtained by drying of the precipitate separated by filtration of the nickel stearate hydrosols. The powder was characterized by FTIR spectroscopy, X-ray diffraction, differential scanning calorimetry, thermal gravimetry, hot-stage microscopy, and colorimetry. The nickel stearate had a solid–solid phase transition at 65°C and become liquid at 100°C. Hot-stage microscopy indicated the melting occurred in a wide temperature range. Spherical liquid droplets were observed in molten nickel stearate. TG analysis indicated NiO and Ni formed as a solid residue and distearyl ketone as the volatile fraction on heating nickel stearate up to 800°C and keeping the temperature constant at 800°C for 20 min. The hydrophilic cellulose surface of a filter paper was changed to hydrophobic surface by wetting it with nickel stearate hydrosol and then drying. The water contact angle at initial time was increased from 30° to 120° and 115° in one surface-coated and dip-coated filter papers, respectively.

12.1 INTRODUCTION

Nickel stearate ($NiSt_2$) is a metal soap used in production of carbon nanotubes,[6,7] mesoporous alumina,[13] and superhydrophobic aluminum surfaces.[16] Carbon nanotubes can be obtained by pyrolysis of nickel stearate at 800–1000°C with high yields in contrast to the electrical arc-discharge technique, and no external C source is required in contrast to chemical vapor deposition process. The solid state stearate provides a medium in which the dispersion of Ni atoms is uniform at a molecular level. Thermal decomposition of the molecules would not only produce Ni nanoparticles as catalytic seeds but also provide sufficient carbon atoms as feedstock for the growth of carbon nanostructures.[6] A new route for the highly convenient scalable production of carbon nanofibers on a sodium chloride support by chemical vapor deposition has been developed by Geng et al.[7] Since the support was nontoxic and soluble in water, it can be easily removed without damage to the nanofibers and the environment. Nickel incorporated alumina catalyst with 3.6 nm pore diameter could be directly prepared using nickel stearate, which was acting as a chemical template and a metal source.[13]

Nickel stearate can be prepared by the precipitation or fusion method used for producing other metal soaps such as zinc stearate, calcium stearate, magnesium stearate, cobalt stearate, and copper stearate ($CuSt_2$).[9,10–12]

Making hydrophilic cellulose hydrophobic is an important process. For example, superhydrophobic cellulose filter papers are used for oil–water separation.[5] Water-resistant cellulosic filters were made for aerosol entrapment in dry and moist medium.[4] Bagasse fibers were made hydrophobic by adsorption of core–shell latex of PSt-co-AH/BA-co-DMAEMA-co-BMA. The hydrophobic fibers could be easily dispersed in hydrophobic polymers such as polyethylene and polypropylene for composite material production.[14] A facile and low-cost superhydrophobic nanocomposite coating on paper surface was fabricated through one-step by simply spraying dispersion, using hydrophobic silica nanoparticles as a filter (SiNPs) and polyvinylidene fluoride (PVDF) as a film-forming material. Such superhydrophobic nanocomposite coatings on paper surface showed little adhesive property with water. The approach to fabricate superhydrophobic coatings on paper surface may have potential application in the fields of anticontamination and protection against moisture.[15]

In the present study, the preparation of nickel stearate sols and powders, and its applications in making cellulose hydrophobic and carbon nanofibers were aimed to be investigated. Nickel stearate was obtained by metathesis reaction and coated on filter paper by dip coating or spreading of the nickel stearate sols.

12.2 MATERIALS AND METHODS

12.2.1 MATERIALS

Sodium stearate, (NaSt) $C_{17}H_{35}COONa$, (commercial product, Dalan Kimya A.S., Turkey) was used in synthesis of metal soaps. The acid value of stearic acid, used in NaSt synthesis, was 208.2 mg KOH/g of stearic acid and it consists of C16–C18 alkyl chain and with 47.7% and 52.3% in weight, respectively. Nickel chloride hexahydrate (Sigma-Aldrich) was used as the metal salt.

12.2.2 PRECIPITATION PROCESS

5.000 g (0.016 mol) of sodium stearate (NaSt) was dissolved in 200 cm³ of deionized water in the reactor at 75°C. Since the NaSt is partly soluble at low temperatures, temperature of dissolution was selected as 75°C. Nickel chloride solutions (10 cm³) containing 50% excess of, 20% less, and 50% less than the equivalent amount of nickel cations to sodium stearate were added to added to 200 cm³ sodium soap solutions at 75°C. The mixtures were stirred at a rate of 160 rpm in a thermostated shaker for 30 min. The suspensions or sols obtained were placed in 250 cm³ graduated cylinders, and the settling rates were measured for 30 min. The suspension with 50% excess salt could be easily filtered. Stable hydrosols were formed for 20% and 50% deficient salt cases. They were passed through filter paper since their particles were smaller than the pore size of the filter paper. Thus, the reaction media with 50% excess salt was filtered by using Buchner funnel and flask at 2×10^4 Pa pressure. Wet metal soap was washed with 200 cm³ deionized water once to remove the by-product NaCl and then, wet metal soap cake was dried in a vacuum oven at 105°C under 2×10^4 Pa pressure to obtain nickel stearate powder.

12.2.3 CHARACTERIZATION OF NICKEL SOAP AND ITS THERMAL DEGRADATION PRODUCTS

The crystalline structure and purity of the nickel soap sample was determined by means of X-ray powder diffractometer (Philips Xpert-Pro) with CuKα radiation at 45 kV and 40 mA. The X-ray scans were performed in the 5–60° 2θ range. Scanning electron microscopy (SEM), (Philips XL30 SFEG) with energy dispersive X-ray (EDX) analyses were used for the identification of particle size and chemical composition. The samples were coated with gold and palladium metals by using the sputtering technique. Fourier transform infrared (FTIR) spectrophotometer (Shimadzu 8601) was used to obtain the FTIR spectra of the products with KBr disc method. In total, 4.0 mg of metal soap and 196 mg of KBr were mixed and pressed into 1 cm discs under 10 t load. EDX analysis was used to determine the product constituents and to check the removal of by-product NaCl. Melting behavior of metal soaps was monitored by using an optical microscope (Olympus CH40) equipped with a hotstage (Olympus CH40 INSTEC—Hotstage). The thermal behavior of the nickel stearate was investigated with thermal gravimetric analysis (TGA) and differential scanning calorimeter (DSC). Setaram Labsys TGA was used for recording sample mass versus temperature and time by heating the samples at 10°C/min rate from room temperature up to 800°C and keeping samples at 800°C for 20 min.

The color of the samples in graduated cylinders was measured using ImageJ software in red green blue (RGB) units.

12.3.4   HYDROPHOBIC CELLULOSE

Whatman black ribbon filter paper, which is made out of pure cellulose with 12–25 μm filtering range, was coated on one surface by spreading and dip coating with nickel stearate hydrosol prepared with 20% less than equivalent metal salt than sodium salt. The filter paper was dried at room temperature under vacuum. Then, the contact angle of water droplets on the surface of the coated paper was measured with 10 s intervals using Attention contact angle measurement device, Theta Lite.

12.3   RESULTS AND DISCUSSION

12.3.1   PRECIPITATION REACTION

When nickel salt solution is added to sodium stearate solution insoluble nickel stearate forms as shown in eq 12.1.

$$Ni^{2+} (aq) + C_{17}H_{35} COO^-(aq) \rightarrow Ni(C_{17}H_{35} COO)_2 (s) \qquad (12.1)$$

The experiments were performed for 50% more, 20% and 50% less than equivalent amount of nickel ions to stearate ions in solution. The mixtures were placed in measuring vessels to measure the settling rates. For the case of 50% more than equivalent nickel ions, the precipitated nickel stearate was in an aqueous solution in which unreacted nickel ions, sodium ions, and chloride ions were present. The presence of the free ions causes flocculation and settling of the nickel stearate particles as seen in Figure 12.1a. The hydrosols with 20% and 50% deficient nickel ions were very stable due to adsorbed water-soluble unreacted sodium stearate on the surface of the nickel stearate particles and they settled at a very slow rate.

The particle size of the flocculated particles was determined from their rate of settling in water.

The terminal rate of fall of a spherical particle in a fluid is given by Stokes' law shown in eq 12.2.

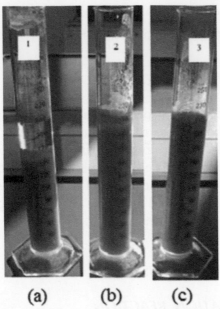

FIGURE 12.1   The nickel stearate suspensions or sols as formed prepared with nickel ions (a) 50% excess, (b) 20% less, and (c) 50% less than the equivalent amount to sodium stearate.

$$v = \frac{2 \times r^2 \times (\rho p - \rho f) \times g}{9 \times \eta}$$

(12.2)

where v: sedimentation rate in units of distance/time (m/s)

r: radius of the spherical particle (m)

ρp and ρf: density of the particle and surrounding fluid, respectively (kg/m³)

g: gravitational constant (m/s)

η: viscosity of the fluid (kg/m s)

The slopes of the graphs in Figure 12.2 showing the height of the clear boundary between water and water with nickel stearate particles in measuring vessels versus time was used as the sedimentation rates in eq 12.2. Taking ρf = 998 kg/m³, ρp = 1130 kg/m³, μ = 0.001002 Pa s at 20°C, and g = 9.81 m/s² in eq 12.2, the radius of the particles in 30% excess nickel salt case was 11 μm, and for the 20% deficient case, it was 3 μm.[17] For the 50% deficient case, the particles were too small to settle. A stable hydrosol was formed in this case.

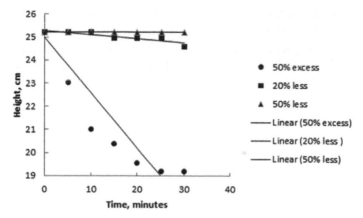

FIGURE 12.2   The height of the boundary between the clear solution and dispersed nickel stearate particles in water.

12.3.2   THE COLOR OF NICKEL STEARATE

The nickel stearates obtained for the three cases had the same green color which can be expressed as 73; 86; 39 in RGB units as measured by ImageJ program.

12.3.3   MORPHOLOGY AND ORDER IN NICKEL STEARATE

The morphology of nickel stearate is seen in Figure 12.3a. As seen in the figure, the nanoparticles of nickel stearate with nearly 100 nm size were agglomerated to form larger particles in micron size. The sedimentation rate in the experiment indicated that the agglomerated size of the particles was 11 μm in water.

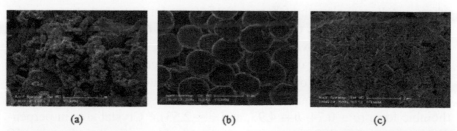

FIGURE 12.3   SEM micrographs of (a) NiSt2, (b) condensed wax in TG analysis, and (c) ash in TG analysis.

The X-ray diffraction diagram of nickel stearate seen in Figure 12.4 was analyzed for bilayer distance.

From the third, fourth, fifth, and seventh order diffraction peaks of the bilayer distance seen in Figure 12.4 and Table 12.1, the bilayer distance was found as 5.1 nm from Bragg's law.

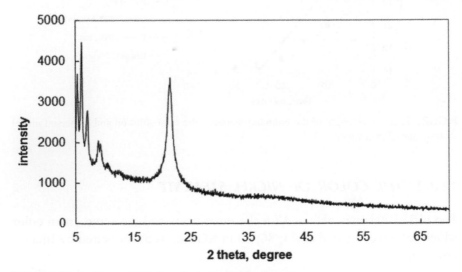

FIGURE 12.4　X-ray diffraction diagram of nickel stearate.

TABLE 12.1　Bilayer Distance form X-ray Diffraction Diagram.

Order of diffraction, n	3	4	5	6	7
2θ value of diffraction.	5.3	6.9	8.85	10.1	12.1
d value of diffraction	5.0	5.1	5.0	5.1	

$$n\lambda = 2d\mathrm{Sin}\theta \tag{12.3}$$

where λ is the wavelength of X-rays, 0.1546 nm for CuK_α radiation, n is the order of diffraction, d is the bilayer distance, and θ is the diffraction peak angle. The alkyl groups were oriented perpendicular to metal layer in nickel stearate since the bilayer distance (i.e., 5 nm) was equal to the length of two stearate groups. Alkyl groups were perpendicular to the planes of the metal ions layer. The alkyl groups of the nickel stearate were arranged in ortho-rhombic form ($a = 0.74$, $b = 4.93$, and $c = 2.53$).[20] Crystal size in perpendicular direction to 200 planes of the orthorhombic unit cells is found as 50 nm from the breadth of the diffraction peak of 200 and Scherrer formula.

$$L = kl \,/\, \beta Cos\theta \qquad\qquad (12.4)$$

where β is the measured full width at half maximum (radians), θ is the Bragg peak angle of the peak, λ is the X-ray diffraction wavelength (in this case, it is 0.1546 nm), L is the effective particle size, and K is a constant = 0.9. This indicated that the particles observed in SEM micrographs were agglomerates of the very small crystals of the metal soaps.

12.3.4 THE MODE OF BINDING OF METAL IONS AND CARBOXYLATE GROUPS

The FTIR spectrum of nickel stearate in Figure 12.5 had vibration peaks at 2916 and 2848 cm^{-1} for asymmetric and symmetric streching vibrations, and at 1465 and 717 cm^{-1} for bending vibrations of CH$_2$ groups. There are different coordination types between metals and carboxylates. The chelating and ionic forms of COO$^-$ have peak maxima at 1540 and 1576 cm^{-1} respectively, in their IR spectra.[1] The asymmetric and symmetric vibrations of carboxylate group are observed at 1540 and 1417 cm^{-1} and indicated that the mode of binding of the carboxylate groups to metal ions was in chelating form.

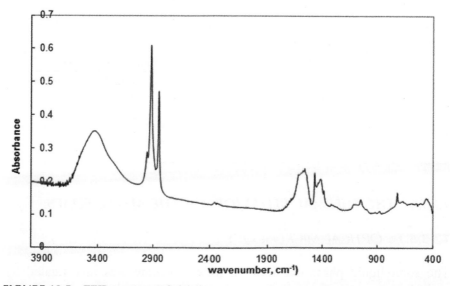

FIGURE 12.5   FTIR spectrum of nickel stearate.

12.3.5 COMPOSITION OF NICKEL STEARATE

The composition of nickel stearate powder was determined by EDX analysis on hydrogen-free basis. The EDX spectrum of nickel stearate is seen in Figure 12.6. As reported in Table 12.2, it contained 8.18% Ni, 7.14 % O, and 84.78 % C.

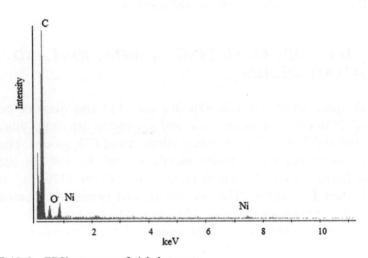

FIGURE 12.6   EDX spectrum of nickel stearate.

TABLE 12.2   Compositions of Nickel Stearate, Volatile Fraction and Nonvolatile Fraction in Mass % and the Emprical Molar Composition.

	Mass, %			Empirical formula		
	Nickel stearate	Volatile fraction	Residue	Nickel stearate	Volatile fraction	Residue
C	84.78	95.65	5.26	51	29	2
O	7.14	4.35	10.83	3	1	3
Ni	8.18	0.00	83.91	1	0	7

12.3.6 THE THERMAL BEHAVIOR OF THE METAL SOAPS

12.3.6.1 OPTICAL MICROSCOPY

The solid–liquid phase transition in nickel stearate was investigated by optical microscopy. The optical micrographs of the soaps were recorded at different time periods during heating them at 2°C/min rate. In general, metal soaps do not transmit light when in solid state and they appear as dark regions in the micrographs. When they are liquefied, the edges of the particles are rounded and the particles become transparent.[12] For cobalt stearate, copper stearate, and magnesium stearate, a sharp transition from solid to liquid was observed at 159°C, 117°C, and 160°C, respectively.[12] However for nickel stearate, the transition occurred in a wider range of temperature; it was not possible to observe a sharp physical change. In Figure 12.7, the optical micrographs of nickel stearate at 42.1°C, 103.7°C, and 149.9°C are seen. At 42.1°C the solid particles and at 149.9°C entirely spherical liquid droplets are observed. As understood from Figure 12.7, melting occurred in a wide range of temperature to give a sharp change in the appearance of nickel stearate.

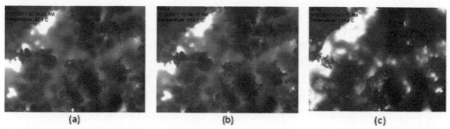

(a) (b) (c)

FIGURE 12.7 Optical micrograph of nickel stearate powder at (a) 42.1°C, (b) 103.7°C, and (c) 149.9°C.

12.3.6.2 DSC ANALYSIS

The DSC analysis gives information about energy changes during heating of the substances. The DSC curve of nickel stearate is shown in Figure 12.8.

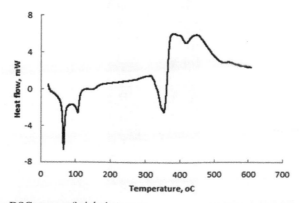

FIGURE 12.8 DSC curve of nickel stearate.

There is a solid phase transition at 65°C and 100°C in DSC curve of nickel stearate due to phase changes. The first peak could be attributed to a solid–solid phase transition as observed for magnesium stearate by Gonen et al.,[12] and 100°C peak is attributed to solid to liquid change. The melting point of nickel stearate is reported as 82–86°C and 100°C in the literature. At 349°C and 409°C there are two endothermic peaks due to thermal degradation.[18,19]

12.3.6.3 TGA AND THERMAL DECOMPOSITION OF NICKEL STEARATE

On heating nickel stearate in Labsys TGA at 10°C/min rate from room temperature up to 800°C and then further heating for 20 min at 800°C, a very interesting behavior was observed. There was a green solid residue in the alumina pan and there were white spherical particles condensed in the space between the alumina pan and alumina pan holder of the TGA instrument shown in Figure 12.9. These particles are thought to be generated from the decomposition of nickel stearate by heat. They were in liquid form and floating in nitrogen flow during the heating period and solidified and collected on the sample holder on cooling.

FIGURE 12.9 LABSYS TGA sample pan holder with the alumina pan placed on top.

No change in mass was observed up to 295°C confirming that the endo-thermic peaks at 65°C and 100°C in DSC curve were related to phase changes only. The mass loss starts at 295°C as seen in the TG curve of nickel stea-rate in Figure 12.10. The mass remained was 99%, 98%, 96%, and 93% at 390°C, 441°C, 555°C, and 800°C, respectively. Further heating at 800°C isothermally resulted in mass loss versus time trend shown in Figure 12.11. The remained mass was 89% at 20 min heating period. The solid residue in the alumina pan and the volatilized fraction that was condensed as spherical particles on the nitrogen flow path were characterized by SEM, EDX, and FTIR.

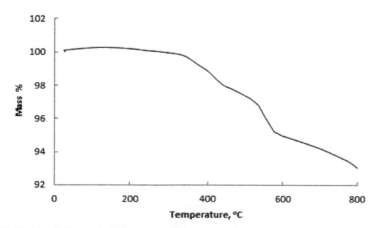

FIGURE 12.10 TG curve of nickel stearate.

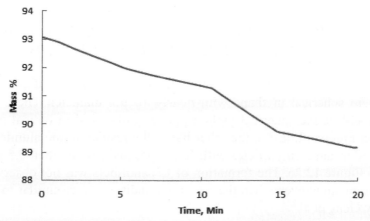

FIGURE 12.11 Change of the remaining mass of the nickel stearate with time on heating at 800°C.

Metal oxides are formed on heating metal soaps at high temperatures.[2,3] The following reaction shown by eq 12.5 was expected to occur during their heating at high temperatures;

$$Ni\,(C_{17}H_{35}\,COO)_2 \rightarrow NiO + C_{17}H_{35}\,COC_{17}H_{35} + CO_2 \qquad (12.5)$$

Nickel oxide, distearyl ketone, and carbon dioxide are possible reaction products. Distearyl ketone will be further pyrolyzed to volatile compounds and carbon-rich nonvolatile compounds.

TGA of nickel stearate in nitrogen atmosphere, by dynamic heating at 10°C/min rate was made by Geng et al.[7] The decomposition of nickel stearate had a rapid start but proceeded slowly at the later stage, and thus displayed a long tail. There was a middle step in the temperature range of 300–400°C, presumably caused by formation of an intermediate substance. The whole process had finally finished at a temperature of 650°C, where the measured residue mass was close to that of the Ni content in the stearate (calculated: 9.4 wt%). The probable chemical decomposition process of nickel stearate at 300–600°C is described below.

$$nNi(CH_3(CH_2)_{16}COO)_2 \rightarrow nNi + \text{gases species}$$
$$(H_2, CO, CO_2, H_2O) + C \qquad (12.6)$$

The decomposition of nickel stearate salts was driven by self-redox reactions, in which a nickel atom obtained two electrons from the two carboxyl ligands to be reduced into metallic nickel.[7]

12.3.6.4   THE MORPHOLOGY OF THE HEATING RESIDUE AND THE PARTICLES CONDENSED IN NITROGEN FLOW PATH

As seen in micrograph in Figure 12.3b, the particles condensed in the flow path were spherical in shape with nearly 10 μm diameter. These particles should be the spherical molten particles observed in optical micrograph in Figure 12.6c. On the other hand, the residue in the alumina pan consisted of lamellar particles with flat edges and smaller than 0.5 μm as seen in Figure 12.3c. The formation of C nanofibers was not observed in the present study even when the reaction conditions were similar to those used by Geng et al.[6]

12.3.6.4.1   The Elemental Composition of the Heating Residue and the Particles Condensed in Nitrogen Flow Path

The EDX spectra of the residue in the alumina pan and the particles condensed in nitrogen flow are shown in Figures 12.12 and 12.13, respectively. The residue contained C, O, and nickel elements. On the other hand, the particles that were evaporated from the alumina pan during heating contained only C

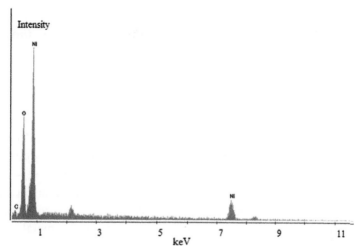

FIGURE 12.12   EDX spectrum of the residue in the alumina pan after TGA.

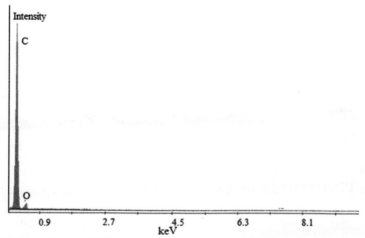

FIGURE 12.13   EDX spectrum of the spherical particles condensed in nitrogen flow path.

and O elements. The average elemental composition of the residue and the volatilized particles in mass %, and their empirical formula are reported in Table 12.2. The elemental composition of the residue indicated for seven Ni atoms, there were three O atoms meaning that not only NiO but also elemental Ni was formed during heating of nickel stearate. Thus, both reactions 12.5 and 12.6 occurred during pyrolysis in nitrogen atmosphere. The volatile fraction had 29 C atoms for 1 oxygen atom. This showed the occurrence of reaction leading to distearyl ketone in eq 12.5.

12.3.6.4.2 The Functional Groups in the residue and the Condensed Particles

The functional groups present in the heating residue and condensed particles were determined by FTIR spectroscopy. There are CH_2 asymmetric streching, CH_2 symmetric streching and CH_2 wagging vibrations at 2916, 2848, and 1462 cm^{-1} with low intensity in the FTIR spectrum of the residue in Figure 12.14, and with high intensity in the FTIR spectrum of volatile particles in Figure 12.15. The peak at 1639 cm^{-1} is attributed to C=C groups and the peak at 1053 cm^{-1} is attributed to C–O group vibrations in the FTIR spectrum of the solid residue. The sharp peak at 2357 cm^{-1} could be due to vibrations of nitrile groups[8] formed by the reaction of the metal soap with

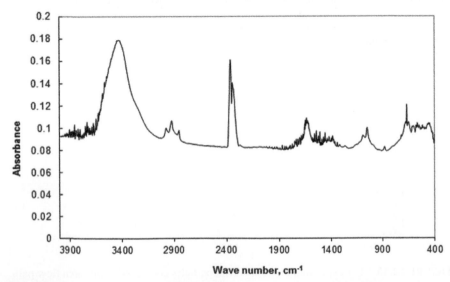

FIGURE 12.14   FTIR spectrum of the residue in the alumina pan after TGA.

flowing nitrogen gas at high temperature. On the other hand, in the spectrum of the condensed particles in Figure 12.15, there is a peak which belongs to C=O stretching vibrations at 1700 cm^{-1}. Thus, the occurrence of reaction shown in eq 12.5 and formation of ketones was also confirmed by FTIR spectroscopy.

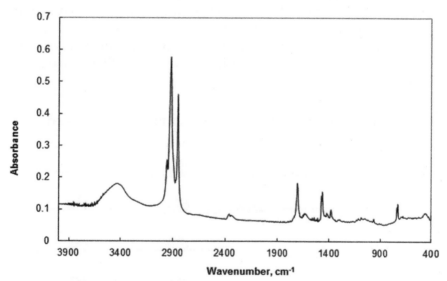

FIGURE 12.15   FTIR spectrum of the particles volatalized from nickel stearate and condensed in nitrogen flow.

12.3.7   HYDROPHOBIC CELLULOSE

Hydrophilic cellulose filter papers could be made hydrophobic by spray coating,[15] by immersion,[5] or by adsorption[14] of water repellent substances. Nickel stearate hydrosol prepared by using 50% deficient nickel ions was used in making cellulosic filter paper hydrophobic in the present study. The paper was coated with nickel stearate by placing a drop of the hydrosol on one face and by immersing the paper in the hydrosol for few seconds, then withdrawing and drying. The photographs of the water droplets placed on the surfaces of filter papers which are uncoated, coated with nickel stearate in one surface, and in two surfaces are shown in Figure 12.16. The contact angle of water droplets in Figure 12.16 was measured and shown in Figure 12.17. As seen in Figure 12.17, water droplet immediately wets the surface of uncoated cellulose with 30° contact angle, and it flows at a fast rate through the pores of the filter paper. The water droplets did not wet the coated surfaces. The initial water contact angle was 120° and 115° for one surface coated and dip coated papers, respectively, and decreased with time as seen in Figure 12.17. The water droplets remained on one surface coated paper up to 80 s, and at 90 s there was still a small amount of water on dip coated paper surface. The coating of filter paper with nickel stearate made it hydrophobic, and slowed down the flow of water through the pores of the paper. The hydrophobic filter paper obtained in the present study could be used in separation of water in oil mixtures by allowing oil to pass through the pores at a fast rate, and leaving water phase on top of the paper as realized by Gao et al.[5]

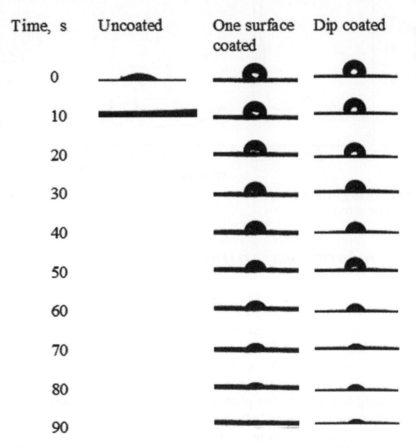

FIGURE 12.16   The photographs of the water droplets on the surfaces of filter paper without coating, one surface coated, and dip coated with nickel stearate taken at 10 s intervals.

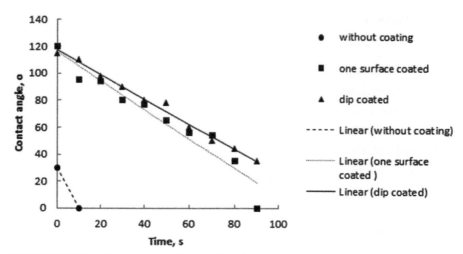

FIGURE 12.17 Contact angle of water droplets versus time on the surface of filter paper without coating, one surface coated, and dip coated with nickel stearate versus time.

12.4 CONCLUSION

Nickel stearate powder and hydrosols were prepared from nickel chloride and sodium stearate by controlling the nickel ion/stearate ion ratio. The powder was crystalline with 50 nm crystal size and 11 μm particle size. The carboxylate groups were in chelating form as indicated by FTIR spectroscopy. There is a solid phase transition at 65°C and melting occurs at 100°C on heating nickel stearate. It decomposed to distearyl ketones, nickel oxide and nickel on heating up to 800°C. Spherical waxy particles were condensed on the nitrogen flow path when it was heated up to 800°C with 10°C/min rate and kept at 800°C for 20 min. Spherical particles were also observed in liquid state when the powder was heated in optical microscope at above 149.9°C. In the solid residue obtained in the same process, NiO and Ni were present in mixture. No carbon nanofiber formation was observed even though it was expected to form. The stable nickel stearate hydrosol was used for making hydrophilic filter paper hydrophobic. The water contact angle at initial time was increased from 30° to 120° and 115° in one surface coated and dip coated filter papers, respectively.

ACKNOWLEDGMENT

The authors thank Dr. Yılmaz Ocak for contact angle measurements.

KEYWORDS

- **nickel stearate**
- **nanoparticles**
- **Stokes' law**
- **cellulose fibers**
- **hydrophilic**
- **hydrophobic**

REFERENCES

1. Benavides, R.; Edge, M.; Allen, N. S. The Mode of Action of Metal Stearate Stabilisers in Poly(Vinyl Chloride). I. Influence of Pre-heating on Melt Complexation. *Polym. Degrad. Stab.* **1994**, *44*, 375–378.

2. Bronstein, L. M.; Atkinson, J. E.; Malyutin, A. G.; Kidwai, F.; Stein, B. D.; Morgan, D. G.; Perry, J. M.; Karty, J. A. Nanoparticles by Decomposition of Long Chain Iron Carboxylates: from Spheres to Stars and Cubes. *Langmuir* **2011**, *27*, 3044–3050.

3. Che, C. J.; Lai, H. Y.; Lin, C. C.; Wang, J. S.; Chiang, R.-K. Preparation of Monodisperse Iron Oxide Nanoparticles via the Synthesis and Decomposition of Iron Fatty Acid Complexes. *Nanoscale Res. Lett.* **2009**, *4*(11), 1343–1350.

4. Heydarifard, S.; Nazhad, M. M.; Xiao, H.; Shipin, O.; Olson, J. Water-resistant Cellulosic Filter for Aerosol Entrapment and Water Purification, Part I: Production of Water-resistant Cellulosic Filter. *Environ. Technol.* **2016**, *37*(13), 1716–1722.

5. Gao, Z. I.; Zhai, X. L.; Liu, F.; Zhang, M.; Zang, D.; Wang, C. Fabrication of TiO_2/EP Super-hydrophobic Thin Film on Filter Paper Surface. *Carbohydr. Polym.* **2015**, *128*, 24–31.

6. Geng, J. F.; Jefferson, D.; Brian, F. G.; Johnson, B. F. Direct Conversion of Nickel Stearate into Carbon Nanotubes or Pure-phase Metallic Ni Nanoparticles Encapsulated in Polyhedral Graphite Cages. *J. Mater. Chem.* **2005a**, *15*(8), 844–849.

7. Geng, J. F.; Kinloch, I. A.; Singh, C.; Golovko, V. B.; Brian, F. G.; Johnson, B. F. G.; Shaffer, M. S. P.; Li, Y.; Windle, A. H. Production of Carbon Nanofibers in High Yields Using a Sodium Chloride Support. *J. Phys. Chem. B* **2005b**, *109*(35), 16665–16670.

8. Gladney, A.; Qin, C.; Tamboue, H. Sensitivity of the C–N Vibration to Solvation in Dicyanobenzenes: A DFT Study. *Open J. Phys. Chem.* **2012**, *2*, 117–122

9. Gonen, M. D.; Balkose, D.; Inal, F.; Ulku, S. Zinc Stearate Production by Precipitation and Fusion Processes. *Ind. Eng. Chem. Res.* **2005**, *44*(6), 1627–1633.

10. Gonen, M.; Balkose, D.; Inal, F.; Ulku, S. The Effect of Zinc Stearate on Thermal Degradation of Paraffin Wax. *J. Therm. Anal. Calorim.* **2008**, *94*(3), 737–742.

11. Gonen, M.; Ozturk, S.; Balkose, D.; Okur, S.; Ulku, S. Preparation and Characterization of Calcium Stearate Powders and Films Prepared by Precipitation and Langmuir–Blodgett Techniques. *Ind. Eng. Chem. Res.* **2010**, *49*(4), 1732–1736.

12. Gonen, M.; Egbuchunam, T. O.; Balkose, D.; Inal, F.; Ulku, S. Preparation and Characterization of Magnesium Stearate, Cobalt Stearate, and Copper Stearate and Their Effects on Poly(Vinyl Chloride) Dehydrochlorination. *J. Vinyl Addit. Technol.* **2015,** *21*(4), 235–244.

13. Kim, Y.; Kim, P.; Kim, C.; Yi, J. Comparison of Mesoporous Aluminas Synthesized Using Stearic Acid and Its Salts. *Korean J. Chem. Eng.* **2005,** *22*(2), 321–327.

14. Pan, Y. F.; Wang, F. T.; Wei, T.; Zhang, C.; Xiao, H. Hydrophobic Modification of Bagasse Cellulose Fibers with Cationic Latex: Adsorption Kinetics and Mechanism. *Chem. Eng. J.* **2016,** *302*, 33–43.

15. Shi, Y.; Xiao, X. Facile Spray-coating for Fabrication of Superhydrophobic SiO_2/PVDF Nanocomposite Coating on Paper Surface. *J. Dispers. Sci. Technol.* **2016,** *37*, 640–645.

16. Xu, N.; Sarkar, D. K.; Chen, X. G.; Tong, W. P. Corrosion Performance of Superhydrophobic Nickel Stearate/Nickel Hydroxide Thin Films on Aluminum Alloy by a Simple One-step Electrodeposition Process. *Surf. Coat. Technol.* **2016,** *302*, 173–184.

17. WEB 1. http://www.chemicalbook.com/ChemicalProductProperty_US_CB5116872.aspx (accessed Nov 8, 2016).

18. WEB 2. http://www.chemicalbook.com/ChemicalProductProperty_US_CB5116872.aspx (accessed Nov 8, 2016).

19. WEB 3. https://www.scbt.com/scbt/product/nickel-ii-stearate-2223-95-2/ (accessed Nov 8, 2016)

20. White, J. L.; Cho, D. *Polyolefines: Processing, Structure Developments and Properties;* Hanser Verlag: Munich, 2001.

CHAPTER 13

NANO ZINC BORATE AS A LUBRICANT ADDITIVE

SEVDIYE ATAKUL SAVRIK, BURCU ALP, FATMA USTUN, and DEVRIM BALKÖSE*

Department of Chemical Engineering, İzmir Institute of Technology, Gulbahce Urla, Izmir, Turkey

*Corresponding author. E-mail: devrimbalkose@gmail.com

CONTENTS

ABSTRACT

Nano zinc borate particles were prepared in the present study by using aqueous solutions with low initial zinc and borate concentrations and at low temperature to prevent particle growth. Span 60 and PEG 4000 were used as templates to control the particle size. However, they increased the aggregation state of the zinc borate nanoparticles in their mother liquor. The particles have H_2O and B_3–O vibrations in their FTIR spectra. All the samples contained very similar amount of water, around 18 %, of which 2% is free water, 13% loosely bound water, and 3 % tightly bound water. The empirical formula of the nanoparticles was found as $3ZnO \ 2B_2O_3 \ 4H_2O$ from EDX analysis. X-ray diffraction diagram indicated that the particles were in amorphous state. When the nanoparticles were added to light neutral oil, the wear scar diameter and friction coefficient were lowered 50% and 20%, respectively. The visible spectrum of the lubricant before and after four ball tests indicated that the yellow color of the oil was darkened during the tests that were carried out at 75°C.

13.1   INTRODUCTION

Lubrication is an art that has been practiced for thousands of years from the early days of the human civilization. The study of lubrication as a science began in the 17th century with the development of bearings and axles. In the early 21st century, the advent of automobiles and steam engines caused the development of modern complex lubricants consisting of base oils and chemical additives.[8] Such additives include dispersants, surfactants, oxidation inhibitors, and antiwear agents.[24] Friction and surface damage can be reduced by applying extreme pressure and antiwear additives. These tend to be sulfur-, chlorine-, and phosphorous-containing compounds designed to react chemically with the metal surfaces, forming easily sheared layers of sulfides, chlorines, or phosphides, and thereby preventing severe wear and seizure. As an environmental protection measure, the use of chlorine- and phosphorous-containing compounds as lubricant additives has been restricted, and so developing new additives that pollute less has, therefore, become the target of researchers. Therefore, organic and inorganic boron-based additives have been of much attention, as they possess a good combination of properties, such as wear resistance, friction-reducing ability, oxidation inhibition, low toxicity, pleasant odor, and compatibility with frictional pairs.[1,34] Hexagonal boron nitride is the most known boron compound which is used as a solid lubricant. It has a lamellar crystalline structure, in which the covalent bonding between molecules within each layer is strong, while the binding between layers is almost entirely by means of weak van der Waals forces. This structure is similar to that of graphite and molybdenum disulfide which are highly successful solid lubricants, and the mechanism behind their effective lubricating performance is understood to be owing to easy shearing along the basal plane of the hexagonal crystalline structures.[12,16] In addition, the employment of hexagonal boron nitride in lubricating oils, metal borates, which are extraordinary ceramic and functional materials have found use as an antiwear and anticorrosion material. In literature, numerous studies have been carried out in recent years on the effects of various metal borate particles as lubricating oil additives on wear and friction.[3,9,10] Their effectiveness can be related to the formation of a borate glass as a tribochemical film or the deposition of particles on the rubbing surface.[29] The friction reduction and antiwear behaviors are dependent on the characteristics of nanoparticles, such as size, shape, and concentration.

Zinc borate is a synthetic hydrated metal borate. There are various kinds of crystalline hydrated zinc borate. These have compositions such as $4ZnO \cdot B_2O_3 \cdot H_2O$, $ZnO \cdot B_2O_3 \cdot 1 \cdot 12H_2O$, $ZnO \cdot B_2O_3 \cdot 2H_2O$, $6ZnO \cdot 5B_2O_3 \cdot 3H_2O$, $2ZnO \cdot 3B_2O_3 \cdot 7H_2O$, and $2ZnO \cdot 3B_2O_3 \cdot 3H_2O$. In these products, B_2O_3/ZnO mole ratios change from 0.25 to 5, which determine the characteristics of product.[20] The production techniques of zinc borate generally include the reaction between zinc source materials (zinc oxide, zinc salts, and zinc hydroxide) and the boron source materials (boric acid and borax).[4,15,22]

13.1.1 NANO ZINC BORATE

Nanoparticles of zinc borate are a requirement for their use as a lubricant additive. Nanoparticle formation by supercritical carbon dioxide (CO_2) drying of zinc borate species was investigated to evaluate possible chemical modifications in the product during the drying.[6] Zinc borates obtained either from borax decahydrate and zinc nitrate ($2ZnO \cdot 3B_2O_3 \cdot 7H_2O$) or from zinc oxide and boric acid ($2ZnO \cdot 3B_2O_3 \cdot 3H_2O$) were dried by both conventional and supercritical carbon dioxide drying methods. It was found that while zinc borate obtained from zinc oxide and boric acid did not have any chemical interaction with CO_2, however, carbonates were formed on the surface of zinc borate obtained from borax decahydrate and zinc nitrate.[6]

Nanometer crystal zinc borate with a particle size of 20–50 nm was prepared using the ethanol supercritical fluid drying technique.[3] Supercritical ethanol drying of zinc borate species to obtain nanoparticles was also investigated by Gonen et al.[7] Zinc borates obtained either from zinc oxide and boric acid ($2ZnO \cdot 3B_2O_3 \cdot 3H_2O$) or from borax decahydrate and zinc nitrate hexahydrate ($ZnO \cdot B_2O_3 \cdot 2H_2O$) were dried by both conventional and supercritical ethanol drying methods. It was found that both zinc borates ($2ZnO \cdot 3B_2O_3 \cdot 3H_2O$ and $ZnO \cdot B_2O_3 \cdot 2H_2O$) had undergone chemical changes that resulted in formation of zinc oxide and boric acid during the supercritical ethanol drying which was carried out at 250°C and 6.4 MPa.[7]

Cu^{2+} doped zinc borate ($Zn_3(BO_3)_2$) nanoparticles were formed by coprecipitation from aqueous zinc nitrate, copper nitrate, and borax decahdyrate solutions and by calcining the precipitate at 700°C.[17] The same coprecipitation method has been used for the synthesis of Co^{2+} and Ni^{2+} doped zinc borate nanopowders.[23]

A nanoflake-like zinc borate $2ZnO \cdot 2B_2O_3 \cdot 3H_2O$ was prepared via coordination homogeneous precipitation method using ammonia, zinc nitrate, and borax as raw materials.[28] $4ZnO \cdot B_2O \cdot 3H_2O$ nanorods were synthesized by a hydrothermal route with surfactant of PEG-300 as a template.[21] These nanorods were rectangular and about 70 nm in thickness and 150–800 nm in width. PEG-300 played a critical role in the formation of $4ZnO \cdot B_2O_3 \cdot H_2O$ nanorods. The pH value and synthesis temperature have important influence on the composition of the products, while the temperature and time have an effect on the morphology of the products.[21] The functional headgroups of the surfactant have coordination bond or strong interaction with nanoparticles and thus kinetically control the growth rates of various faces of crystals, which could control the morphology. As a nonionic polymer, PEG has hydrophilic –O– and hydrophobic –CH_2–CH_2– on the long chains. Therefore, water is a good solvent for PEG, favoring the full extension of the polymer chains. The atom O in PEG has coordination abilities with metal ions. The more or less extended order of the polymer chain could provide a new sort of organization to metal atoms along the polymer backbone, which affords the diverse assembling ways for the 1D growth of the final products.[21]

The cluster-like $4ZnO \cdot B_2O_3 \cdot H_2O$ nanostructure and $4ZnO \cdot B_2O_3 \cdot H_2O$ nanoribbons have been prepared under hydrothermal conditions at 120°C for 12 h.[30] Aqueous solutions of zinc sulfate having sodium bis(2-ethylhexyl) sulfosuccinate (AOT) and excess KNO_3 and borax solutions were mixed and heated in an autoclave at 120°C. $4ZnO \cdot B_2O_3 \cdot H_2O$ nanoribbons were firstly formed through conventional nucleation and subsequently through crystal growth process.

$$[B_4O_5(OH)_4]_2^- + H_2O \rightarrow 2H_3BO_3 + 2[B(OH)_4]^- \tag{13.1}$$

$$2Zn^{2+} + [B(OH)_4]^- \rightarrow Zn_2(OH)BO_3 + 3H^+ \qquad (13.2)$$

Zinc borate nanowhiskers $4ZnO \cdot B_2O_3 \cdot H_2O$ were in situ successfully synthesized via one-step precipitation reaction.[5] Netlike zinc borate was produced from ZnO, boric acid, and ammonia by coordination amorphous precipitation.[27] The mechanism of the coordination homogeneous precipitation method is as follows: first, zinc ions react with ammonia, forming a complex solution coexisting with precipitator boric acid. Secondly, the complex dissociates to release the metal ions via diluting the complex solution with water and heating the complex solution to remove ammonia from aqueous solution. When the zinc ions reach a certain amount, the zinc borate sediments are formed in the solution. Because the metal ions and the precipitator are dispersed in the solution homogeneously, the precipitation reaction of metal ions and precipitator can reach molecular level, which ensures the sedimentation of desired nanomaterials yielding and separating out homogeneously from the solution.[27] Zinc borate ($4ZnO \cdot B_2O_3 \cdot H_2O$) nanowhiskers were synthesized via one-step precipitation reaction in aqueous solution of sodium borate and zinc nitrate with phosphate ester as the modifying agent. The zinc borate nanowhiskers have the monoclinic crystal structure with diameters of 50–100 nm, lengths of about 1 μm, and the aspect ratios close to 10–20.[33] Nano-sized zinc borate powder with a formula of $4ZnO \cdot B_2O_3 \cdot H_2O$ was synthesized using $2ZnO \cdot 3B_2O_3 \cdot 3.0–3.5H_2O$ as a starting chemical which was produced using a wet chemical method.[14] After dissolving $2ZnO \cdot 3B_2O_3 \cdot 3.0–3.5H_2O$ in an ammonia solution, the clear solution was boiled until a white powder formed. The resultant powder was $4ZnO \cdot B_2O_3 \cdot H_2O$.[14]

The crystal and hydrophobic zinc borate ($Zn_2B_6O_{11} \cdot H_2O$) nanodiscs and nanoplatelets were successfully prepared by a wet method using $Na_2B_4O_7 \cdot 10H_2O$ and $ZnSO4 \cdot 7H_2O$ as raw materials in situ aqueous solution, and oleic acid as the modifying agent. It had been found that the as-prepared materials displayed nanodisc morphology with average diameters from 100 to 500 nm and the thicknesses about 30 nm.[25,26] Measurements of the relative water contact angle and the active ratio indicated that $Zn_2B_6O_{11} \cdot 3H_2O$ samples were hydrophobic.

It is also possible to obtain nanoparticles of zinc borate by mixing two inverse emulsions (water-in-oil) with sorbitan monostearate (Span 60) as a surfactant and light neutral oil as a continuous phase. The first and second emulsions contained aqueous solutions of borax decahydrate ($Na_2B_4O_7 \cdot 10H_2O$) and zinc nitrate ($Zn(NO_3)_2 \cdot 6H_2O$), respectively, as the dispersed phases. The produced particles were zinc borate crystals having both rodlike and spherical morphologies, and the diameters of spherical particles were changing between 20 and 30 nm.[19]

13.1.2 TRIBOLOGICAL PROPERTIES

Tribological properties of 500 SN oil-containing nanometer zinc borate particles prepared by supercritical ethanol drying method were measured by Dong and Hu.[3] Results indicated that compared with the base oil the wear resistance and load-carrying capacity of the oil were improved and the friction coefficient was decreased. Diboron trioxide was formed and tribochemical boronization took place in friction. Nanometer zinc borate took effect by deposition of diboron trioxide on the rubbing surface and tribochemical boronization.[3] It was shown that the friction coefficient of the base oil was decreased by the addition of hydrophobic zinc borate nano-discs.[26] Zinc borate produced by precipitation method decreased the wear scar diameter from 1.402 to 0.639 mm and the friction coefficient from 0.099 to 0.064. The inverse emulsion was effective in decreasing wear scar diameter and the friction coefficient by lowering them to 0.596 mm and 0.089, respectively.[19]

The surface modification of zinc borate ultrafine powders (ZB UFPs) on their tribological properties as lubricant additives in liquid paraffin (LP) was investigated by Zhao et al.[31] ZB UFPs were successfully modified by hexadecyltrimethoxysilane (HDTMOS) and oleic acid (OA). HDTMOS-modified ZB UFP (HDTMOS-ZB UFP) had a small conglomerate size, good stability in the organic solvent, and sound antiwear property. It has been observed that a continuous and tenacious tribofilm on the worn surface, generated from HDTMOS-modified ZB UFP as a lubricant additive in LP, plays an important role in the outstanding antiwear property.[31]

Vegetable oils can be used as base oils in lubricants. The tribological properties of ZB UFP employed as a lubricant additive in sunflower oil were investigated by Zhao et al.[32] Tribofilms with dark colors were generated on the worn surfaces and showed good contrast with the substrate. The combination of sunflower oil with 0.5% ZB UFP delivered the most balanced performance in friction and wear reduction. The study has demonstrated the possibility of application of this industrially applicable solid lubricant additive (zinc borate) with a decomposable vegetable-based lubricant oil.[32]

The production and characterization of nanoparticles of hydrated zinc borate for lubrication were aimed at in the present study. Dilute solutions of borax and zinc nitrate were mixed instantly at room temperature to avoid particle growth, and hydrosols were formed. The particles were separated by centrifugation and washed with ethanol and dried at 25°C under vacuum.

13.2 EXPERIMENTAL

13.2.1 MATERIALS

Anhydrous borax (Sigma-Aldrich), zinc nitrate hexahydrate ($Zn(NO_3)_2 \cdot 6H_2O$) (Sigma-Aldrich), light neutral oil (TUPRAS A.S.), sorbitan monostearate (Span 60, Sigma-Aldrich), and PEG 4000 (Merck) were used in the preparation of hydrated zinc borate nanoparticles and lubricants.

13.2.2 PREPARATION OF NANOPARTICLES OF HYDRATED ZINC BORATES

In total, 50 cm³ of 0.1 mol dm⁻³ sodium borate solution was added instantly to 50 cm³ of 0.1 mol dm⁻³ zinc nitrate solution and mixed at 600 rpm for 2 h at ambient temperature of 23°C. Experiments were repeated by adding 1 cm³ of 0.002 M Span 60 and 1 cm³ of 0.4 g PEG 4000 in 100 cm³ to the mixtures. While hydrosols were being mixed, their temperature and pH values were recorded. Since the particles passed through the Whatman black ribbon filter paper, they were separated from the aqueous phase by centrifugation. They were separated by centrifugation using Rotofix 32 centrifuge at 2000 rpm for 10 min, washed with ethanol, and centrifuged again at 2000 rpm for 10 min. The gelatinous precipitates were dried under vacuum at 10 kPa for 18 h at 25°C to obtain nanoparticles. Since the nanoparticles were in an agglomerated state, they were ground in a porcelain mortar and pestle before use as lubricant additive.

13.2.3 CHARACTERIZATION OF HYDRATED ZINC BORATE POWDERS

The morphologies of the samples were examined using QUANTA 250F scanning electron microscope (SEM). EDX analysis was carried out using the same instrument. The particle size of the powders was measured by Malvern 2000 Zetasizer. X-ray diffraction (XRD) diagrams were obtained by Philips X'Pert Pro X-ray diffractometer employing Ni-filtered Cu K_α radiation.

FTIR spectra of the samples were taken with SHIMADZU FTIR-8400S using KBr disc technique. TG analysis was performed by using SETARAM Labsys TGA to determine changes in weight with heating under nitrogen flow at 40 cm^3 min^{-1} rate and at a heating rate of 10°C min^{-1} up to 600°C.

13.2.4 PREPARATION AND TESTING OF LUBRICANTS

25 cm^3 of light neutral oil, 0.25 g of sorbitan monostearate, and 0.25 g of the nanoparticles of hydrated zinc borate obtained with span 60 template were mixed thoroughly at 600 rpm rate and at 160°C for 1 h on a magnetic hot plate (Ika RH Digital KT/C) and left to cool down to room temperature by continuous stirring. A four ball tester (Ducom) was used to determine the friction coefficient and wear scar diameter. The tests were performed according to ASTM D 4172-94 at 392 N and the test duration was 1 h at 75°C. The upper ball was rotated at 1200 rpm. Test balls were made from AISI standard steel No. E-52100 and had 12.7 mm diameter. Microphotographs of the wear scars of the three fixed and one rotating test balls were taken by using Olympos BX 60 equipped with Canon Powershot 590IS camera.

The visible spectrum of base oil separated by centrifugation from base oil was taken by using Perkin Elmer UV-Vis spectrophotometer by using base oil without any additive as the reference.

13.3 RESULTS AND DISCUSSION

13.3.1 REACTION OF AQUEOUS BORAX AND ZINC NITRATE SOLUTIONS

The zinc borate precipitation reaction is expected to occur as given in eq 13.3.

$$(y/2)B_4O_7^{2-} \text{ (aq)} + xZn^{2+}\text{(aq)} + zH_2O \rightarrow xZnO \cdot yB_2O_3 \cdot zH_2O \text{ (s)} \quad (13.3)$$

There are other simultaneous reactions in the reaction medium, such as

$$2Zn^{2+}\text{(aq)} + 3OH^-\text{(aq)} + NO_3^-\text{(aq)} \rightarrow Zn_2(OH)_3(NO)_3\text{(s)} \quad (13.4)$$

$$Zn^{2+}\text{(aq)} + 2OH^-\text{(aq)} \rightarrow Zn(OH)_2\text{(s)} \quad (13.5)$$

The pH changes during reaction of aqueous borax and zinc nitrate solutions. The measured initial temperature and pH values of the solutions are shown in Table 13.1. As seen in the table, while the borax solution was at pH 9.02, the zinc nitrate solution is at pH 4.75.

TABLE 13.1 pH and Temperature Values Before and After Mixing Solutions.

Solution	pH		Temperature (°C)	
	Before	**After**	**Before**	**After 120 min**
$Zn(NO_3)_2(aq)$	4.75	–	23.2	
$Na_2B_4O_7(aq)$	9.02	–	23.2	
Mixture without any additive	–	7.13	–	26.6
Mixture with span 60	–	7.35	–	25.8
Mixture with PEG 4000	–	7.63	–	27.0

The reason of zinc nitrate giving an acidic solution is due to hydrolysis of Zn^{2+} ions in water giving $Zn(OH)_2$.

$$Zn^{2+} + 2H_2O \rightarrow Zn(OH)_2 + 2 H^+ \qquad (13.6)$$

where 0.008% of the Zn^{2+} should be present as $Zn(OH)_2$ in the initial solution.

The change of pH and temperature of the reaction medium with time after mixing the reactants are shown in Figure 13.1.

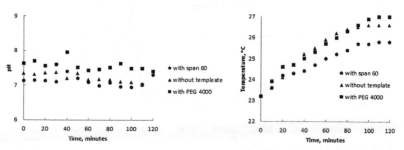

FIGURE 13.1 The change of (a) pH and (b) temperature after mixing borax and zinc nitrate solutions.

The pH was 7.35, 7.13, and 7.63 immediately after mixing the reactants; it was 7.4, 7.30, and 7.46 for mixtures without template, with Span 60, and with PEG 4000, respectively, at 120 min after reaction. The change of the pH values could be attributed to the precipitation of $Zn_2(OH)_3(NO)_3$ or $Zn(OH)_2$

by reaction of Zn^{2+} ions with OH ions initially present in borax solution.[2] However, the change in Zn^{2+} concentration due these reactions could be up to only $3.5 \times 10^{-7}\%$ due to low concentrations of OH^- groups in initial solutions. Thus, the occurrence of reaction shown in eq 13.3 is the most possible reaction. The reaction was an exothermic and slow since a temperature rise of $3.4°C$, $3.2°C$, and $3.8°C$ was observed for mixtures without any template, with Span 60 and with PEG 4000, respectively, from the initial temperature of $23.2°C$ in 120 min as seen in Table 13.1 and Figure 13.1b.

13.3.2   FUNCTIONAL GROUPS BY FTIR SPECTROSCOPY

The FTIR spectra of the samples without any template and with templates Span 60 and PEG 4000 are very similar to each other as seen in Figure 13.2. There is a broad peak at 3358 cm^{-1} due to hydrogen-bonded O–H group vibrations. Asymmetric B_3–O vibrations were observed at 1383 and at 1350 cm^{-1}. H–O–H bending vibration was observed at 1624 cm^{-1}. At 1014 cm^{-1}, a peak for symmetric B_3–O vibration was present. Out of plane bending vibration of B_3–O gave a small peak at 694 cm^{-1}.[11,35]

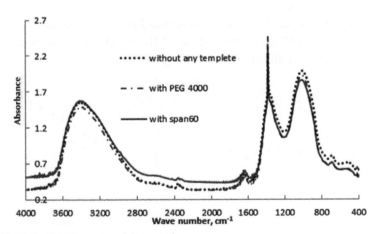

FIGURE 13.2   FTIR spectra of the samples.

13.3.3   TG ANALYSIS AND WATER CONTENT OF THE PRECIPITATED PRODUCTS

TG curves of the samples were very similar to each other as depicted in Figure 13.3. On heating the samples, only water is eliminated from the samples. Free water, loosely bound water, and tightly bond water were removed as water vapor; thus the mass decreased with increasing temperature. The temperature ranges are 25–85°C, 85–285°C, and 285–1000°C for external, loosely bound, and tightly bound water, respectively.[13] All the samples contained very similar amount of water, around 18 % of which was 2% free water, 13% loosely bound water, and 3% tightly bound water as reported in Table 13.2. The sample without any template contained 0.6% more free water than samples with PEG 4000 and Span 60. This was attributed to hydrophilic nature of the sample without any template. The total amount of water was less than the zinc borate samples prepared from higher initial borax and zinc nitrate concentrations. Savrik[18] has determined that the samples prepared at 70°C from 1 mol dm^{-3} initial concentration contained around 20.6% water with the same FTIR spectra of the samples in the present study.

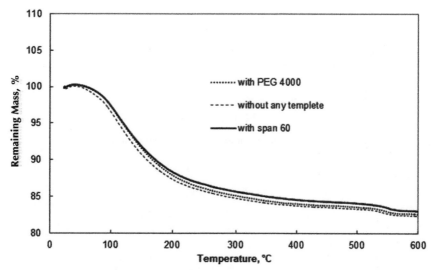

FIGURE 13.3   TG analysis curves of the samples.

TABLE 13.2   Remaining Mass %, Free, Bound, and Tightly Bound Water % in Zinc Borates.

Sample	Mass loss, %			Free water, %	Loosely bound water; %	Tigtly bound water, %	Total water, %
	At 85°C	At 285°C	At 600°C				
Without any template	98.09	85	82.36	1.91	13.09	2.64	17.64
With span 60	98.77	85.9	83.03	1.23	12.87	2.87	16.97
With PEG 4000	98.73	85.3	82.6	1.27	13.43	2.7	17.4

13.3.4 CHEMICAL ANALYSIS BY ENERGY DISPERSIVE X-RAY SPECTROSCOPY

The energy dispersive spectroscopy was used to determine the elemental composition of zinc borates. The energy dispersive X-ray spectroscopy (EDX) spectrum of the samples without any template and with Span 60 and PEG 4000 are seen in Figures 13.4, 13.5, and 13.6, respectively. As seen in the figures, they are very similar to each other.

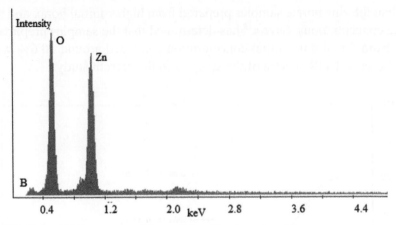

FIGURE 13.4 EDX spectrum of the sample without any templete.

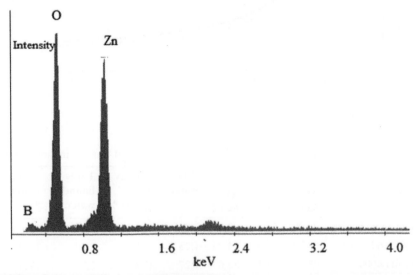

FIGURE 13.5 EDX spectrum of the sample with span 60.

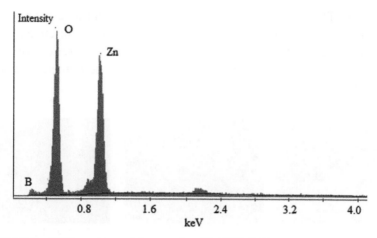

FIGURE 13.6   EDX spectrum of the sample with PEG 4000.

The chemical compositions of the samples are similar to each other. The H_2O content calculated from material balance in EDX analysis was lower than the content determined by TG analysis. Since EDX operates under high vacuum, the free water and some of the lightly bound water should have been removed from the sample during the test. The empirical formula of the nanoparticles was approximately $3ZnO \cdot 2B_2O_3 \cdot 4H_2O$ from EDX analysis. However, it could be a mixture of zinc borate and zinc hydroxide due to parallel reactions.

TABLE 13.3   Elemental Composition and Empirical Formula of the Samples.

Sample	Mass, %						
	Zn	B	O	ZnO	B_2O_3	H_2O	
						from EDX	from TG analysis
Without any template	54.33	13.37	32.3	45.88	33.06	2.37	17.64
With span 60	53.58	12.78	33.64	46.52	32.56	4.99	16.97
With PEG 4000	54.51	11.82	33.69	48.84	30.47	7.12	17.40

13.3.5   MORPHOLOGIES OF THE POWDERS

The SEM micrographs of the dried samples are shown in Figure 13.7. The dried powders consisted of agglomerates of primary particles of 40, 50, and 80 nm for samples without template, with Span 60, and with PEG 4000, respectively. The particles attracted to each other due to capillary forces during drying. Water in the wet samples was replaced with ethanol which has a lower surface tension to decrease the attractive forces between the particles due to capillarity. However, they were also agglomerated even by drying of the ethanolic hydrogel.

FIGURE 13.7   SEM micrographs of zinc borates: (a) without any templete, (b) with span 60, and (c) with PEG 4000.

13.3.6   PARTICLE SIZE DISTRIBUTION OF THE INITIALLY FORMED HYDRATED ZINC BORATES IN SOLUTION

The particle size distributions of zinc borates in aqueous solution are shown in Figure 13.8. The size distribution of particles were monodisperse with number average sizes of 35.3 and 234 nm, respectively, for zinc borates

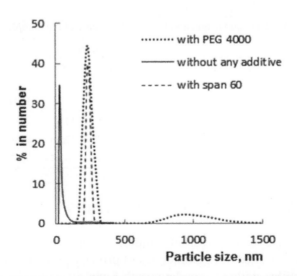

FIGURE 13.8   Particle size distribution of zinc borates as determined by zeta sizer.

without any template and with Span 60 (Table 13.4). However, the particle size distribution of particles with PEG 4000 was bidisperse and the average size was 228.5 and 969.7 nm for 94.6% and 5.4% of the particles (Table 13.4). While the primary particles were dispersed in zinc borate without any template, these particles were agglomerated in the presence of surface active agents Span 60 and PEG 4000.

TABLE 13.4   Particle Size Distribution.

Sample	Peak number	Area	Average particle size, nm	Width of distribution, nm
Without any template	1	99	35.3	16.9
With span 60	1	100	234.8	39.8
With PEG 4000	1	94.6	228.5	8.9
	2	5.4	969.7	39.3

13.3.7 XRD ANALYSIS

The XRD diagrams of the samples are shown in Figure 13.9. As seen in the figure, there are not any sharp diffraction peaks. Only a broad peak is observed at 2θ value of 28° indicating that all the samples were amorphous. The small size of the crystals formed in the present study caused broadening of the diffraction lines. The crystal growth occurs in time at high temperature

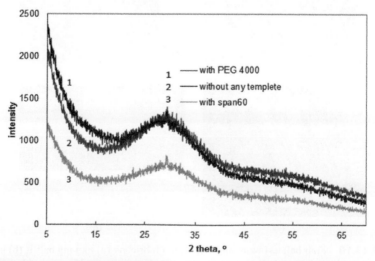

FIGURE 13.9   X-ray diffraction diagrams of the samples.

and at high initial concentration, and X-ray diagram with sharp diffraction peaks is obtained. Savrik[18] obtained sharp diffraction peaks for zinc borate from 1 mol dm^{-3} initial concentrations at 70°C in 2 h. The major peaks in XRD pattern of zinc borate were observed at 13.07°, 17.58°, 19.66°, 25.54°, 26.38°, 27.04°, 29.42°, and 36.97° 2θ values. When these values were compared with those in the database, it was implied that all the diffraction peaks were perfectly indexed with those of $Zn(B_3O_3(OH)_5)H_2O$ (JPDS PDF File Number 721789). In literature, this type of zinc borate was also defined as $2ZnO_3 \cdot B_2O_3 \cdot 7H_2O$.[19]

13.3.8 THE PRECIPITATES AS LUBRICANT ADDITIVES

Previous studies also had shown the lowering of the friction coefficient when zinc borate nanoparticles were added to the base oil.[3,26] The results of the four ball tests are shown in Figure 10 for the lubricating oil prepared in the present study. The wear scar diameter for ball 1, 2, and 3 are 662, 704, and 701 μm, respectively as seen in Figure 13.10. The average wear scar diameter was 689 μm. The change of the friction coefficient with time during the

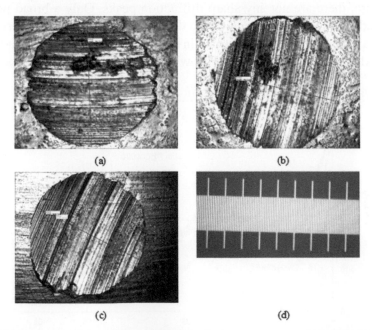

FIGURE 13.10   Four ball test wear surfaces for the lubricant: (a) moving ball 1, (b) moving ball 2, (c) movibg ball 3, and (d) 1 mm scale taken at the same magnification with the balls.

test is shown in Figure 13.11 and it is 0.079 as reported in Table 13.5. The wear scar diameter was lowered from 1402 to 689 μm and the friction coefficient was lowered from 0.099 to 0.079 by adding nano zinc borate to light neutral oil. The wear scar diameter and friction coefficient was lowered to 50% and 20%, respectively. Compared to inverse emulsion case, the friction coefficient was 11% lower, but the wear scar diameter was 15.6% higher for the nano zinc borate case.

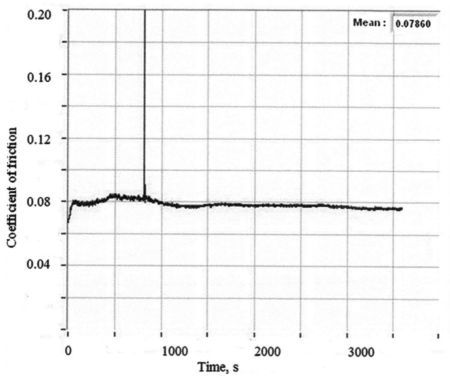

FIGURE 13.11   Change of friction coefficient with time.

TABLE 13.5   Friction Coefficient and Wear Scar Diameter of Light Neutral Oil, Zinc Borate Prepared by İnverse Emulsion, Zinc Borate Prepared by İnverse Emulsion and Zinc Borate Prepared in the Present Study.

Lubricant types	Friction coefficient	Wear scar diameter, μm	Ref.
Light neutal oil	0.099	1402	Savrik et al (2011)
Inverse emulsion	0.089	596	Savrik et al (2011)
Nano zinc borate	0.079	689	Present study

The visible spectrum of the lubricant before and after four ball tests indicated that the yellow color of the oil was darkened during the tests that were carried out at 75°C. As seen in Figure 13.12, the absorbance of the lubricating oil at 414 nm increased from 0.06 to 0.84 after the test. This discoloration was due to oxidation and cross-linking reactions in base oil. Thus, it is necessary to add antioxidants other than zinc borates to the lubricant to avoid oxidation.

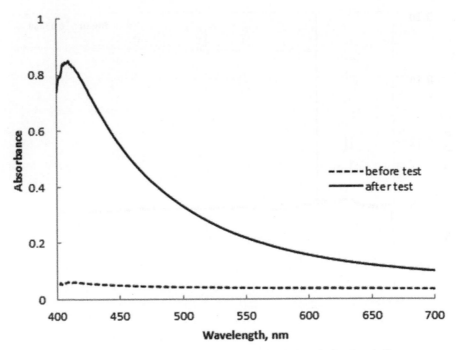

FIGURE 13.12   Visible spectra of the lubricating oil before and after four ball test.

13.4   CONCLUSIONS

The nano zinc borate particles were prepared in the present study by using low initial zinc and borate concentrations and low temperature to prevent particle growth. The templates Span 60 and PEG 4000 increases the aggregation state of nanoparticles of zinc borate in their mother liquor. The particles were separated from mother liquor by centrifugation since they were too small and passed through filter paper in filtration. They were washed in ethanol which has lower surface tension to reduce the capillary forces that attracts the particles to each other. However, they were also agglomerated even by drying of the particles washed with ethanol. Thus, they were ground in a porcelain mortar and pestle before using in lubricating oil. The particles have H_2O and B_3–O vibrations in their FTIR spectra. The zinc borates obtained in the present study contained around 18 % water of which 2% free water, 13% loosely bound water, and 3% tightly bound water. EDX analysis indicated the presence of around 55% ZnO and 33% B_2O_3. The empirical formula of the nanoparticles was approximately $3ZnO \cdot 2B_2O_3 \cdot 4H_2O$ from EDX analysis. However, it could be a mixture of zinc borate and zinc hydroxide. XRD diagram indicated the particles were in amorphous state. When the nanoparticles were added to light neutral oil, the wear scar diameter and friction coefficient were lowered to 50% and 20%, respectively. However, the oil color was darker after the four ball tests indicating addition of antioxidants is necessary. Further studies should be made in synthesis, characterization of zinc borate nanoparticles, and their use in nanoparticle state as lubricant additives and formulating lubricants.

ACKNOWLEDGMENT

The authors thank OPET A.Ş. for four ball tests.

KEYWORDS

- lubrication
- zinc borate
- nanoparticles
- four ball test
- boric acid

REFERENCES

1. Battez, A. H.; Rico, J. E. F.; Arias, A. N.; Rodriguez, J. L. V.; Rodriguez, R. C.; Díaz Fernández, J. M. The Tribological Behaviour of ZnO Nanoparticles as an Additive to PAO6. *Wear* **2006**, *261*, 256–263.

2. Biswick, T.; Jones, W.; Pacu, A.; Ewa Serwicka, E. Synthesis, Characterisation and Anion Exchange Properties of Copper, Magnesium, Zinc and Nickel Hydroxy Nitrates. *J. Solid State Chem.* **2006,** *179*(1), 49–55.

3. Dong, J. X.; Hu, Z. S. A Study of the Anti-wear and Friction-reducing Properties of the Lubricant Additive, Nanometer Zinc Borate. *Tribol. Int.* **1998,** *31*(5), 219–223.

4. Eltepe, H. E.; Balkose, D.; Ulku, S. Effect of Temperature and Time on Zinc Borate Species Formed from Zinc Oxide and Boric Acid in Aqueous Medium. *Ind. Eng. Chem. Res.* **2007,** *46*(8), 2367–2371.

5. Gao, P. Q.; Zhang, Y. Synthesis and Characterization of Zinc Borate Nanowhiskers and Their Inflaming Retarding Effect in Polystyrene. *J. Nanomater.* **2015,** *2015*, (Article ID 925060).

6. Gonen, M.; Balkose, D.; Gupta, R.; Ulku, S. Supercritical Carbon Dioxide Drying of Methanol–Zinc Borate Mixtures. *Ind. Eng. Chem. Res.* **2009,** *48*(14), 6869–6876.

7. Gonen, M.; Balkose, D.; Ulku, S. Supercritical Ethanol Drying of Zinc Borates of $2ZnO.3B_2O_3$. $3H_2O$ and ZnO. B_2O_3. $2H_2O$. *J. Supercrit. Fluids* **2011,** *59*, 43–52.

8. Hsu, S. M. Molecular Basis of Lubrication. *Tribol. Int.* **2004,** *37*, 553–559.

9. Hu, Z. S.; Dong, J. X. Study on Antiwear and Reducing Friction Additive of Nanometer Titanium Borate. *Wear* **1998,** *216*, 87–91.

10. Hu, Z. S.; Dong, J. X.; Chen, G. X.; He, J. Z. Preparation and Tribological Properties of Nanometer Lanthanum Borate. *Wear* **2000,** *243*, 43–47.

11. Jun, L.; Shupping, X.; Shiyang, G. FT-IR and Raman Spectroscopic Study of Hydrated Borates. *Spectrochim. Acta Part A: Mol. Biomol. Spectrosc.* **1995,** *51*, 519–532.

12. Kimura, Y.; Wakabayashi, T.; Okada, K.; Wada, T.; Nishikawa, H. Boron Nitride as a Lubricant Additive. *Wear* **1999,** *232*, 199–206.

13. Knowlton, G. D.; White, T. R. Thermal Study of Types of Water Associated with Clinoptilolite. *Clays Clay Miner.* **1981,** *29*(5), 404–411.

14. Mergen, A; Ipek, Y; Bolek, H; Oksuz, M. Production of Nano Zinc Borate (4ZnO. B_2O_3. H_2O) and Its Effect on PVC. *J. Eur. Ceram. Soc.* **2012,** *32*(9), 2001–2005.

15. Nies, N.; Beach, L.; Hulbert, R. W. Zinc Borate of Low Hydration and Method for Preparing Same. U.S. Patent 3,649,172, 1972.

16. Pawlak, Z.; Kaldonski, T.; Pai, R.; Bayraktar, E.; Oloyede, A. A Comparative Study on the Tribological Behaviour of Hexagonal Boron Nitride (h-BN) as Lubricating Microparticles—An Additive in Porous Sliding Bearings for a Car Clutch. *Wear* **2009,** *267*, 1198–1202.

17. Reddy, C. V.; Sarma, G.; Ravikumar, R. V. S. S. N.; Shima, J. Synthesis and Spectroscopic Characterizations of Copper Ions Doped Zinc Borate Nanoparticles *Optik* **2016,** *127*(10), 4536–4540.

18. Savrik, S. A. Enhancement of Tribological Properties of Mineral Oil by Addition of Sorbitan Monostearate and Zinc Borate. Ph.D. Dissertation, Graduate School of Engineering and Sciences of İzmir Institute of Technology, 2010.

19. Savrik, S. A.; Balkose, D.; Ulku S. Synthesis of Zinc Borate by Inverse Emulsion Technique for Lubrication. *J. Therm. Anal. Calorim.* **2011,** *104*(2), 605–612.

20. Schubert, D. M.; Alam, F.; Visi, M. Z.; Knobler, C. B. Structural Characterization and Chemistry of Industrially Important Zinc Borate $Zn[B_3O_4(OH)_3]$. *Chem. Mater.* **2003,** *15*, 866–871.

21. Shi, X. X.; Li, M.; Yang, H.; Chen, S.; Yuan, L.; Zhang, K.; Sun, J. PEG-300 Assisted Hydrothermal Synthesis of $4ZnO$ B_2O_3 H_2O Nanorods. *Mater. Res. Bull.* **2007,** *42*(9), 1649–1656.

22. Shi, X.; Xiao, Y.; Yuan, L.; Sun, J. Hydrothermal Synthesis and Characterizations of 2D and 3D $4ZnO \cdot B_2O_3 \cdot H_2O$ Nano/Microstructures with Different Morphologies. *Powder Technol.* **2009,** *189*, 462–465.

23. Shim, J.; Reddy, C. V.; Sarma, G. V. S. S.; Murthy, P. N.; Ravikumar, R. V. S. S. N. Effect of Co(2+) and Ni(2+)-doped Zinc Borate Nano Crystalline Powders by Co-precipitation Method. *Spectrochim. Acta Part A Mol. Biomol. Spectrosc.* **2015,** *142*, 279–285.

24. Smiechowski, M. F.; Lvovich, V. F. Electrochemical Monitoring of Water-surfactant Interactions in Industrial Lubricants. *J. Electroanal. Chem.* **2002,** *534*, 171–180.

25. Tian, Y. M. Y.; He, Y.; Yu, L.; Deng, Y.; Zheng, Y.; Sun, F.; Liu, Z.; Wang, Z. In Situ and One-step Synthesis of Hydrophobic Zinc Borate Nanoplatelets. *Coll. Surf. A Physicochem. Eng. Asp.* **2008,** *312*(2–3), 99–103.

26. Tian, Y. M.; Guo, Y. P.; Jiang, M.; Sheng, Y.; Hari, B.; Zhang, G.; Jiang, Y.; Bing Zhou, Z.; Zhu, Y.; Wang, Z. Synthesis of Hydrophobic Zinc Borate Nanodiscs for Lubrication. *Mater. Lett.* **2006,** *60*(20), 2511–2515.

27. Ting, C.; Jian-Cheng, D.; Long-Shuo, W.; Fan, Y.; Gang, F. Synthesis of a New Netlike Nano Zinc Borate. *Mater. Lett.* **2008,** *62*(14), 2057–2059.

28. Ting, C. D.; Jian-Cheng, D.; Long-Shuo, W.; Gang, F. Preparation and Characterization of Nano-zinc Borate by a New Method. *J. Mater. Process. Technol.* **2009,** *209*(8), 4076–4079.

29. Varlot, K.; Kasrai, M.; Bancroft, G. M.; Yamaguchi, E. S.; Ryason, P. R.; Igarashi, J. X-ray Absorption Study of Antiwear Films Generated from ZDDP and Borate Micelles. *Wear* **2001,** *249*, 1029–1035.

30. Zhang, Y. Y.; Xue, L.; Liu, Z-H. Preparation of Cluster-like Nanostructure and Nanoribbon for $4ZnO \cdot B_2O_3 \cdot H_2O$ and the Evaluation of Their Flame Retardant Properties by a Thermal Analysis Method. *Thermochim. Acta* **2010,** *506*(1–2), 52–56.

31. Zhao, C. L.; Jiao, Y.; Chen, Y. K.; Ren, G. The Tribological Properties of Zinc Borate Ultrafine Powder as a Lubricant Additive in Sunflower Oil. *Tribol. Trans.* **2014,** *57*(3), 425–434.

32. Zhao, C.; Chen, Y. K.; Jiao, Y.; Loya, A.; Ren, G. G. The Preparation and Tribological Properties of Surface Modified Zinc Borate Ultrafine Powder as a Lubricant Additive in Liquid Paraffin. *Tribol. Int.* **2014,** *70*, 155–164.

33. Zheng, Y. H.; Tian, Y. M.; Ma, H.; Qu, Y.; Wang, Z.; An, D.; Guan, S.; Gao, X. Synthesis and Performance Study of Zinc Borate Nanowhiskers. *Coll. Surf. A Physicochem. Eng. Asp.* **2009,** *339*(1–3), 178–184.

34. Zheng, Z.; Shen, G.; Wan, Y.; Cao, L.; Xu, X.; Yue, Q.; Sun, T. Synthesis, Hydrolytic Stability and Tribological Properties of Novel Borate Esters Containing Nitrogen as Lubricant Additives. *Wear* **1998,** *222*, 135–144.

35. Zhihong, L.; Bo, G.; Mancheng, H.; Shuni, L.; Shuping, X. FT-IR and Raman Spectroscopic Analysis of Hydrated Cesium Borates and Their Saturated Aqueous Solution. *Specrochim. Acta Part A* **2003,** *59*, 2741–2745.

CHAPTER 14

DIFFUSION OF COPPER NITRATE IN AQUEOUS SOLUTIONS AT 298.15 K

LUIS M. P. VERISSIMO[1,2,*], ANA M. T. D. P. V. CABRAL[2,3],
FRANCISCO J. B. VEIGA[3,4], MIGUEL A. ESTESO[1,2], and
ANA C. F. RIBEIRO[2]

[1]U. D. Química Física, Facultad de Farmacia, Universidad de Alcalá, 28871 Alcalá de Henares, Madrid, Spain

[2]Department of Chemistry, Coimbra Chemistry Centre, University of Coimbra, 3004-535 Coimbra, Portugal

[3]Faculty of Pharmacy, University of Coimbra, Azinhaga Sta. Comba, 3000-548 Coimbra, Portugal

[4]Faculty of Pharmacy, REQUIMTE/LAQV, Pharmaceutical Technology, University of Coimbra, Coimbra, Portugal

*Corresponding author. E-mail: luis.verissimo@uc.pt

CONTENTS

ABSTRACT

Mutual diffusion coefficients have been measured for copper(II) nitrate in water at 298.15 K and at concentrations between 0.001 and 0.020 mol kg^{-1}. The diffusion coefficients were measured using the Taylor dispersion method. New experimental data are compared with those computed using Onsager–Fuoss and Pikal theories. The Nernst diffusion coefficients, both derived from our diffusion data (1.266 $\times$ 10^{-9} m^2 s^{-1}) and calculated (1.261 $\times$ 10^{-9} m^2 s^{-1}) at 298.15 K, are discussed. Also, the conductance of ion pairs and the degrees of dissociation in these solutions have been estimated using these results.

14.1 INTRODUCTION

The knowledge of electrolyte diffusion data is of great interest not only for fundamental purposes but also in order to be used in many applicational fields, for example, corrosion studies. Among the electrolytes, copper(II) nitrate is widely used in nitration processes, while fundamental research on its properties continues to produce interesting results.[1,2] Recent work found that Cu(NO$_3$)$_2$ electrochemical properties could be promisingly useful in the battery construction field. As Shu et al.[3-6] show that copper(II) nitrate has good storage performance as anode material for lithium- and sodium-ion batteries. However, as far as the authors have researched, data on mutual diffusion coefficients, D, of aqueous solutions of Cu(NO$_3$)$_2$ are not available. We have measured these parameters in the concentration range from 0.001 to 0.020 mol·kg^{-1} using the Taylor dispersion method.[7-12] The results, thus, obtained were analyzed with the help of the Onsager–Fuoss[13] and Pikal[14] models, considering that copper(II) species may be present as complexes or pair ions. In addition, by applying our data to the semiempirical model suggested by Harned,[15,16] the conductance of ion pairs Cu(NO$_3$)$_x^{2-x}$ and the respective degrees of dissociation were obtained.

14.2 EXPERIMENTAL

14.2.1 REAGENTS

Copper(II) nitrate hemi(pentahydrate) (Riedel-de Haën, pro analysis >98%) was used without further purification (Table 14.1). The solutions were prepared in molality units, m, using ultrapure water (Millipore, 18.2 Mohm/

cm at 298.15 K). Water content in copper(II) nitrate was accounted upon solution preparation. The solutions were freshly prepared and deaerated for about 30 min before each set of runs.

TABLE 14.1 Sample Description.

Chemical name	Source	Purity
$Cu(NO_3)_2$-2.5-hydrate	Riedel-de Haën	Mass fraction > 0.98

14.2.2 DIFFUSION MEASUREMENTS

The Taylor dispersion method was used for measuring diffusion coefficients. It is based on the dispersion of a very small amount of solution injected into a laminar carrier stream of solvent or solution of different composition flowing through a long capillary tube [with length and radius of 3.2799 ($\pm$0.0001) $\times$ 10^3 cm and 0.05570 ($\pm$0.00003) cm, respectively].[7-12] At the start of each run, a 6-port Teflon injection valve (Rheodyne, model 5020) was used to introduce 0.063 cm^3 of solution into the laminar carrier stream of slightly different composition. A flow rate of 0.23 cm^3 min^{-1} (corresponding to 3.5 rotations per min of the peristaltic pump head) had been used, and was controlled by a metering pump (Gilson model Minipuls 3) to give retention times of about 8 $\times$ 10^3 s. The dispersion tube and the injection valve were kept at 298.15 K ($\pm$0.01 K) in an air thermostat.

Dispersion of the injected samples was monitored using a differential refractometer (Waters model 2410) at the outlet of the dispersion tube. Detector voltages, $V(t)$, were measured at accurately timed 5 s intervals with a digital voltmeter (Agilent 34401 A); the data being stored in a computer having an IEEE-488 interface. Binary diffusion coefficients were evaluated by fitting the detector voltages to the dispersion equation.

$$V(t) = V_0 + V_1 t + V_{max} \left(t_R / t\right)^{1/2} \exp[-12D\left(t - t_R\right)^2 / r^2 t] \quad (14.1)$$

The additional fitting parameters were the mean sample retention time, t_R; peak height, V_{max}; baseline voltage, V_0; and baseline slope, V_1.

14.3 RESULTS AND DISCUSSION

Mutual diffusion coefficients, D, of $Cu(NO_3)_2$ in aqueous solutions at 298.15 K and the standard deviations of the means are shown in Table 14.2, where D

is the mean value of, at least, three independent measurements. On the basis of previous papers reporting data obtained with this technique,[7-12] we believe that our uncertainty is less than 3%.

TABLE 14.2 Diffusion Coefficients, D, of $Cu(NO_3)_2$ in Aqueous Solutions at Various Concentrations, c, and the Standard Deviations of the Means, S_D at 298.15 K and at 101.3 kPa.

m (mol kg^{-1})	D (10^{-9} m^2 s^{-1})^a	S_D (10^{-9} m^2 s^{-1})^b
0.001	1.173	0.010
0.002	1.140	0.009
0.005	1.081	0.002
0.008	1.070	0.002
0.010	1.060	0.002
0.020	1.044	0.002

^{a}D is the mean diffusion coefficient for 3 experiments.
^{b}S_D is the standard deviation of that mean. Standard uncertainties u with 0.68 level of confidence are u $(c) = 0.03$, $u(T) = 0.01$ K, and $u(p) = 2.03$ kPa.

The following polynomial in $m^{1/2}$ (eq 14.2) resulted from a good fit the data by a least squares procedure, where m represents the molality of the solutions.

$$D/(m^2s^{-1}) = 1.266 - 3.411m^{1/2} + 13.112m \quad (R^2 = 0.990) \quad (14.2)$$

The goodness of the fit (obtained with a confidence interval of 98%) can be assessed by the excellent correlation coefficients, R^2, and the low standard deviation (<1%).

As a first approach, the experimental mutual diffusion coefficients at 298.15 K were compared with those estimated by using Onsager–Fuoss equation[13] which are applicable for dilute solutions (eq 14.3). For these solutions, the electrophoretic effects are described while the phenomena such as hydrolysis,[17] complexation, and/or ion association[18-21] are not taken into consideration, that is, the experimental data can be compared with the calculated D on the basis of eq 14.3.

$$D = F_M \times F_T = 2000RT \frac{\overline{M}}{c}\left(1 + c\frac{\partial \ln y_\pm}{\partial c}\right) \quad (14.3)$$

where

$$F_T = \left(1 + c\frac{\partial \ln y_\pm}{\partial c}\right) \tag{14.4}$$

$$F_M = \left(D^0 + \Delta_1 + \Delta_2\right) = 2000RT\frac{\overline{M}}{c} \tag{14.5}$$

D^0 is the Nernst limiting value of the diffusion coefficient given by:

$$D^0 = \frac{RT}{F^2}\frac{|Z_c| + |Z_a|}{|Z_c \times Z_a|}\frac{\lambda_c^0\lambda_a^0}{\lambda_c^0 + \lambda_a^0} \tag{14.6}$$

and

$$\frac{\overline{M}}{c} = 1.0741 \times 10^{-20}\frac{\lambda_1^0\lambda_2^0}{|z_1|v_1\Lambda^0} + \frac{\Delta\overline{M}'}{c} + \frac{\Delta\overline{M}''}{c} \tag{14.7}$$

In eq 14.7, the first- and second-order electrophoretic terms are given by:

$$\frac{\Delta\overline{M}'}{c} = \frac{\left(|z_2|\lambda_1^0 - |z_1|\lambda_2^0\right)^2}{|z_1 z_2|2\left(\Lambda^0\right)^2}\frac{3.132 \times 10^{-19}}{\eta_0(\varepsilon T)^{1/2}}\frac{c\sqrt{\tau}}{(1+ka)} \tag{14.8}$$

and

$$\frac{\Delta\overline{M}''}{c} = \frac{\left(z_2^2\lambda_1^0 - z_1^2\lambda_2^0\right)^2}{\left(\Lambda^0\right)^2}\frac{9.304 \times 10^{-13}c^2}{\eta_0(\varepsilon T)}\varphi(ka) \tag{14.9}$$

where D is the mutual diffusion coefficient of the electrolyte in $m^2\ s^{-1}$, R is the gas constant in $J\ K^{-1}\ mol^{-1}$, T is the absolute temperature, z_1 and z_2 are the algebraic valences of a cation and of an anion, respectively, and the last term in parenthesis is the activity factor, with $y_\pm$ being the mean molar activity coefficient, c is the concentration in $mol\ m^{-3}$, η_0 is the viscosity of the water in $N\ s\ m^{-2}$, N_A is the Avogadro's constant, e_0 is the proton charge in Coulombs, z_1 and z_2 are the stoichiometric coefficients, λ_1^0 and λ_2^0 are the limiting molar conductivities of the cation Cu^{2+} and anion NO_3^-, respectively, given in the literature[22] as $56.60 \times 10^{-4}\ \Omega^{-1}\ m^2\ mol^{-1}$ and $71.42 \times 10^{-4}\ \Omega^{-1}\ m^2$ mol^{-1}, respectively, k is the "reciprocal average radius of ionic atmosphere" in m^{-1} (e.g., Ref. [16]), a is the mean distance of closest approach of ions in m, $\varphi(ka) = \left| e^{2ka} E_i(2ka) / (1+ka) \right|$ has been tabulated by Harned and Owen,[15] and the other letters represent well-known quantities.[15]

Table 14.3 shows the values resulting from the application of this theory for the concentrations of 0.000–0.020 mol kg^{-1}. By comparison between these values and our experimental data, we can well see that there are significant deviations among them. This is not surprising if we take into account the formation of complexes of the type $Cu(NO_3)_x^{2-x}$ and the variety of ion pairs eventually formed, the change with concentration, factors not taken into account in the Onsager–Fuoss model. Based on previously published work[23] and assuming that eq 14.10 is valid for the present case, we may attempt to estimate the degrees of dissociation, α, for each concentration and the equivalent conductance of the ion pair, λ_{IP}^0. Our data D are taken as D_{IP} (i.e., diffusion coefficients of copper nitrate for different concentrations, taking into account the ionic species involved, Cu^{2+} and NO_3^- ions, and the respective ion pairs).

TABLE 14.3 Diffusion Coefficients of $Cu(NO_3)_2$ Calculated from Onsager-Fuoss and Pikal Theories, D_{OF}, and D_{Pikal} at 298.15 K.

m (mol kg^{-1})	D'_{OF} (10^{-9} m^2 s^{-1})[a]	$\Delta D/D'_{OF}$ (%)[b]	D_{Pikal} (10^{-9} m^2 s^{-1})[a]	$\Delta D/D_{Pikal}$ %[b]
0.000	1.261	0.4	1.261	0.4
0.001	1.207	−2.8	1.190	−1.4
0.002	1.191	−4.3	1.166	−2.2
0.005	1.168	−7.4	1.114	−2.9
0.008	1.158	−7.6	1.074	−0.4
0.010	1.156	−8.3	1.048	+1.1
0.020	1.138	−8.3	1.008	+4.2

[a]D_{OF} and D_{Pikal} represent the values of D calculated by eqs 14.3 and 14.15, respectively.

[b]D/D_{OF} and $\Delta D/D_{Pikal}$ represent the relative deviations between our D (Table 14.3) and D_{OF} and D_{Pikal} values, respectively, using the Kielland's value ($a = 4.8 \times 10^{-10}$ m [21]) for the mean distance of closest approach of ions.

$$\frac{D_{IP}}{D_{OF}} = 1 + (1-\alpha)\left[\lambda_{IP}^0 \left(\frac{1}{\lambda_1^0} + \frac{1}{\lambda_2^0} \right) - 1 \right] \qquad (14.10)$$

Having in mind that we do not have data for all parameters involved in this eq 14.10, it seems first appropriate to estimate λ_{IP}^0 (equivalent conductance of the ion pair) from the D value at specific concentration, and then use it to estimate α at other concentrations. For example, considering $\alpha = 0.5$ for 0.005 mol kg^{-1} Cu(NO$_3$)$_2$ from eq 14.10, we have obtained $\lambda_{IP}^0 = 26.87$ s cm^2. Then, with this parameter and also from eq 14.10, calculations led to different values of α (Table 14.4). Despite limitations, in general, these calculations led to acceptable values of α, decreasing from 0.001 to 0.020 mol kg^{-1}, evidencing the possible presence of ion pairs and led us to conclude that the Onsager–Fuoss theory is inadequate as a method to calculate diffusion coefficients of the Cu(NO$_3$)$_2$.

TABLE 14.4 Estimations of the Conductance of Ion Pair, $\ddot{e}_P^0$, and of the Degrees of Dissociation, α.

m (mol kg^{-1})	α ($\lambda_{IP}^0 =$ 29.80 S cm^2)[a]	α ($\lambda_{IP}^0 =$ 28.87 S cm^2)[b]	α ($\lambda_{IP}^0 =$ 26.87 S cm^2)[c]	α ($\lambda_{IP}^0 =$ 26.78 S cm^2)[d]	α ($\lambda_{IP}^0 =$ 26.43 S cm^2)[e]	α ($\lambda_{IP}^0 =$ 26.78 S cm^2)[f]
0.001	0.50	0.67	0.82	0.81	0.83	0.83
0.002	g	0.50	0.71	0.72	0.74	0.74
0.005	g	0.13	0.50	0.51	0.54	0.55
0.008	g	0.11	0.49	0.50	0.53	0.54
0.010	g	0.05	0.45	0.46	0.50	0.51
0.020	g	0.04	0.45	0.46	0.49	0.50

[a]Value calculated with eq 14.10, assuming $\alpha = 0.50$ for $m = 0.001$ mol kg^{-1}.
[b]Value calculated with eq 14.10, assuming $\alpha = 0.50$ for $m = 0.002$ mol kg^{-1}.
[c]Value calculated with eq 14.10, assuming $\alpha = 0.50$ for $m = 0.005$ mol kg^{-1}.
[d]Value calculated with eq 14.10, assuming $\alpha = 0.50$ for $m = 0.008$ mol kg^{-1}.
[e]Value calculated with eq 14.10, assuming $\alpha = 0.50$ for $m = 0.01$ mol kg^{-1}.
[f]Value calculated with eq 14.10, assuming $\alpha = 0.50$ for $m = 0.001$ mol kg^{-1}.
[g]Value without physical meaning.

The model of mutual diffusion in binary electrolytes developed by Pikal[14] includes, in addition to Onsager–Fuoss equation,[13] new terms resulting from the application of the Boltzmann exponential function for the study of diffusion. As a result of this procedure, a term representing the effect of ion pair formation appears in the theory as a natural consequence of the electrostatic interactions. The electrophoretic correction is the sum of two terms

$$\Delta \upsilon_j = \Delta \upsilon_j{}^L + \Delta \upsilon_j{}^S \qquad (14.11)$$

where $\Delta \upsilon_j{}^L$ represents the effect of long-range electrostatic interactions, and $\Delta \upsilon_j{}^S$ represents the short range ones.

Designating $M = 10^{12} L/c$ as the solute thermodynamic mobility, where L is the thermodynamic diffusion coefficient, then ΔM can be represented by the eq 14.12:

$$(1/M) = (1/M^0)(\Delta M/M^0) \qquad (14.12)$$

where M^0 is the value of M at infinitesimal concentration and

$$\Delta M = \Delta M^{0F} + \Delta M_1 + v M_2 + \Delta M_A + \Delta M_{H1} + \Delta M_{H2} + \Delta M_{H3} \qquad (14.13)$$

The first term in eq 14.13, ΔM^{0F}, represents the Onsager–Fuoss term for the effect of the concentration in the solute thermodynamic mobility, M; the second term, ΔM_1, is a consequence of the approximation applied on the ionic thermodynamic force; the other terms result from the application of the Boltzmann exponential function.

The relation between the solute thermodynamic mobility and the mutual diffusion coefficient is given by eq 14.14:

$$D = (L/c)10^3 RT\upsilon [1 + c(\mathrm{d}\ln\gamma_\pm/\mathrm{d}c)] \qquad (14.14)$$

From eqs 14.13 and 14.11, Pikal's equation (eq 14.15) can be deducted

$$D = \frac{1}{\dfrac{1}{M^0}\left(1 - \dfrac{\Delta M}{M^0}\right)}(10^3\, R\, T\mathrm{v})[1 + c\,(\mathrm{d}\ln\gamma\pm/\mathrm{d}c) \qquad (14.15)$$

The results predicted from the above model, D_{Pikal}, are in a good agreement with the experimental values by 2% as shown in Table 14.3. This is not surprising if we take into account that ion pairs eventually present in these solutions are now taken into account in this model, unlike in the Onsager–Fuoss approach.

The deviation between the limiting D^0 value calculated by extrapolating experimental data to $c \to 0$ (Table 14.2) and the Nernst value (eq 14.6) (Table 14.3) is also acceptable (0.4%).

The decrease of the diffusion coefficient of this electrolyte, when the concentration increases, may be interpreted on the basis of species resulting from the short-range electrostatic interactions (leading to the formation of ion pairs).

For $m = 0.020$ mol kg^{-1}, the theoretical values from the above models differ considerably from experimental observation. This is to be expected from the limitations of the above models if we take into account the variation with concentration of parameters such as of hydration, viscosity, and others, now relevant for higher concentrations.

14.4 CONCLUSIONS

For the considered concentration range (0.001–0.020 mol kg^{-1}), the results from the Onsager–Fuoss model differ considerably from experimental observation. However, our experimental data and the Pikal modeled values are in good agreement (deviations, in general, are less than 2%). Possibly, these differences can be attributed to the formation of ion pairs of the type $Cu(NO_3)_x^{2-x}$ not contemplated in Onsager's development, contrarily to the Pikal's theory. This work also established reasonable values for the conductance of these ion pairs and the respective degrees of dissociation, allowing us to have a better knowledge on the structure of these systems.

ACKNOWLEDGMENTS

The authors are grateful for funding from "The Coimbra Chemistry Centre" which is supported by the Fundação para a Ciência e a Tecnologia (FCT), Portuguese Agency for Scientific Research, through the projects UID/QUI/UI0313/2013 and COMPETE.

KEYWORDS

- **mutual diffusion**
- **copper(II) nitrate**
- **aqueous solutions**
- **transport properties**
- **degrees of dissociation**

REFERENCES

1. Gao, M.; Li, Y.; Gan, Y.; Xu, B. *Angew. Chem. Int. Ed. Engl.* **2005**, *54*, 8795–8799.
2. Willenberg, B.; Ryll, H.; Kiefer, K.; Tennant, D. A.; Groitl, F.; Rolfs, K.; Manuel, P.; Khalyavin, D.; Rule, K. C.; Wolter, A. U. B.; Süllow, S. *Phys. Rev. B* **2015**, *91*, 060407-1-5.
3. Zheng, X.; Wu, K.; Mao, J.; Jiang, X.; Shao, L.; Lin, X.; Li, P.; Shui, M.; Shu, J. *Electrochim. Acta* **2014**, *147*, 765–772.
4. Jiang, X.; Wu, K.; Shao, L.; Shui, M.; Lin, X.; Lao, M.; Long, N.; Ren, Y.; Shu, J. *J. Power Sources* **2014**, *260*, 218–224.
5. Wu, K.; Wang, D.; Shao, L.; Shui, M.; Ma, R.; Lao, M.; Long, N.; Ren, Y.; Shu, J. *J. Power Sources* **2014**, *248*, 205–211.
6. Li, P.; Wang, P.; Zheng, X.; Yu, H.; Qian, S.; Shu, M.; Lin, X.; Long, N.; Shu, J. *J. Electroanal. Chem.* **2015**, *755*, 92–99.
7. Tyrrel, H. J. V. *J. Chem. Educ.* **1964**, *41*, 397–400.
8. Callendar, R.; Leaist, D. G. *J. Solut. Chem.* **2006**, *35*, 353–379.
9. Barthel, J.; Gores, H. J.; Lohr, C. M.; Seidl, J. J. *J. Solut. Chem.* **1996**, *25*, 921–935.
10. Lopes, M. L. S. M.; Nieto de Castro, C. A.; Sengers, J. V. *Int. J. Thermophys.* **1992**, *13*, 283–294.
11. Loh, W. *Quim. Nova* **1997**, *20*, 541–545.
12. Alizadeh, A.; Nieto de Castro, C. A.; Wakeham, W. A. *Int. J. Thermophys.* **1980**, *1*, 243–284.
13. Onsager, L.; Fuoss, R. M. *J. Phys. Chem.* **1932**, *36*, 2689–2778.
14. Pikal, M. J. *J. Phys. Chem.* **1971**, *75*, 663–675.
15. Harned, H. S.; Owen, B. B. *The Physical Chemistry of Electrolytic Solutions*, 3rd ed.; Reinhold Publishing Corporation: New York, 1964.
16. Robinson, R. A.; Stokes, R. H. *Electrolyte Solutions*, 2nd ed.; Butterworths: London, 1959.
17. Baes, C. F., Jr.; Mesmer, R. E. *The Hydrolysis of Cations;* John Wiley & Sons: New York, 1976.
18. Burguess, J. *Metal Ions in Solution;* John Wiley & Sons: Chichester, Sussex, England, 1978.
19. Lobo, V. M. M.; Ribeiro, A. C. F. *Port. Electrochim. Acta* **1994**, *12*, 29–41.
20. Lobo, V. M. M.; Ribeiro, A. C. F.; Andrade, S. G. C. S. *Ber. Bunsen. Gesel. Phys. Chem. Chem. Phys.* **1995**, *99,* 713–720.
21. Lobo, V. M. M.; Ribeiro, A. C. F.; Andrade, S. G. C. S. *Port. Electrochim. Acta* **1996**, *14*, 45–124.
22. Dobos, D. *Electrochemical Data. A Handbook for Electrochemists in Industry and Universities;* Elsevier: New York, 1975.
23. Lobo, V. M. M.; Quaresma, J. L. *Electrochim. Acta* **1990**, *35*, 1433–1436.

CHAPTER 15

A CRITICAL OVERVIEW OF APPLICATION OF EVOLUTIONARY COMPUTATION IN DESIGNING PETROLEUM REFINING UNITS: A VISION FOR THE FUTURE

SUKANCHAN PALIT*

Department of Chemical Engineering, University of Petroleum and Energy Studies, Post Office Bidholi via Premnagar, Dehradun 248007, Uttarakhand, India

*Corresponding author. *E-mail: sukanchan68@gmail.com; sukanchan92@gmail.com*

CONTENTS

ABSTRACT

Petroleum refining and petrochemical processes are moving toward a newer scientific paradigm and a newer visionary future. Science, technology, and engineering are witnessing drastic changes in this century of human civilization. Energy and environmental sustainability are at a state of immense distress. Scientific vision and deep scientific introspection are unfolding new chapters in human history and time. At this threshold of a new century and a newer scientific revolution, the need for chemical process engineering tools has widened the scientific vision and unfolded the murky depths of modeling, simulation, optimization, and control. Depletion of fossil fuel resources is plundering and threatening the true scientific vision of petroleum engineering science. Global status of energy sustainability is at a disastrous state. The author delineates the concept of modeling, simulation, control, and optimization of petrochemical processes and petroleum refining with deep comprehension and cogent insight. Vision of science and engineering is at its helm at each step of scientific endeavor in the present century. At this crucial juxtaposition of human civilization and human scientific pursuit, the immediate need of science is to reenvision the areas and domain of petroleum refining and the field of petrochemicals. Chemical process modeling and simulation are the pallbearers and the hallmarks of a greater emancipation of scientific innovation in decades to come. This treatise brings forward to the scientific horizon the wide and versatile world of process modeling, simulation, control, and optimization of the fluidized catalytic cracking unit in a petroleum refinery and other chemical engineering systems. Chemical engineering systems are robust, and deep introspection is necessary in comprehending and unraveling the murky depths of the visionary world of chemical process modeling and simulation. Vision of science and engineering, the defiant urge to excel, and the progress of human scientific endeavor will go a long way in visionary emancipation of chemical process engineering and petroleum engineering science.

15.1 INTRODUCTION

Petroleum engineering in today's world is moving toward a visionary frontier. Depletion of fossil fuel resources has changed the dimension of future scientific thought and scientific vision. The challenge of human civilization in today's world is appalling and surpassing vast and versatile boundaries. The need of the hour is to gain ground in research frontiers in the field of design of petroleum refining unit. Chemical process modeling and simulation are veritably ushering in a new era in the field of scientific cognizance. Energy sustainability in today's world has an unsevered umbilical cord with petroleum refining and the wide world of petrochemicals. Science and engineering needs to be reenvisioned in the field of chemical process modeling and process optimization. The author willfully rewrites the history of modeling, simulation, optimization, and control of petrochemical processes and petroleum refining as a whole. With deep insight, the author gleans through the nitty-gritty's of multiobjective optimization and application of genetic algorithm (GA) in designing petroleum refining units. The science of evolutionary multiobjective optimization (EMO) and bio-inspired optimization stands as a major hallmark of this critical overview. Petroleum refining is today in a state of immense catastrophe. The dire straits in designing petroleum refining units and petrochemical processes need to be reenvisioned. At such a critical juncture, scientific research and the wide world of scientific judgement and vision will go a long and visionary way in true realization of global energy sustainability. Man's vision, a scientist's prowess, and the scientific urge to excel are the veritable pallbearers of tomorrow's technological vision. Petrochemical processes, in a similar manner, are witnessing drastic and dramatic challenges. The world of difficulties and barriers in application of mathematical modeling and process optimization needs to gain ground in its computational skills and relevant scientific rigor.[15]

Science of optimization is today ushering in a new eon of scientific rejuvenation and deep scientific vision. The author repeatedly stresses upon the immense success, the wide potential, and the futuristic vision of the mathematics of multiobjective optimization.

15.2   VISION OF THE TREATISE

In this treatise, the author brings forward in the scientific horizon the vast and versatile world of multiobjective optimization and application of GA. The challenge, the vision, and the scientific rigor in the field of bio-inspired optimization are immense and far-reaching. Human scientific endeavor and the immense scientific urge to excel are changing the scenario of research pursuit. Petroleum refining and petrochemical processes are undergoing immense challenges and veritable difficulties. Vision of science and engineering is moving fast surpassing far-reaching boundaries. The author with deep introspection and cogent insight gleans through the scientific imagination and scientific fortitude behind EMO. The groundbreaking goals of chemical process modeling and process optimization are delineated in details in this well-informed and well-observed treatise. The world of optimization and the intricate details of process modeling and simulation are the hallmarks of the newer scientific realm and newer scientific regeneration. Human civilization needs to be reenvisioned with respect to energy sustainability. Holistic sustainable development and the immediate need of the hour with respect to application of applied mathematics in chemical process engineering and petroleum engineering science are ushering in a new eon of science and technology.[15]

15.3 SCOPE OF THE STUDY

The scope of the study is immense and versatile. Scientific rigor in the field of chemical process engineering and petroleum engineering science is changing the face of energy and environmental sustainability. Rigors of science and engineering are appalling nowadays. The fate of environment and devastating ecological disbalance are of immense concern. Human civilization is at a devastating state. At this crucial juxtaposition, petroleum refining needs to be reenvisioned. The scope of this study encompasses the immense scientific rigor behind the modeling, simulation, control, and optimization of chemical engineering systems especially the design of Fluidized Catalytic Cracking Unit (FCCU) in a petroleum refinery. Vision of science, the wide world of technological advancement, and the progress of scientific rigor will lead a long and visionary way in the true realization of energy sustainability. Global energy crisis today stands in the midst of difficult crisis. This study focuses on the vast and varied issues of design of FCC unit, the immense challenges behind design of chemical engineering systems, and the success of today's petrochemical processes. Petrochemical processes and petroleum refining need to be reenvisioned with every step of human civilization and human scientific pursuit. The author skillfully delineates the advancements, the progress, and the future trends in the century-long pursuit in the field of process optimization of petroleum refining units. Bio-inspired optimization, application of GA, and the imminent and immediate success of EMO are gaining ground in the scientific scenario. The challenge and vision of application of process optimization are opening up new vistas in the field of petroleum engineering science and chemical process modeling. Scientific research pursuit, the progress of petroleum engineering science, and the success of chemical process modeling, simulation, and control are reshaping the entire gamut of human scientific rigor. In a similar vein, academic rigor and the immense scientific judgement are reshaping the world of bio-inspired optimization and GA. Scope of this treatise is vast and versatile. The author treads through a weary path in reenvisioning the treatise on bio-inspired optimization. GA and its world of challenges are unraveling the scientific vision and deep scientific understanding. Petroleum refining and the world of petrochemicals are witnessing widespread and remarkable challenges. Human scientific endeavor and the progress of technology are gaining immense grounds in today's scientific panorama. The challenge needs to be envisaged at each step of scientific grit and determination.[15]

15.4   THE NEED AND THE RATIONALE BEHIND THIS STUDY

The scientific challenge in this human mankind and research pursuit is to preserve ecological balance and to realize energy sustainability. Human scientific endeavor and the contribution of civil society all will lead a long way in the true realization of energy sustainability and the true success of petroleum engineering science. The need and the rationale behind this study encompass vast and versatile boundaries. Science of modeling and simulation of a petroleum refining unit is difficult and arduous. This study focuses on the wide principle and concept of optimization in designing the riser reactor of an FCCU. The challenge and the vision are unimaginable as the author treads the weary path of multiobjective optimization and the arduous aisles of GA. Bio-inspired optimization stands as a veritable backbone of the entire study. This study delineates the immense pros and cons in the design of a riser reactor in an FCCU. EMO is gearing toward newer challenges as science and engineering move toward a visionary eon and a newer realm. The challenge of human civilization today lies in the successful realization of energy and environmental sustainability. Science and technology today have a wide and versatile vision. This study pointedly focuses on the optimization and control of petrochemical processes and the greater vision of a petroleum refining unit such as an FCCU.[15]

15.5   SCIENTIFIC DOCTRINE BEHIND PETROCHEMICAL PROCESSES AND THE WORLD OF PETROLEUM REFINING

Petrochemical processes and petroleum refining in today's visionary world are ushering in a new eon in the wide world of scientific rigor, scientific validation, and scientific judgement. Science and engineering are revolutionizing the present state of human scientific pursuit and deep and innovative area of scientific validation. The author targets this treatise toward this visionary direction. Petroleum refining and petrochemical processes are witnessing a new dawn of human mankind. Energy sustainability and global energy crisis are changing the face of scientific vision, scientific introspection, and scientific truth. The scientific rigor, the innovative academic pursuit, and the visionary challenges will all lead a long way in the true realization of energy sustainability. Academic rigor in today's human civilization is in the path of a new regeneration. The world of petroleum refining and petrochemical processes is witnessing new test of our times. Scientific acuity, the progress of technology, and the future road toward a visionary future are all the pallbearers toward a newer scientific horizon and a newer arena of scientific understanding.[15]

15.6 PETROLEUM REFINING AND THE SCIENTIFIC DOCTRINE OF FLUIDIZED CATALYTIC CRACKING

Petroleum refining today is witnessing drastic challenges. A scientist's vision, technological advancements, and the vast scientific rigor will all lead a long way in the true realization of petroleum refining. Petroleum refining today stands in the midst of scientific vision and wide scientific introspection. Immense scientific challenges such as depletion of fossil fuel sources have plunged the scientific domain to delve deep into newer vision and newer innovations. Fluidized catalytic cracking (FCC) stands at the forefront of the vast body of scientific endeavor in design of petroleum refining units. This technology is robust and far-reaching. Today, scientific profundity, scientific forbearance, and scientific knowledge are tested in the field of scientific endeavor in the field of petroleum refining. Technology needs to be redefined and rebuilt as global concerns for depletion of fossil fuel resources gain immense heights. The author rigorously defines and discusses with cogent insight the wide path toward scientific vision in the field of design of FCC unit in a petroleum refinery.[15]

15.7 WHAT IS FCC?

FCC is one of the most important conversion processes used in petroleum refineries. It is vastly used to convert the high-boiling and high molecular weight hydrocarbon fractions of petroleum crude oils to more valuable gasoline, olefinic gases, and other products. Cracking of petroleum hydrocarbons was originally done by thermal cracking, which has been almost completely replaced by catalytic cracking because it produces more gasoline with a higher octane rating. It also produces by-product gases that are more olefinic, and hence more valuable, than those produced by thermal cracking. The feedstock to an FCC is usually that portion of the crude oil that has an initial boiling point of 340°C or higher at atmospheric pressure and an average molecular weight ranging from 200 to 600 or higher. This portion of crude oil is often referred to as heavy gas oil or vacuum gas oil (HVGO). The FCC process vaporizes and breaks the long-chain molecules of the high-boiling hydrocarbon liquids into much shorter molecules by contacting the feedstock, at high temperature and moderate pressure, with a fluidized powered catalyst.[15] Technology and science are reemphasized and reemboldened as human scientific endeavor ushers in newer and visionary innovations.[15]

15.7.1 CRACKING KINETICS

Refineries vary by complexity and intricacy; more complex refineries have more secondary conversion capability, meaning they can produce different types of petroleum products. FCC, a type of secondary unit operation, is primarily used in producing additional gasoline in the refining process. Unlike atmospheric distillation and vacuum distillation, which are physical separation processes, FCC is a chemical process that uses a catalyst to create new and smaller molecules from larger molecules to make gasoline and distillate fuels. The catalyst is a solid sand-like material that is made fluid by the hot vapor and liquid fed into the FCC (much as water makes sand into quicksand). Because the catalyst is fluid, it can circulate around the FCC, moving between reactor and regenerator vessels.[15] The FCC uses the catalyst and heat to break apart the large molecules of gas oil into the smaller molecules that make up gasoline, distillate, and other high-value products such as butane and propane. After the gas oil is cracked through contact with the catalyst, the resulting effluent is processed in fractionators, which separate the effluent based on various boiling points into several intermediate products, including butane and lighter hydrocarbons, gasoline, light gas oil, heavy gas oil, and clarified slurry oil.[15]

15.7.2   SCIENTIFIC ENDEAVOR IN THE FIELD OF MATHEMATICAL MODELING OF RISER AND REGENERATOR IN FCCU

Mathematical modeling and simulation of petroleum engineering systems are the primary focus of this well-researched treatise. The author rigorously does a well-researched review on the recent scientific advances in the field of modeling, simulation, and optimization of FCCU. Frontiers of technology and scientific motivation need to be reemphasized and reenvisioned as scientific research pursuit enters a newer phase in the world of science and technology.[15]

Chang et al.[1] discussed with cogent insight computational fluid dynamics investigation of hydrodynamics, heat transfer, and cracking reaction in a heavy oil riser with bottom airlift loop mixer. The bottom mixer causes the hot and cool catalysts to mix well. There exist optimal product distributions in the riser. The product yields are more sensitive to injection angle than to droplet size of feedstock and reaction temperature. By extending a validated gas–solid flow model to incorporate feedstock vaporization and 12-lump kinetic model, a three-phase flow and reaction model is established.[1,15]

FCC is one of the most conversion processes in petroleum refineries. It is widely used to convert the high-boiling and high molecular weight hydrocarbon fractions of crude oils to more valuable gasoline, olefinic gases, and other products. About 45% gasoline and one-third propylene worldwide comes from FCC process. Especially in China, due to lack of hydrocracking and hydro-conversion units, FCC remains the most important and profitable heavy oil conversion process. During its history of 60 years, FCC technology has been under tremendous scientific rejuvenation and scientific revamping. The present technology drivers include processing increased feedstocks and abiding to environmental regulations. Science and engineering are in the path of new scientific destiny as petroleum engineering science ushers in a new eon of scientific vision. To meet the cause of technology drivers and scientific/technology understanding, many nobel advancements such as short contact time units, double riser technology, and the recycling of deactivated catalyst mixed with regenerated catalyst have been developed. The technology that recycles deactivated catalyst and mixes it with regenerated catalyst could be dated back to "RxCat" process by UOP, USA, in which the deactivated catalyst of low temperature was recycled from FCC stripper to the bottom of riser reactor to process heavy oil. Later, this technology is improved and applied in flexible dual-riser fluid catalytic cracking (FDFCC) process. The challenge and wide vision of science are opening up new avenues of scientific innovation and scientific understanding in the design of petroleum refining units. Application of multiobjective optimization and GA in modeling and simulation of petroleum refining has reached its visionary zenith with the passage of scientific history, scientific forbearance, and wide scientific vision. In FDCC process, the spent catalyst from naphtha riser, which is in lower temperature and still retains an inherently high residue activity level, is recycled to an airlift loop mixer located at the base of heavy feedstock riser and mixed with the regenerated catalyst. The mixed catalyst is then applied to refine heavy feedstock. In addition, there generally exist catalyst coolers in FCC units to remove surplus heat. The cooled catalysts of low temperature and high activity could also be used for this purpose.[1,15]

Petroleum refining is moving toward immense scientific regeneration. This technology is a wide eye-opener to future scientific research pursuit. Engineering science and the wide world of petroleum engineering are moving very fast crossing visionary frontiers. Energy sustainability and holistic sustainable development are the utmost need of the hour. In such a crucial juncture of sustainability concerns, deep thought and scientific vision will lead a long way in the true emancipation of science and engineering.

Chen et al.[2] discussed with cogent foresight evaluation of role of intra-particle mass and heat transfers in a commercial FCC riser in a widely researched mesoscale study. This chapter provides new insights into the fundamental mechanism of catalytic cracking from the mesoscale viewpoint.[2] A comprehensive single-particle model was developed to characterize detailed chemical and physical phenomena occurring within catalyst particles in a commercial FCC riser from an industrial-scale refinery. This model integrates the mass, energy, and momentum balances as well as the equations for gas-state, lumped-species reaction kinetics, the multicomponent diffusion, and convective heat transfer.[2] The march of technology and the wide scientific motivation are the challenges of human civilization and scientific endeavor today.[2] This model envisions prediction of temperature, pressure, species mass fraction distributions, as well as the reaction rate and the effective diffusivity coefficient within the particles as a function of catalyst position in the riser. FCC scientific endeavor is reaching new scientific pinnacles and scientific progeny. The world of challenges in energy sustainability and petroleum engineering science are opening up new future directions and newer visionary innovations in the path toward furtherance of engineering science. The authors of this treatise provide deep foresight into the world of FCC modeling and simulation.[2]

Zhang et al.[3] comprehended deeply the success of modeling and simulation of FCC risers. The authors in a widely researched paper tackled modeling of FCC risers with special pseudo-components (SPCs).[3] Technology and science of petroleum engineering science are today wide and varied. The challenge and vision of FCC modeling and simulation are surpassing scientific vision and scientific splendor. Science needs to be more reenvisioned and rejuvenated as energy sustainability gains immense importance. SPCs are proposed to be the compositional entities for characterizing the complicated mixtures of stock and product oils involved in FCC risers.[3] SPCs have invariant and definite physicochemical properties and are defined in pairs of light and heavy oil cuts of narrow boiling range. A steady-state model for a prototype riser with a side feed stream is formulated where material and heat balance is strictly observed, the hydraulic behavior is considered, and the kinetic model of a previous visionary research widely adopted. Science and engineering of petroleum refining are redefining itself with the passage of scientific history and time. Human scientific endeavor's immense prowess, petroleum refining's immense scientific potential, and the concerns of energy sustainability are the visionary forerunners toward a newer eon of petroleum engineering science. It is of great significance to optimize the design and operation of industrial FCC units because of its immense throughputs. Therefore, modeling and simulation of FCC processes have been considered to be immensely important in research since 1940s. One of the challenging areas of research in modeling an FCCU is to describe the chemical reactions of feed and product oils which are complicated mixtures of numerous hydrocarbons and nonhydrocarbons. This area of scientific profundity and scientific splendor needs to explored and investigated in details. Research today is profound with the rise of concern for energy sustainability and the success of petroleum engineering science. The authors also reviewed the various scientific forays into the science of kinetic modeling of FCC and the design of an FCC riser. Technology today is vast and versatile as depletion of fossil fuel resources ushers in a new era in the field of petroleum refining. The authors also delineated the kinetic reactions of catalytic cracking, characterization based on SPCs, and an exploration into a steady-state model for a prototype riser.

Koratiya et al.[4] discussed lucidly modeling, simulation, and optimization of FCC downer reactor. Downer reactor, in which gas and solids move downward cocurrently, has unique features such as it accommodates high-severity operation at the initial stage with the benefit of near-plug flow reactor. In the present research work, mathematical model for downer reactor has been developed, in which a five-lump model is used to characterize the feed and the products, where gas oil cracks to give lighter fractions and coke.[4] Optimization study of FCC downer reactor to maximize its profitability and satisfy real-life constraints and nondominated sorting genetic algorithm (NSGA-II) used to solve a two-objective optimization problem are investigated in this chapter.[4]

Souza et al.[5] comprehended in lucid details a two-dimensional model for simulation, control, and optimization of FCC risers. Technological vision is slowly evolving as scientific research pursuit enters into a newer world of scientific vision and deep scientific understanding. A simplified two-dimensional (2D) model formulation for FCC risers has been developed.[5] The model approximates the mixture of gas oil, steam, and solid catalyst that flows inside the riser reactor by an equivalent fluid with average properties.[5] Scientific validation of FCC modeling is the focal point of this well-researched review chapter. Science, technology, and engineering are gearing forward today toward scientific vision and scientific rejuvenation at a rapid pace. Scientific endeavor in the field of petroleum refining needs to be reemphasized and rebuilt at the utmost and a veritable need of the hour. The authors reiterate the success of innovation in the field of modeling and simulation of FCC risers.[5]

15.8   WHAT DO YOU MEAN BY EVOLUTIONARY COMPUTATION?

In computer science, evolutionary computation is a subfield of artificial intelligence (more particularly soft computing) that can be defined by the type of algorithms it is concerned with. These algorithms, called evolutionary algorithms (EAs), are based on adopting Darwinian principles, hence the name. Technically, they belong to the family of trial and error problem solvers and can be considered global optimization methods with a metaheuristic or stochastic optimization character, distinguished by the use of a population of candidate solutions (rather than just iterating over one point in the search pace). The application of recombination and evolutionary strategies makes them less prone to get stuck in local optima than alternative methods. Technology of evolutionary computation is witnessing a newer revolutionary and paradigmatic shift. Artificial intelligence and evolutionary computation are the two opposite sides of the visionary scientific coin. Evolutionary computation uses iterative progress, such as growth and development in a population. This population is then selected in a guided random approach using parallel processing to achieve the desired end. Such processes are often inspired by biological mechanisms of evolution.[6,15]

The use of Darwinian principles for automated problem-solving originated in the 1950s. It was not until 1960s that three distinct interpretations of this idea started to be developed in three different places. Evolutionary programing was introduced by Lawrence Vogel in the US, while John Henry Holland called his method a GA.[15] Technology moved at a rapid pace after the first invention of evolutionary programming. Today, scientific vision and scientific legacy are witnessing immense challenges. The success of engineering science and computer science are redefining scientific history. These areas, after the first discovery, developed separately for about 15 years. From the early 90s, they are unified as different representatives (dialects) of one technology called evolutionary computing. Evolutionary computing techniques mostly involve metaheuristic optimization algorithms. They are widely divided into: (1) ant colony optimization algorithm, (2) artificial bee colony algorithm, (3) cultural algorithms, (4) differential evolution, (5) EA, (6) particle swarm optimization, and (7) swarm intelligence, and many classifications.[15]

15.9   WHAT IS GA?

In computer science and operations research, a GA is a metaheuristic inspired by the process of natural selection that belongs to the larger class of EAs. GAs are commonly used to generate high-quality solutions to optimization and search problems by relying on bio-inspired operators such as mutation, crossover, and selection. Technology and science of GA are highly advanced today and surpassing wide scientific landscapes. The author revisits and reemphasizes the success of application of GA and multiobjective optimization in designing chemical engineering systems such as petroleum refinery.[15]

In a GA, a population of candidate solutions (called individuals, creatures, or phenotypes) to an optimization problem is evolved toward better solutions. Science and engineering of GAs are moving at a rapid pace. The author pointedly focuses on the immense scientific potential and deep scientific introspection behind application of GA in engineering systems. Each candidate solution has a set of properties (its chromosomes or genotype) which can be mutated and altered; traditionally, solutions are represented in binary as strings of 0s and 1s, but other encodings are also possible. The evolution usually starts from a population of randomly generated individuals and is an iterative process, with the population in each iteration called generation. In each generation, the fitness of every individual in the population is evaluated; the fitness is usually the value of the objective function in the optimization problem being solved. The more fit individuals are stochastically selected from the current population, and each individual's genome is modified (recombined and possibly randomly mutated) to form a new generation. The new generation of candidate solutions is then used in the next iteration of the algorithm. The intricacies of GA are widely observed in this treatise.[15]

15.9.1   RECENT SCIENTIFIC ENDEAVOR IN THE FIELD OF GA

Scientific research pursuit in the field of GA is undergoing drastic challenges. Scientific vision, scientific forbearance, and the vast scientific urge to excel are the torchbearers toward a greater scientific legacy today. In this chapter, the author pointedly focuses on the success of application domain of evolutionary computation toward the furtherance of science.

Deb et al.[6] deeply comprehended and investigated in the domain of a fast and elitist multiobjective GA which is termed as NSGA-II. Multiobjective evolutionary algorithms (MOEAs) that use nondominated sorting and sharing have been criticized mainly for their: (1) computational complexity, (2) nonelitism approach, and (3) the need for specifying a sharing parameter. NSGA-II is a highly advanced form of GA. The challenge, the vision, and the potential of NSGA-II are immense. Over the years, the main criticisms of NSGA are:

• high computational complexity of nondominated sorting,
• lack of elitism,
• need for specifying the sharing parameter.

This chapter widely observes all of these issues and proposes an improved version of NSGA called NSGA-II. Scientific endeavor needs to be reenvisioned and reenvisaged with the passage of scientific history and scientific research pursuit. From the simulation results on a number of difficult test problems, the authors target two more strategies: (1) Pareto archived evolution strategy and (2) strength Pareto EA. They also lucidly discussed simple constraint-handling strategy with NSGA-II that suits well for any EA.

Bandyopadhayay et al.[7] in a visionary and phenomenal paper discussed a simulated annealing-based multiobjective optimization algorithm of AMOSA. This chapter widely observes a simulated annealing (SA)-based multiobjective optimization algorithm that incorporates the concept of archive in order to provide a set of trade-off solutions for the problem under consideration. The multiobjective optimization problem has a rather different perspective compared with one having a single objective. In the single-objective optimization, there is only one global optimum, but in multiobjective optimization, there is a set of solutions, called Pareto-optimal (PO) set, which are considered to be equally important; all of them constitute global optimum solutions. Technology and engineering science are today surpassing vast and versatile visionary boundaries. The challenge of human civilization is toward efficient processes in chemical engineering and petroleum engineering systems. The success and the vision of AMOSA are immense and versatile. Over the past decade, a number of MOEAs have been suggested. The main reason for the popularity of EAs for solving multiobjective optimization is their population-based nature and ability to find multiple optima simultaneously. SA, another popular search algorithm, utilizes the principles of statistical mechanics regarding the behavior of a large number of atoms at low temperature, for finding minimal cost solutions to large optimization problems by minimizing the associated energy. Technological vision and scientific objectives are enhanced as a scientist's immense vision and prowess enters a newer paradigmatic era. The author in this chapter rigorously points out the success and vision behind the application of multiobjective optimization science. In statistical mechanics, investigating the ground states or low-energy states of matter is of fundamental importance. These states are achieved at very low temperatures. However, it is not sufficient to lower the temperature alone since this results in unstable states. In the annealing process, the temperature is first raised and then decreased gradually to a very low value, while ensuring that one spends sufficient time at each temperature value. This process yields stable low-energy states. Technology and science of SA are so vast and versatile today. SA and its interconnected area of multiobjective simulated annealing (MOSA) are one of the diverse fountainheads of multiobjective optimization. This area of scientific research pursuit is replete with scientific vision and scientific profundity. SA has shown the path toward newer vision and newer innovation in the wide world of GA. Researchers have proved that SA, if annealed sufficiently slowly, converges to the global optimum. Being strongly visionary, SA has applications in wide and versatile areas by optimizing a single criterion.

Deb et al.[8] comprehended with deep foresight in a report multiobjective optimization using EAs. This treatise is a comprehensive report and an innovative futuristic vision. As the name suggests, multiobjective optimization involves optimizing a number of objectives simultaneously. The technology and science of optimization are today reaching wide scientific frontiers and crossing visionary boundaries. The problem becomes challenging when the objectives are of conflict to each other, that is, the optimal solution of an objective function is different from that of the other. This scientific conflict and immense scientific vision result in parameterized procedure. The author rigorously pursues the path of multiobjective optimization with a brief introduction, its operating principles, thus outlining the current research and applications of EMO.

15.10   WHAT IS MULTIOBJECTIVE OPTIMIZATION?

Multiobjective optimization (also known as multiobjective programming, vector optimization, multicriteria optimization, multiattribute optimization, or Pareto optimization) is an area of multiple-criteria decision-making. Scientific vision, scientific prowess, and wide scientific cognizance are challenged today in the field of modeling, simulation, and optimization of petroleum refining units. Multicriteria optimization is today evolving into a next-generation technology for modeling and simulation of petroleum refining units primarily FCCU. Today, science is a huge colossus with a definite vision and will of its own. This is an area of multiple-criteria decision-making that is concerned with mathematical optimization problems involving more than one objective function to be optimized simultaneously. Multiobjective has been applied in many fields of science, including engineering, economics, and logistics where optimal decisions need to be taken in the presence of trade-offs between two or more conflicting objectives. Minimizing cost while maximizing comfort when buying a car and maximizing performance while minimizing fuel consumption and emission of pollutants of a vehicle are examples of multiobjective optimization problems involving two or three objectives, respectively. In practical problems, there can be more than three objectives.

15.10.1   RECENT SCIENTIFIC RESEARCH PURSUIT IN THE FIELD OF MODELING, SIMULATION, AND MULTIOBJECTIVE OPTIMIZATION OF PETROLEUM REFINING UNITS (FCC)

FCCU stands as a major unit in a petroleum refinery. Today, scientific vision, scientific understanding, and deep scientific profundity are the forerunners and pallbearers toward a greater emancipation of environmental sustainability. Technological advancements in the field of petroleum refining and petroleum engineering science need to be reenvisioned in the path toward immense scientific emancipation. Research endeavor in the field of modeling, simulation, and optimization of FCCU is surpassing scientific vision and wide scientific frontiers. The author rigorously comprehends the research trends—past and future in FCC modeling.

Dagde et al.[9] lucidly discuss with cogent insight modeling and simulation of industrial FCC unit and the analysis is based on five-lump kinetic scheme for gas–oil cracking. Scientific vision and technological objectives are changing with the path of scientific endeavor. Models which describe the performance of the riser and reactor regenerator reactors of FCC unit are elucidated in details. The riser reactor is modeled as a plug flow reactor operating adiabatically, using five-lump kinetics for the cracking reactions. The regenerator reactor is divided into a dilute region and dense region, with the dense region divided into bubble phase and an emulsion phase. The models are validated using plant data obtained from a functional FCC unit. It is observed that predictions of models compare well with plant data for both reactors. The challenge of modeling and simulation in FCC reactors and other petroleum refining units are evolving tremendously toward greater scientific vision and scientific forbearance. The success of petroleum refining is immense and is the utmost need of the hour this decade.

Bhende et al.[10] well observed in a comprehensive treatise modeling and simulation for FCC unit for the estimation of gasoline production. FCC is a definitely important scientific endeavor in the petroleum refining industry for the conversion of heavy gas oil to gasoline and diesel. Technology is widely expanding with every step of scientific life. Furthermore to the focal point, valuable gases such as ethylene, propylene, and isobutylene are produced. This work develops a mathematical model that simulates the behavior of the FCC unit, which consists of feed and preheat system, riser, stripper, reactor, regenerator, and the main fractionators.

Ahari et al.[11] discussed in deep details a mathematical modeling of the riser reactor in industrial FCC unit. For improvement of petroleum process technology, availability of a modeling program for FCC is of utmost importance in unit optimization and scale-up or reducing the problems of FCC unit. The success of this treatise is that a one-dimensional adiabatic model for riser reactor of FCC is developed. Simulation studies are performed to explore the effect of changing process variables, such as input catalyst temperature and catalyst-to-oil ratio (COR).

Bispo et al.[12] discussed with cogent insight modeling, optimization, and control of FCC unit with the help of neural network and evolutionary methods. This chapter presents a simulation study of the use of artificial neural network (ANN) model for control and optimization of a fluidized-bed catalytic cracking reactor–regenerator system (FCC). This case study, whose phenomenological model was validated with industrial data, is a multivariable and nonlinear process with strong interactions among the operational variables. In order to obtain a dynamic model of the FCC system, a feed-forward ANN model was identified. Technology of ANN is highly advanced and scientifically progressive. Technological challenges and scientific vision are the forerunners toward mankind's immense scientific prowess and emancipation of basic needs of civilization. At such a critical juncture of scientific history and advancements, depletion of fossil fuel resources assumes vital importance. ANN is an innovative technology for modeling and simulation of FCC unit.

Rao et al.[13] comprehended lucidly industrial experience with object-oriented modeling in a well-defined and well-observed FCC case study. In activities such as design, optimization, and control of processes, realistic process models, which incorporate physics and chemistry of the process in adequate detail, are of vital importance. This work presents the framework of one such multipurpose process simulator, MPROSIM, an object-oriented modeling and simulation environment. Technological vision and scientific advancements are gearing toward a newer eon in the modeling and simulation of FCC unit. The success and the path toward scientific effectivity in model simulation are ushering in a newer futuristic vision. The immense utility of the framework is illustrated by exploring how the model for FCCU could be fine-tuned both structurally and parametrically with the vision toward furtherance of modeling and simulation science.

15.11 WHAT IS MOSA?

SA is a probabilistic technique for approximating the global optimum of a given function. Specifically, it is a metaheuristic to approximate global optimization in a large search space. It is often used when the search space is discrete. For problems where finding an approximate global optimum is more important than finding a precise local optimum in a fixed amount of time, SA may be preferable to alternatives such as gradient descent. The name and inspiration come from annealing in metallurgy, a technique involving heating and controlled cooling of a material to increase the size of crystals and reduce their defects.[15]

15.11.1   RECENT RESEARCH THRUST AREAS IN THE FIELD OF MOSA

MOSA is one of the new and innovative branches of scientific endeavor in the field of optimization science. Technology is today moving extremely fast surpassing immense hurdles and wide scientific boundaries. This field is replete with scientific vision and deep scientific profundity and forbearance. Chemical process engineering and petroleum engineering science are today linked with optimization science and applied mathematics as an unsevered umbilical cord. The author rigorously ponders on the areas of scientific vision in multiobjective optimization and widely explores nontraditional algorithms to solve optimization problems. Visionary issues of design of petroleum refining units are challenged at step of science and engineering.

Ziane et al.[14] discussed lucidly SA optimization for multiobjective economic dispatch solution. This chapter presents a MOSA optimization to solve a dynamic generation dispatch problem. Science of optimization and evolutionary computation are witnessing immense scientific and academic rigor today. Human civilization and human scientific endeavor are moving toward a newer scientific regeneration and a newer scientific vision. In this work, the problem is formulated as a multiobjective one with two competing functions, namely economic cost and emission functions, subject to different constraints. The inequality constraints considered are the generating unit capacity limits while the equality constraint is generation-demand balance. Experimental results show a proficiency of SA over other existing techniques in terms of effectiveness.

Bandyopadhaya et al.[7] deeply comprehended with deep and cogent insight a SA-based multiobjective optimization algorithm which is termed as AMOSA. Science and engineering are at its helm as optimization surpasses visionary frontiers. SA, another popular search algorithm, utilizes the principles of statistical mechanics regarding the behavior of a large number of atoms at low temperature, for finding the minimal cost solutions to large optimization problems. SA is a wide and visionary direction in the field of optimization science. Engineering science is moving at a rapid pace. Mankind's scientific rigor, the wide academic rigor, and the passage of scientific history are all leading a long and visionary way in the true emancipation of both SA and multiobjective optimization. This is a visionary treatise focusing on the research trends and the wide directions of optimization science.

15.12 SCIENTIFIC VISION, SCIENTIFIC INGENUITY, AND SCIENTIFIC INTROSPECTION IN THE FIELD OF MULTIOBJECTIVE OPTIMIZATION AND GA

GA is the innovative avenue of scientific endeavor today. Multiobjective optimization, SA, and application of GA will all lead a long and visionary way in the true emancipation of chemical process engineering and petroleum engineering science. A scientist's wide vision, the vast scientific horizons, and the technological objectives are the torchbearers toward a greater visionary era in the field of optimization science and applied mathematics. Applied mathematics and optimization science are veritably ushering in a new era in the field of modeling, simulation, and optimization of petroleum refining units and petrochemical processes. Science and engineering of petroleum refining are today crossing wide and vast visionary boundaries. The author in this treatise widely observes the technological intricacies in the design of petroleum refining units primarily the FCCU. FCCU is an important unit in petroleum refinery. Upgradation of crude petroleum is the focal point in FCC. Technological vision and scientific objectives need to be enhanced as human scientific endeavor ushers in a new world of petroleum refining.

15.13 TECHNOLOGICAL OBJECTIVES AND RESEARCH FRONTIERS IN OPTIMIZATION SCIENCE

Optimization science is today in the path of newer scientific regeneration and visionary scientific rejuvenation. Multiobjective optimization and modeling and simulation of petroleum refining are the two opposite sides of the visionary coin today. The vast challenges and the definite vision of science need to be reenvisioned and reemphasized as science of petroleum refining crosses veritable scientific frontiers. Design of chemical engineering systems and petroleum engineering systems are today dependent on evolutionary computation and GA. Technology and engineering science are today crossing visionary scientific landscapes. A scientist's immense vision, human scientific endeavor's wide and visionary prowess, and the vast academic and scientific rigor will all lead a long way in the true realization of energy sustainability. Petroleum refining and effective energy sustainability are today linked by an unsevered umbilical cord. Sustainable development whether it is energy or environment is the crucial need of the hour.

15.14 THE WIDE CHALLENGES AND THE VISION BEHIND MULTIOBJECTIVE OPTIMIZATION AND MODELING AND SIMULATION OF PETROLEUM REFINING UNIT

Multiobjective optimization is a proven tool in the field of modeling and simulation of a petroleum refining unit. The technology is not new yet needs wide scientific introspection and deep comprehension. Depletion of fossil fuel resources today is an enigma of science and engineering. The deep vision and the true challenges of modeling, simulation, and optimization of petrochemical processes need to be reenvisioned and reenshrined with the passage of scientific history and time. Technology and engineering of chemical process engineering are immensely challenged today. Challenges and barriers in the field of wide domain in petroleum engineering and chemical process engineering are immense and groundbreaking. Science is advancing extremely fast and scientific fortitude is redefined. Modeling and simulation of a petroleum refining unit is of immense and vital importance on the path toward furtherance of science, and broadly chemical process engineering and petroleum engineering science.[15]

15.15 MATHEMATICAL MODELING AND THE VISION OF FUTURE CHEMICAL PROCESS ENGINEERING

Mathematical modeling and the wide world of chemical process engineering are on the path of immense scientific judgement and wide scientific vision. Scientific forbearance, scientific candor, and the futuristic vision of technology are in a state of unimaginable distress with the depletion of fossil fuel resources. Science is a huge colossus without a will of its own. Environmental sustainability and the challenges of nuclear engineering are changing the face of scientific endeavor and innovative scientific research pursuit. Applied mathematics and the world of chemical process modeling are undergoing the test of its times in the futuristic world of scientific vision and scientific fortitude.

Mathematical modeling is a veritable process engineering tool in today's path of scientific research pursuit. Modeling, simulation, control, and optimization are the hallmarks toward a greater scientific vision and greater scientific cognizance in the design of FCCU in a petroleum refinery. Energy sustainability needs to be redefined today with the evergrowing march of science. Sustainable development, whether it is energy or environmental, is of immense global concern today. The march and forays into science of optimization are of utmost need at this critical juncture. Scientific success, mankind's scientific prowess, and scientific rejuvenation will all lead a long way on the path toward true emancipation of optimization science and science of modeling today.

15.16 OPTIMAL PROBLEM FORMULATION

Many scientists and technologists in industries and academics face difficulty in understanding the role of optimization in engineering design. For many in the scientific research domain, optimization is an esoteric technique used only in mathematics and operations research-related activities. Science of operations research is surpassing visionary boundaries. With the advent of computers, optimization has become a part of computer-aided design activities.[15]

There are two distinct types of optimization algorithms which are in use today. First, there are algorithms which are deterministic, with specific rules for moving from one solution to the other. Second, there are algorithms which are stochastic in nature, with probabilistic transition rules. These algorithms are comparatively new and are gaining popularity due to certain properties which the deterministic algorithms do not have.[15]

15.17 ENGINEERING OPTIMIZATION PROBLEMS

Engineering optimization problems are redefining and restructuring the visionary world of modeling, simulation, and control of a petroleum refining unit. Vision of science and validation of technology are opening up new vistas in the field of GA and bio-inspired optimization. This treatise gleans with deep scientific comprehension the success of endeavor in the field of modeling and simulation of an FCC unit in a petroleum refinery. FCC unit stands as a major component in the operation of a petroleum refinery. The challenge, the vision, and immense scientific urge to excel are the redefining parameters in the arduous task of design of a petroleum refining unit. Today, science of multiobjective optimization and GA needs to be reenvisioned and redefined with immense perspicuity. The challenge of solving engineering optimization problems lies in the heart of science of chemical process engineering and applied mathematics.[15]

15.17.1 SINGLE-VARIABLE OPTIMIZATION PROBLEMS

Single-variable optimization problems and its application are in a state of immense mathematical distress. The world of multivariable optimization problems is the immediate need of the hour. Science of EMO and bio-inspired optimization are veritably changing the scientific landscape. Single-variable optimization problems and its scientific genre are defunct as science moves toward multivariable optimization problems. Technology and science are moving at a rapid pace surpassing one visionary boundary over another. Single-variable optimization is less relevant these days with the progress of science and forays into research pursuit. Mathematics' ultimate acuity and the visionary world of challenges in the field of chemical process modeling and optimization are transforming the global scientific paradigm.[15]

Since single-variable functions involve only one variable, the optimization procedures are simple and easier to understand. Moreover, these algorithms are repeatedly used as a subtask of many multivariable optimization methods. Therefore, a clear understanding of these algorithms will help readers learn complex algorithms. Technological vision, the wide world of optimization, and the progress of bio-inspired optimization are leading a long way in the true emancipation of the research pursuit in applied mathematics and chemical process modeling. In this age of technological revolution and scientific perspicuity, the immediate need of human civilization is to target toward effectiveness of a mathematical procedure. In such a crucial situation, the hallmark of science of optimization is to target toward multiobjective optimization. The algorithms described in this section can be used to solve minimization problems of the following type:

$$\text{Minimize } f(x),$$

where $f(x)$ is the objective function and x is a real variable. The purpose of an optimization algorithm is to find a solution x, for which the function $f(x)$ is minimum. In this section, two distinct types of algorithm are presented. Direct search methods use only objective function values to locate the minimum point, and gradient-based methods use the first- and/or the second-order derivatives of the objective function to locate the minimum point. Gradient-based optimization methods do not really require the function $f(x)$ to be differentiable or continuous. Gradients are computed numerically wherever required. Although the optimization methods mentioned here are for minimization problems, they can also be used to solve maximization problems.

15.17.1.1 MULTIVARIABLE OPTIMIZATION PROBLEMS

Multivariable optimization problems are changing the global process modeling and simulation scenario. Human scientific vision, scientific candor, and scientific astuteness are today's hallmarks of a newer visionary boundary of technology and engineering. The author deeply contemplates the vast, visionary, and versatile domains of science and engineering of mathematical modeling of a petroleum refining unit in particular. Petroleum engineering science today is at a state of deepest disaster and deep introspection. Petroleum engineering, environmental engineering science, and the world of chemical process engineering are in today's human civilization on the path of immense scientific revelation. In a similar vein, mathematical modeling and science of optimization are opening up new chapters of scientific innovation in years to come.[15]

15.17.1.2 CONSTRAINED OPTIMIZATION ALGORITHMS

Science and technology of optimization are scaling new heights in today's world. Bio-inspired optimization and the visionary gamut of GA in today's scientific scenario are surpassing wide and versatile frontiers. The wide vision, the true challenges, and the visionary frontiers of constrained optimization algorithms are the hallmarks of a new generation and a new paradigm of GA. The technology of EMO is immensely challenging and scaling visionary heights. Constrained optimization algorithms and the interlinked application of GAs are the torchbearers toward a greater and widespread foray into the science of chemical process modeling, simulation, and optimization. The author with deep and cogent insight gleans through the nitty-gritties and nuances of EMO along with its instinctive and innovative applications in designing of a petroleum refining unit. Today's scientific and engineering scenario is of constrained optimization.[15]

15.17.1.3 SPECIALIZED ALGORITHMS

Specialized algorithms in the field of optimization science are overcoming immense scientific sagacity and scientific boundaries. The urge for scientific validation, the imminent and innovative need of scientific cognizance, and the future progress of human mankind are the parameters of a greater visionary future in the field of modeling, simulation, and optimization of a petroleum refining unit.

15.17.1.4 NONTRADITIONAL OPTIMIZATION ALGORITHMS

Nontraditional optimization algorithms are changing the face of scientific wisdom in the field of EMO. The vicious challenges, the subtle scientific astuteness, and the wide world of bio-inspired optimization are unfolding the hidden scientific truths. Nontraditional optimization is witnessing a new challenge of the day. Scientific wisdom and scientific acuity are changing the face of scientific research pursuit in energy sustainability. Nontraditional optimization and bio-inspired optimization in today's scientific scenario are surpassing visionary barriers.

15.18 CRACKING KINETICS AND THE WIDE WORLD OF FCC

Cracking kinetics stands as a heart of chemical reaction engineering in the vast world of FCCU in a petroleum refinery. Vision of science, progress of technology, and the scientific urge to excel are the visionary parameters in the path toward realization of energy sustainability. Human scientific vision, scientific candor, and a nation's economic growth will lead a long way in true emancipation of a nation's energy sustainability program. Cracking kinetics and scientific research endeavor in this direction are unfolding newer visionary realm and overcoming veritable scientific boundaries. Human scientific vision in today's world is witnessing immense challenges. Man's scientific astuteness and science's immense acuity will lead a long way in redefining the science of petroleum refining and the world of petrochemicals.[15]

15.19 DEACTIVATION PHENOMENON AND THE VISIONARY AREA OF SCIENTIFIC VALIDATION AND FAR-REACHING SCIENTIFIC VISION

Deactivation phenomenon in chemical reaction engineering and the subsequent endeavor in chemical process engineering are the hallmarks toward a greater vision in the vast and versatile world of chemical process modeling and simulation. Mankind's immense prowess, a scientist's definitive vision, and the wide progress and forays into technological advancement are the torchbearers toward a greater visionary tomorrow. The road toward scientific challenge is arduous and exceedingly inspiring. In today's scientific civilization, scientific cognizance and scientific sagacity are in a state of deep distress. Scientific acuity and scientific astuteness are witnessing veritable challenges in the path toward technological vision.

15.20 SCIENTIFIC VISION, ADVANCEMENT IN CHEMICAL PROCESS MODELING, AND THE FUTURE OF PETROLEUM REFINING

Advancement in chemical process modeling and the future of petroleum refining today stands in the midst of unimaginable crisis and deep introspection. Science of chemical process modeling and simulation and subsequent optimization and control are the visible parameters of the global energy challenge today. The global energy crisis needs to be reenvisioned and readdressed. Chemical process modeling, simulation, and control are in the process of a newer generation, and optimization stands in the heart of advancement of chemical engineering and petroleum engineering science today. Design of an FCC unit in petroleum refining is a prerogative to the success of realization of energy sustainability.

15.21 VISION OF PETROLEUM REFINING, ENERGY SUSTAINABILITY, AND THE FUTURE OF GLOBAL ENERGY SCENARIO

Petroleum refining in today's human civilization and in academic rigor has an umbilical cord with global energy sustainability. Global energy scenario today is in dire straits. The failure of nonrenewable energy scenario has urged the global scientific generation to innovate and apply deep scientific vision. Vision of science and engineering is on the path of immense restructuring. Future of global energy scenario is disastrous and bleak with the advancement of human civilization. Progress of human mankind, the wide path toward energy and environmental sustainability, and the realization of environmental engineering rigor are all the pallbearers of a greater visionary future. Scientific vision, technological validation, and the immense scientific and academic rigor will all lead a long way in the true emancipation of holistic sustainable development. Scientific candor and immense scientific cognizance are the overall upshot of this treatise.

15.22   SCIENTIFIC DOCTRINE AND SCIENTIFIC COGNIZANCE BEHIND MULTIOBJECTIVE OPTIMIZATION AND GA

GA and bio-inspired optimization are the pallbearers toward a newer and visionary scientific regeneration and scientific forbearance. Applied mathematics and chemical process modeling and simulation are moving toward a newer scientific journey and a newer visionary future. The vision of scientific rigor needs to be challenged and reenvisioned. The scientific doctrine of GA is witnessing inspiring challenges. Design of a petroleum refining unit and of petrochemical process is the veritable backbone of this treatise. The science of design and the vision of technology to move forward are the pallbearers of a greater emancipation of application of multiobjective optimization and GA. Human scientific endeavor in today's human civilization is surpassing wide and versatile visionary frontiers. Technological validation is the other side of the visionary coin of GA and bio-inspired optimization.

15.23   GA AND MULTIOBJECTIVE OPTIMIZATION: ITS SCIENTIFIC UNDERSTANDING AND THE FUTURISTIC VISION

Scientific understanding and scientific introspection in today's visionary horizon are in the path of immense glory and scientific fortitude. Applied mathematics and application of chemical process modeling are gaining ground in the wide scientific horizon of chemical and petroleum engineering science. Scientific vision and the avenues of science in GA and bio-inspired optimization are witnessing new challenges and vicious difficulties. The futuristic vision needs to be reshaped and readdressed at each step of human scientific endeavor. Petroleum refining and petrochemicals sector are in the path of a new rejuvenation. Technological and scientific validations are the veritable backbones of our times. Scientific understanding and scientific acuity in today's human civilization are moving toward a newer phase and a newer visionary horizon. Futuristic vision in the field of optimization and bio-inspired optimization are gearing toward a newer visionary era. Pragmatic approaches toward modeling and simulation are the necessity of the hour.

15.24 THE DOCTRINE AND CHALLENGES OF THE APPLICATION OF EMO

The challenges and the difficulties in the field of EMO and bio-inspired optimization are enormous and far-reaching. Human scientific research pursuit and the world of scientific validation are crossing today's scientific boundaries. Technological validation, the immense progress of science, and the wide and vast domain of GA will lead a long and visionary way in the true realization of global energy sustainability. Global energy crisis in today's world has an umbilical cord with global energy sustainability. A deep comprehension and immense investigation are the immediate need of the hour in the wide foray into energy sustainability. Scientific doctrine and scientific sagacity are in today's world witnessing immense and drastic challenges.

15.25 GA, BIO-INSPIRED OPTIMIZATION, AND THE FUTURE OF CHEMICAL PROCESS OPTIMIZATION AND CONTROL

GA, bio-inspired optimization, and the future of chemical process optimization are witnessing major challenges and leading an arduous path of glory. The future of chemical process optimization lies in the hands of petroleum engineers and science of chemical engineering. The wide scientific endeavor, the progress of chemical engineering science, and the visionary path toward energy sustainability will lead a long and far-reaching way in the true realization of successful sustainable development. Scientific endeavor in the field of chemical and petroleum engineering science is witnessing widespread visionary development. The challenge, the vision, and the world of scientific validation are the hallmarks of a newer visionary tomorrow.

15.26 VISIONARY CHALLENGES AND SCIENTIFIC ENDEAVOR IN THE FIELD OF MULTIOBJECTIVE OPTIMIZATION

Science and engineering in today's world are moving very fast surpassing one paradigm over another. Multiobjective optimization and bio-inspired optimization are transforming the scientific horizon. Scientific endeavor and instinctive scientific introspection are the hallmarks of a newer scientific regeneration and a newer scientific understanding. Vision of science and technological validation today stands as a major hallmark in the true pursuit of science and engineering. Multiobjective optimization and bio-inspired optimization are the challenging domains of science today. The author with deep and cogent insight unravels the murky depths of human scientific challenges and presents before the scientific panorama the vision behind multiobjective optimization, GA, and its vast and versatile application in modeling and simulation of a petroleum refining unit. A scientist's vision and the world of difficult barriers are unfolding the far-reaching chapters of EMO in years to come. Human scientific challenges in today's human civilization are evolving into a newer visionary paradigm.

15.27 MODELING, SIMULATION, AND CONTROL OF PETROCHEMICAL PROCESSES

Petroleum refining and petrochemical processes are in today's world in the path of immense regeneration. Scientific vision and deep scientific introspection are the visionary coin words of today. Energy sustainability and global energy crisis are gaining ground with the passage toward a newer visionary tomorrow. The catastrophe of global energy scenario, the progress and vision of technology, and the wide and versatile technological horizon are all the pallbearers of a new generation of scientific hope and scientific determination. Modeling, simulation, and control of petrochemical processes and petroleum refining are the crux of future scientific endeavor and the future world of visionary research pursuit. Scientific validation and the world of scientific scale-up are the visionary roads of tomorrow. Mankind's prowess is in a state of immense disaster with the passage of history and time. In today's scientific world, optimization and control of a petrochemical process are opening new doors of innovation and new aisles of scientific vision and deep understanding.

15.28 SCIENTIFIC RESEARCH PURSUIT, THE CHALLENGES, AND THE ADVANCEMENT OF SCIENCE IN THE DOMAIN OF GA

Scientific research pursuit and its varied challenges are surpassing visionary boundaries day by day. The vision of science and its unimaginable advancements in the domain of GA are the hallmarks toward a newer realm and a visionary scientific horizon. In today's world, scientific advancements are veritably dependent on scientific validation and scientific scale-up. In such a situation, GA, bio-inspired optimization, and EMO are opening up and unfolding the hidden scientific truths of chemical process modeling and simulation. Advancements and achievements of science are inspiring and far-reaching in today's path of human civilization. Design of a fluidized bed reactor and the design of the riser reactor are the major concerns in the wide world of petroleum engineering science. The author, with deep details, relates the versatile research pursuit in the vast domain of chemical process modeling of riser reactor of an FCC unit with precision and in details.

15.29   RIGOR OF SCIENCE, THE WIDE VISION, AND THE FUTURE OF APPLICATION OF GA AND EMO

Rigor of science and the true vision of technological advancements are the torchbearers toward a newer paradigm and a newer optimism in the domain of EMO. Science and technology in today's world are witnessing unimaginable challenges. Technological vision is at its helm with the passage of history and time. Petroleum refining in a similar vein is crossing innumerable difficulties and challenges. Technological validation is on the other side of the coin with scientific vision on one side. Rigor of science needs to be envisioned at every step of human scientific research pursuit. Global energy crisis is in a state of immense difficulties and unimaginable barriers. The immediate need of the hour is wide vision and rigorous scientific validation in the attainment of energy sustainability. Scientific profundity and scientific discernment today stand in the midst of immense wisdom and deep introspection. Erudition and sagacity are the major torchbearers toward a greater emancipation of science and engineering in today's human scientific pursuit. The wide vision of science in today's scientific horizon is stunted with the passage of history and time. The scientific challenge needs to be revisited at each step of research foray.

Technological advancements, the rigors of science and technology, and the need for energy sustainability are all leading a long way in the true emancipation of energy engineering today. GA and EMO are rigorously changing the face of scientific research pursuit today. The challenge, the vision, and the visionary urge to excel are the torchbearers toward a newer visionary era of optimization science.

15.30   FUTURISTIC VISION AND FUTURE FLOW OF THOUGHTS

Petroleum refining is the backbone and the forerunner of technological vision and scientific grandeur today. Depletion of fossil fuel resources is challenging the wide scientific horizons of petroleum refining and the holistic world of petroleum engineering science. Futuristic vision in scientific endeavor in the field of petroleum engineering should be targeted toward the realization of energy sustainability. Scientific challenges, scientific cognizance, and scientific vision need to be rebuilt and reemphasized with each step of future life. Evolutionary computation today stands in the midst of deep scientific comprehension and wide introspection. The march of technology in the research areas of fossil fuel technology and science needs to be reenvisioned and reenvisaged. The answers to energy and environmental sustainability are vast and versatile with the forays into science and engineering. Evolutionary computation, GA, bio-inspired optimization, and multiobjective optimization are the torchbearers toward a greater visionary eon of applied mathematics and scientific computation. Futuristic vision and future flow of thoughts should be directed toward greater emancipation of petroleum engineering science and energy sustainability. This treatise gives vast and wide glimpses into the latent and unknown world of computation and nonrenewable/renewable technology. Windows of innovation and scientific instinct are wide open as science plunges forward toward unknown and murky depths.

15.31 FUTURE TRENDS IN RESEARCH IN EVOLUTIONARY COMPUTATION

Evolutionary computation is today an enigma of science. Technological barriers, scientific vision, and scientific objectives are the support and backbones of immense research pursuit today. The visionary path of evolutionary computation, the wide scientific urge to excel and devise, and the concerns of depletion of fossil fuel resources will lead a long way in the true emancipation of petroleum refining and petroleum engineering science today. Evolutionary computation has a wide research vision today. Robust modeling and the success of simulation science are rigorously changing the face of scientific validation. SA, GA, and bio-inspired optimization are the newer faces of scientific innovation today. These are the definite visionary endeavors of optimization science. In this treatise, the author widely observes the immense vision behind different and diverse areas of research in optimization science and petroleum engineering today.

15.32 CONCLUSION

Engineering science and computational science are witnessing immense scientific challenges and scientific regeneration. Progress of human civilization and scientific endeavor is reaching immense scientific pinnacles. The world of science and engineering is facing immense pragmatic challenges such as depletion of fossil fuel sources. In such a crucial juncture of scientific history and time, scientific vision in the field of chemical process engineering and petroleum engineering is of utmost need and of immense concern. Petroleum refining science needs to be revamped and rebuilt with the utmost target of furtherance of science and engineering. The author in this widely observed treatise pointedly focuses on the potential and subsequent success in the field of evolutionary computation, multiobjective optimization, and the wide world of GA. The wide windows of modeling and simulation are ushering in a new eon in the field of applied mathematics and the holistic world of chemical process engineering. Today, technology, whether it is mathematical computation or chemical process engineering, is conquering the scientific landscape. The vision of science, technological splendor, and the scientific urge to excel are changing the face of computer science and mathematical computation/evolutionary computation. In this treatise, the author willfully points out the success of research endeavor in the field of evolutionary computation with the sole target toward the furtherance of enigma of petroleum engineering science.

ACKNOWLEDGMENT

The author willfully acknowledges the contribution of the chancellor, vice-chancellor, faculty, and students of University of Petroleum and Energy Studies, Dehradun, India without whom this project would not have been complete. The author also wishes to gratefully acknowledge the immense support of Professor Santosh Kumar Gupta and Professor Vijay Parthasarathy who are the author's academic and doctoral supervisors. In the end, the author respectfully acknowledges the contributions of Subimal Palit, the author's late father, who was an eminent textile engineer and who taught the author the rudiments of chemical engineering.

KEYWORDS

- multiobjective
- optimization
- genetic algorithm
- petroleum
- refining

REFERENCES

1. Chang, J.; Meng, F.; Wang, L.; Zhang, K.; Chen, H.; Yang, Y. CFD Investigation of Hydrodynamics, Heat Transfer, and Cracking Reaction in a Heavy Oil Riser with Bottom Airlift Mixer. *Chem. Eng. Sci.* **2012**, *78*, 128–143.
2. Chen, G. Q.; Luo, Z. H.; Lan, X. Y.; Xu, C. M.; Gao, J. S. Evaluation the Role of Intra-particle Mass and Heat Transfers in a Commercial FCC Riser: A Meso-scale Study. *Chem. Eng. J.* **2013**, *228*, 352–365.
3. Zhang, J.; Wang, Z.; Jiang, H.; Chu, J.; Zhou, J.; Shao, S. Modeling Fluid Catalytic Cracking Risers with Special Pseudo Components. *Chem. Eng. Sci.* **2013**, *102*, 87–98.
4. Koratiya, V. K.; Kumar, S.; Sinha, S. Modeling, Simulation, and Optimization of FCC Downer Reactor. *Pet. Coal* **2010**, *52*, 183–192.
5. Souza, J. A.; Vargas, J. V. C.; Von Meien, O. F.; Martignoni, W.; Amico, S. C. A Two-dimensional Model for Simulation, Control, and Optimization of FCC Risers. *Am. Inst. Chem. Eng. J.* **2006**, *52*(50), 1895–1905.
6. Deb, K.; Pratap, A.; Agarwal, S.; Meyarivan, T. A Fast and Elitist Multiobjective Genetic Algorithm: NSGA-II. *IEEE Trans. Evolut. Comput.* **2002**, *6*(2), 182–197.
7. Bandyopadhyay, S.; Saha, S.; Maulik, U.; Deb, K. A Simulated Annealing-based Multi-objective Optimization Algorithm: AMOSA. *IEEE Trans. Evolut. Comput.* **2008**, *12*(3), 269–283.
8. Deb, K. *Multi-objective Optimization Using Evolutionary Algorithms: An Introduction;* Kanpur Genetic Algorithm Report No. 2011003, 2011.
9. Dagde, K. K.; Puyate, Y. T. Modeling and Simulation of Industrial FCC Unit: Analysis Based on Five-lump Kinetic Scheme for Gas–Oil Cracking. *Int. J. Eng. Res. Appl.* **2012**, *2*(5), 698–714.
10. Bhende, S. G.; Patil, K. D. Modeling and Simulation for FCC Unit for the Estimation of Gasoline Production. *Int. J. Chem. Sci. Appl.* **2014**, *5*(2), 38–45.
11. Ahari, J. S.; Farshi, A.; Forsat, K. A Mathematical Modeling of the Riser Reactor in Industrial FCC Unit. *Pet. Coal* **2008**, *50*(2), 15–24.
12. Bispo, V. D. S.; Silva, E. S. R. L.; Meleiro, L. A. C. Modeling, Optimization, and Control of a FCC Unit Using Neural Networks and Evolutionary Methods. *Engevista* **2014**, *16*(1), 70–90.

13. Rao, R. M.; Rengaswamy, R.; Suresh, A. K.; Balaraman, K. S. Industrial Experience with Object Oriented Modeling: FCC Case Study. *Chem. Eng. Res. Des.* **2004,** *82*(A4), 527–552.
14. Ziane, I.; Benhamida, F.; Graa, A. Simulated Annealing Optimization for Multi-objective Economic Dispatch Solution. *Leonardo J. Sci.* **2014,** (25), 43–56.
15. Elnashaie, S. S. E. H.; Elshishini, S. S. *Modeling, Simulation and Optimization of Industrial Fixed Bed Catalytic Reactors*; Gordon and Breach Science Publishers: USA, 1993.

CHAPTER 16

MATHEMATICAL MODEL OF CENTRIFUGAL SEDIMENTATION OF CORPUSCLES IN A CYCLONIC DEDUSTER

R. R. USMANOVA[1] and G. E. ZAIKOV[2,*]

[1]Department of Science, Ufa State Technical University of Aviation, Ufa 450000, Bashkortostan, Russia

[2]N. M. Emanuel Institute of Biochemical Physics, Russian Academy of Sciences, Moscow 119991, Russia

*Corresponding author. E-mail: GEZaikov@Yahoo.com

CONTENTS

ABSTRACT

In this chapter, developments on mathematical model of centrifugal sedimentation of corpuscles in a cyclonic deduster are presented in detail.

16.1 INTRODUCTION

Modern development of the industry is characterized by the increasing intensification of various processes that is accompanied by considerable emissions. The problem of trapping of dust touches practically all branches of modern manufacturing industry, that is, manufacturing of building materials (cement), metallurgy, a woodworking, the chemical industry, and many other industries. Annually, in an aerosphere of the earth, about 580 million tons of dust, carbon black, and ashes are thrown out.[1] Along with the demand of maintenance of the minimum dust emissions, the dust removal equipment should be quite reliable for a longtime maintenance and should not demand for the big expenses at installation and service.

The inertia dedusters, dry and wet cyclone separators, and mechanical and electric filters are now applied for clearing of gases of dust. In case of a high initial dustiness (more than 10–15 g/m), coarse clearing are usually applied, depending on specificity of local conditions and dry or wet cyclone separators.

The principle act of cyclone separators is based on centrifugal separation of a mixture consisting of dust and gas. The rotary motion is given to a dusty gas stream at an entry in the cyclone separator. A strong field of a centrifugal force of inertia, leading to sedimentation of corpuscles of dust, is created. An inertial force acting on corpuscles in the cyclone separator considerably exceeds gravitational forces; therefore, gabarits of cyclone separators is much less, than, for example, at dusty chambers where sedimentation occurs by gravity. In connection with toughening of demands to clearing of technological air, the trend to replace cyclone separators with fabric filters was outlined. However, thanks to variety of advantages, cyclone separators cannot be completely expelled now other than apparatuses of clearing of gases. Therefore, actually there is a problem in correct sampling of regimes, maintenance of already working cyclone separators, and working out of new models of cyclone separators with the higher technical data accommodated for concrete working conditions.

On a way to create more perfect cyclonic dedusters and assemblies there are certain difficulties caused mainly by absence of exact methods of forecasting of operational parameters of the future apparatuses (taking into account the concrete conditions of their work). Existing design procedures of parameters of cyclone separators are bulky and restricted. Therefore, their refinement and automation is necessary.

16.2 PRINCIPLE OF ACT OF A CYCLONIC DEDUSTER AND ITS KEY PARAMETRES

Wide use of cyclonic dust removal devices is defined by some of their advantages[1,2] with respect to other apparatuses of analogous applications:

1) relative simplicity of manufacturing and low cost (both at manufacturing and at maintenance);
2) possibility of functioning in the conditions of varying heat and pressure without any basic changes in designs;
3) possibility of trapping of abrasive turnings on protection of internal surfaces of cyclone separators by special coverings;
4) conservation of demanded level of fractional efficiency for clearing growth of mass concentration of a firm phase.

Nevertheless, there are the factors complicating extension of cyclonic dust removal devices. Despite active expansion of scope of wet clearing of gas, there are no data about full replacement of such cyclonic apparatuses with traditional dust removal equipment (wet filters and scrubbers) in at least one of the areas of the industry.[1,3]

Many authors[3-5] see principal cause of restricted use of such apparatuses as the absence of authentic methods of calculation of hydrodynamic processes in cyclonic dedusters and also the criteria of scale transition from laboratory models to plants.

The intensification of processes in cyclonic apparatuses leads to essential increase in a twisting of a stream that can cause stability loss for which disintegration of a symmetric whirlwind kernel and origination after area of disintegration of the indignant traffic in the form of single or double spiral whirlwinds is characteristic. Possibility of formation of a whirlwind is defined by value of a Reynolds number and twisting parameter. It is observed that the increase in a twisting of a stream is conducted to accelerate disintegration of a whirlwind kernel, thereby expanding boundary line of unsteady operation of the inertia apparatuses.

In modern exploratory works, in the theory and practice of the equipment of a gas cleaning, it is possible to note an extensive range of researches in this area that speaks about seeming simplicity of implementation of the twirled streams in plants. However in the majority works, the difficult structure of a real current[2] is ignored. Many researchers[4] are restricted to indicate only average characteristics of a stream which do not give representation about the space structures of a real current without which studying, it is impossible to define whether peak efficiency of process of separation is attained to find boundary lines of conversion zones and truly to size up new is constructive circuit solutions.

Thus, it is impossible to create a similarity theory of strongly twirled streams necessary for purposeful perfection of cyclonic apparatuses for a gas cleaning. This fact has led creation of a considerable quantity of inconsistent recommendations for sampling of design data of cyclonic apparatuses (twisting devices and setting sizes), which have been realized in various industrially released apparatuses.

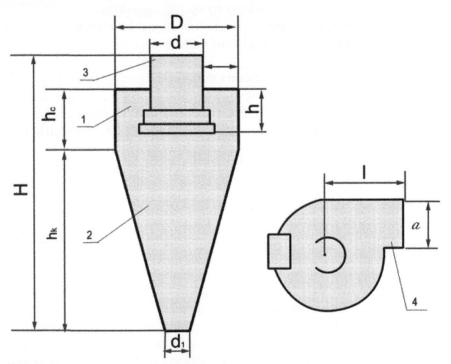

FIGURE 16.1 The constructive circuit design of a cyclonic deduster: (1) a cylindrical shell; (2) a conic shell; (3) a pipe exhaust; and (4) the tangential upstream end.

Recommendations about designing are based, as a rule, only on the experimental research data that narrows a range of their usability.

In the majority of researches of cyclonic apparatuses, agency of factors is sized up indirectly, for example, on clearing factor. Agency of design data of the cyclone separator on its parameters, first of all, on its efficiency is more low observed.

To the basic critical bucklings defining efficiency of a dust separation, the following Figure 16.1 refers to: diameter of a cylindrical part (D or $2R$), its altitude (H_u), altitude of a conic part (H_κ), the square and the inlet opening form (a relationship of altitude of an entry h_{ex} and width b_{entry}), diameter (d) and extent of a depth of the exhaust tube in the cyclone separator (h), and an angle of feeding of gases into the cyclone separator. Sampling of a rational slope of the upstream end allows to raise efficiency of clearing of gases and to lower the aerodynamic resistance of an apparatus a little.

In the capacity of explanatories (Table 16.1) are resulted in drawing the generalized values of the recommended parameters[3] which range of spread testifies to trouble of a current situation.

TABLE 16.1 Recommendations for Sampling of Design Data.

Parameter	Aspect	Value
Relative width of the upstream end	b/D	0.05–0.35
Relative diameter of the upstream end	d/D	0.15–0.75
Relationship of sizes of the upstream end	h/b	1–6
Relationship of the square of an entry and exit	A_{en}/A_{ex}	0.6–2.5
Relative length of the exhaust tube	L_{ex}/D	0.5–1.8
Relative length of the case of the apparatus	l/D	1.5–5.5
The relation of a conic part to length of the case	L_{con}/l	0–1

The similar situation is observed in industrial production where cyclonic dedusters are produced in the conditions of individual exhaustion since for them the manufacturing methods from the sheet metal rolling, widely used in case of simple cyclone separators, are not possible.

Following problems have been put in-process:

1. to build mathematical model of a dust separation in the cyclonic apparatus,
2. to plan and make numerical experiment,
3. on the basis of the numerical data of experiment, to build the functional dependences presenting process of a dust separation.

16.3　SUBSTANTIVE PROVISIONS OF MATHEMATICAL MODEL OF A DUST SEPARATION

Basis of mathematical model of process of a dust separation in the cyclone separator is studying of mechanical trajectories of separate corpuscles of dust and research of dependence of these paths from the most essential factors. Calculation of paths of corpuscles will allow to find conditions of their trapping and quantitatively to define agency of essential factors on dust separation process.

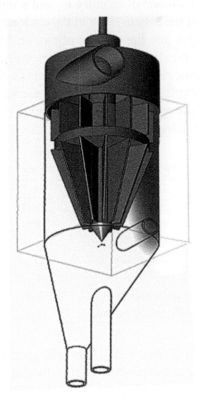

FIGURE 16.2　The geometrical model of a cyclonic deduster.

The principle act of a cyclonic deduster consists in the following. The dusty gas stream is inducted into the cyclone separator so that to give it a rotary motion (Fig. 16.2). It becomes or by means of tangential (simple or screw) an entry, or by means of a profile of the shovels twisting a stream. This twirled stream enters into the cyclone separator, as a rule, develops in it, and gets out through the symmetric exhaust tube. The corpuscles of dust which are in a stream, under the influence of an inertia centrifugal force, are kick to walls of the cyclone separator and on spiral paths slip through an exhaust outlet in the loading pocket.

During construction of mathematical model following basic assumptions have been made:

1.   gas is considered ideal and incompressible liquid,
2.   the gas stream is symmetric concerning an axis and is stationary,
3.   the corpuscle is considered a sphere of small diameter that allows to use the Stokes' formula for calculation of force of resistance,
4.   the density of corpuscles considerably exceeds gas density, therefore, it is possible to neglect Archimedes force,
5.   concentration of dust is low; hence, it is possible not to consider interaction of corpuscles,
6.   the corpuscle which has attained a lateral wall of the case of the cyclone separator is considered trapped.

The forces acting on a corpuscle at a gas rotary motion in a curvilinear stream are presented in Figure 16.3.

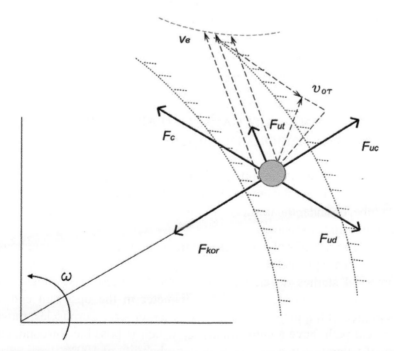

FIGURE 16.3   The forces acting on a corpuscle in a gas stream.

Because of twirl of a cleared stream in the cyclone separator, the field of the inertia forces which leads to separation of a mix of gases and corpuscles is created. Therefore for calculation of mechanical trajectories of corpuscles, it is necessary to know their equations of traffic and aeromechanics of a gas stream in the cyclone separator. According to the assumption of small concentration of dust, agency of corpuscles on a gas stream it is possible to neglect. Hence, it is possible to observe traffic of a separate corpuscle in the field of speed of a gas stream. Therefore, the problem of definition of paths of corpuscles in the cyclone separator decomposes on two:

1. definition of a field of speed of a gas stream,
2. integration of the equations of traffic of a corpuscle with the account of a settlement field of speed of gas.

The assumption about a rotational symmetry of an observed problem (except for an inlet opening) allows consideration of traffic of corpuscles cylindrical axes.

The greatest difficulty is represented by fine dust trapping for which force of resistance with sufficient accuracy is computed by Stokes' formula; these reasons cause the third assumption. At increase in dustiness, the factor of clearing of the cyclone separator grows;[3] therefore, calculation of parameters of the cyclone separator at a small dustiness (according to an assumption 5) secures its minimum efficiency.

16.4   DEFINITION OF MECHANICAL TRAJECTORIES OF CORPUSCLES OF VARIOUS DIAMETERS

At the analysis of results of traffic of corpuscles in a gas bottle, it is necessary to install, whether the corpuscle will be taken out by a gas stream in the exhaust tube, or under the influence of an inertial force will move to walls of the apparatus and is separated.

In Figure 16.4, paths of corpuscles of the various diameter counted on the offered model are presented.

Numerical studies developed for the most adverse case of an initial arrangement of corpuscles of the set diameter in the upstream end that allows to predict trapping of a corpuscle of larger size. Corpuscles, moving on a critical path, have a cutoff diameter which is possible for conducting process of separation in a gas bottle with probability of 100%.

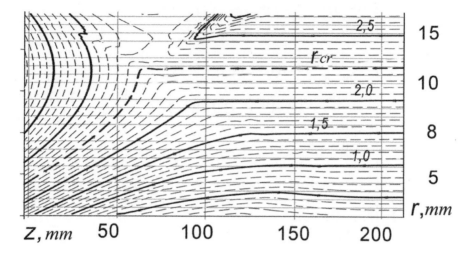

FIGURE 16.4 Paths of corpuscles $d \leq 5$ mic: the critical and counterbalanced.

The system of the equations of traffic of a corpuscle switches on sharply excellent values of derivatives, that is, a rigid system.[7] The aspect of a matrix of system of the equations of the first approach (16.1) is caused by number 10^5:

$$
\begin{pmatrix}
-18C_r - \dfrac{V}{r} & -\dfrac{U}{r} & 0 & -18\dfrac{C_p C_r}{r^2} - \dfrac{UV}{r^2} & 0 \\[2ex]
\dfrac{2U}{r} & -18C_r & 0 & 18C_p\dfrac{dC_z}{dr} - \dfrac{U^2}{r^2} & 18C_{rt}\dfrac{dV_z}{dz} \\[2ex]
0 & 0 & -18C_r & 18C_p\dfrac{dW_z}{dr} & 18C_r\dfrac{dW_z}{dz} \\[2ex]
0 & 1 & 0 & 0 & 0 \\[1ex]
0 & 0 & 1 & 0 & 0
\end{pmatrix}
\tag{16.1}
$$

For the solution of system (16.1), the implicit method of the solution providing wide limits of stability of computing operations has been used. On n integration step value of a variable, we will mark out as x^n, and on $(n + 1)$ step we will mark out x^{n+1}, as a result we will gain difference analogue, which after integration by Euler's implicit method will become:

$$
\begin{cases}
\dfrac{U^{n+1}-U^n}{\Delta t}=18C_p\left(\dfrac{C_r}{r^n}-U^{n+1}\right)-\dfrac{U^{n+1}V^n}{r^n} \\[2ex]
\dfrac{V^{n+1}-V^n}{\Delta t}=18C_p\left(V_r\left(r^n,z^n\right)-V^{n+1}\right)+\dfrac{\left(U^n\right)^2}{r^n} \\[2ex]
\dfrac{W^{n+1}-W^n}{\Delta t}=18C_p\left(W_r\left(r^n,z^n\right)-W^{n+1}\right) \\[2ex]
\dfrac{r^{n+1}-r^n}{\Delta t}=V^{n+1} \\[2ex]
\dfrac{z^{n+1}-z^n}{\Delta t}=W^{n+1} \\[2ex]
\dfrac{\varphi^{n+1}-\varphi^n}{\Delta t}=\dfrac{U^{n+1}}{r^n}
\end{cases}
\tag{16.2}
$$

Because of irregularity of distribution of speed of corpuscles on apparatus altitude, the integration step was chosen automatically. The iterative formula of an implicit method of Euler, taking into account (16.1) for definition of paths of corpuscles, will register as:

$$
\begin{cases}
U^{n+1}=\left(U^n+18C_p\Delta t^n/r^n\right)\big/\left(1+\left(18C_p+V^n/r^n\right)\Delta t^n\right) \\[2ex]
V^n=\left(V^n+\left(18C_pV_r\left(r^n,z^n\right)+\left(U^n\right)^2/r^n\right)\Delta t^n\right)\big/\left(1+18C_p\Delta t^n\right) \\[2ex]
W^n=\left(W^n+18C_pW_r\left(r^n,z^n\right)\Delta t^n\right)\big/\left(1+18C_p\Delta t^n\right) \\[2ex]
r^{n+1}=r^n+V^{n+1}\Delta t^n \\[2ex]
z^{n+1}=z^n+W^{n+1}\Delta t^n \\[2ex]
\varphi^{n+1}=\varphi^n+U^{n+1}\Delta t^n/r^{n+1} \\[2ex]
\Delta t^{n+1}=C\big/\sqrt{\left(V^n\right)^2+\left(W^n\right)^2}
\end{cases}
\tag{16.3}
$$

The calculation program on Euler's method of paths of corpuscles which builds paths of corpuscles depending on the set characteristics of process of a gas cleaning has been developed. Defining paths of corpuscles, it is possible to find efficiency of their trapping in the given apparatus.

The solution lapse was sized up by formula:

$$
\varepsilon_r=\left|\frac{\partial r}{\partial(1/N)}\right|\frac{1}{Nr}+\left|\frac{\partial r}{\partial h}\right|\frac{h}{r}+\left|\frac{\partial r}{\partial \Delta t}\right|\frac{\Delta t}{r}.
\tag{16.4}
$$

$\delta, \text{M}\kappa\text{M}$	0.5	1.0	2.0	3.0	5.0	10
$\Delta t_\delta / \tau = 1/(4\delta)$	1/2	1/4	1/8	1/12	1/20	1/40
ε		0.26	0.13	0.064	0.040	0.025	0.013

For corpuscles with a size of δ from 2 to 10 μm, the lapse of scalings is in limits from 6.4% to 1.3% that is comprehensible to technical calculations. For small corpuscles with δ from 0.5 to 1.0 μm, raise of a relative lapse to 2.6% which if necessary can be reduced is admitted. Advantage of such method of designing consists that the results gained on the counting machine, it is possible to compare to the data of commercial tests of each of maintained gas bottles. It gives the chance to develop the program at each stage of use. It is the most important to inoculate the program so that to achieve more exact forecasting of characteristics of installation of a gas cleaning:

1) to predict efficiency of a deduster for more wide range of conditions, than it is characteristic for industrial practice, and precisely to size up a lapse;
2) to provide possibility of inexpensive optimization of such variables separated by change such as geometry of the apparatus and dustiness of passing gas.

16.5   OPTIMIZATION OF THE CYCLONIC APPARATUS

Sampling of cyclonic apparatuses for gas cleaning systems is influenced by demanded efficiency (it is defined by clearing factor), a water resistance, metal consumption, gabarits, properties of dust (abrasivity, stickiness), etc.

First of all, parameters of the cyclonic apparatus should match to properties of a trapped dust. With other things being equal, diameter of trapped corpuscles d_p will be the less, than the charge G will be more. Therefore, sampling of such regime of maintenance of the cyclonic apparatus at which its greatest possible loading would be provided is desirable. However, at excess of charge $G < 2\,G$ above some threshold, efficiency of trapping ceases to increase.[7] For the cyclonic apparatuses produced in lots, optimum speed of traffic of gases[6] if it is unknown, the maximum speed in cross section should be accepted in the interval of 3.5–4.5 mps for cylindrical apparatuses and 1–2 mps for conic apparatuses is experimentally installed.

Entrance speed of dusty stream it is recommended to choose of 10–30 mps from an interval, smaller values match to highly effective cyclonic apparatuses.

Sampling of sizes of a conic part of the case

By consideration of traffic of corpuscles in a gas bottle, the assumption that the paths of corpuscles of size from 0.5 to 1.0 μm coincide with streamlines of the bearing medium was accepted. Such assumption practically does not influence the form of a curve and only slightly raises a design value d at comprehensible convergence with experiment on integrated values of efficiency of separation.

Let's write down a condition of balance of corpuscles:

$$(\rho_\rho - \rho)\frac{\pi d_\rho^3}{6}\cdot\frac{v_\varphi^2}{r} = \frac{24a}{Re}\cdot\frac{\pi d_\rho^2}{4}\cdot\frac{\rho v_r^2}{2} \qquad (16.5)$$

$$a = \frac{1}{0.848.1g\dfrac{\xi}{0.057}},$$

where α—factor of the form of a corpuscle:[6]

$$Re = \frac{v\cdot d\cdot p}{\mu}$$

Re—a corpuscle Reynolds number;
D—equivalent diameter of a corpuscle.

Let's write down a condition (16.5) in a following aspect:

$$\frac{(\rho_r - \rho)d_\rho^2\cdot\rho_\varphi^2}{18\mu ra} = \frac{v_z}{\rho_+} \qquad (16.6)$$

Let's inject two scale magnitudes: radius $r_я$, in the capacity of scale of length and scale of speed v gas. From 16.6 we will gain

$$\frac{(\rho_r - \rho)d\rho^2\cdot v^2\varphi}{18\mu ra} = \frac{v_\varphi r v_z}{r v_\varphi^2} \qquad (16.7)$$

The left side of an eq 16.7 represents the Stokes' number whose factor is the parameter of a twisting depending in the conditions of self-similarity of speed only from relative sizes of the apparatus.

The right side of an eq 16.7 depends on relative radius of a whirlwind kernel and from relative radial coordinate $r_{min.}$

As a result of the spent analysis, it is possible to draw following leading outs:

To increase in radius of an axial zone and a share of a cylindrical part in the general ratios of the apparatus, there is an increase in efficiency of clearing;

With growth α_0 efficiency of trapping of small corpuscles decreases that especially it is appreciable at $r_0 > 0.2$;

With an elongation of a conic part of the apparatus at constant altitude of its flowing part, efficiency of clearing increases.

Hence, for achievement of peak efficiency of trapping of small corpuscles, the share of a cylindrical part in gas bottle total length should be the least and a share of a conic part be the greatest. The cone angle is recommended to be accepted in limits 15–20 as its further decrease leads to essential increase in altitude of a gas bottle. Other geometrical relationships are recommended to be chosen from a condition $r_0 < 0.2$.

On the developed relationships, the industrial cyclonic deduster has been counted. The affinity of recommended values to the nominal parameters known from references testifies to legitimacy of the stated recommendations.

16.6 CONCLUSIONS

1. The analysis of a current state and prospects of development of the equipment of a gas cleaning has shown that in the chemical industry and manufacture branches contiguous with it, apparatuses of a gas cleaning with the twirled traffic of the disperse medium whose efficiency is completely defined by hydrodynamic perfection of process of separation are used.

2. The algorithm of modeling process of separation of a dispersoid in a gas stream in program complex Ansys CFX has been developed. At the analysis of results of traffic of corpuscles in a deduster it is installed, whether the corpuscle will be taken out by a gas stream in the exhaust tube, or under the influence of an inertial force will move to walls of the apparatus and is separated.

3. Ordering and refinement of recommendations about designing of geometric proportions of the setting of a cyclonic deduster, and also from the point of view of optimization of hydrodynamic conditions of its work, are executed. It is observed that with an elongation of a conic part of the apparatus at constant altitude of the setting, efficiency of clearing increases.

KEYWORDS

- **cyclonic deduster**
- **critical bucklings**
- **gas cleaning**
- **optimization**
- **Ansys CFX**
- **path of corpuscles**

REFERENCES

1. Barahtenko, G. M.; Idelchik, I. E. *Industrial and Sanitation Gas;* Moscow: Chemistry, 1984; pp 34–52.
2. Straus, V. *Industrial Cleaning Gases;* Moscow: Chemistry, 1981; pp 89–95.
3. Tarasova, L. A Hydraulic Calculation of the Resistance of the Vortex Unit. *Chem. Pet. Mech. Eng.* **2004,** *4,* 11–12.
4. Shilyaev, M. I. *Aerodynamics and Heat and Mass Transfer of Gas-dispersion Flow: Studies Allowance;* Tomsk: Industrial Engineering, 2003; pp 23–30.
5. Artyukhov, A. E. Computer Simulation of Vortex Flow. *Hydrodyn. J. Manuf. Ind. Eng.* **2013,** *12* 25–29.
6. Deutsch, M. E.; Filippov, G. A. *Gas Dynamics of Two-phase Media;* Moscow: Energy, 1988; pp 16–24.
7. Usmanova, R. R.; Zaikov, G. E. *Clearing of Industrial Gas Emissions: Theory, Calculation, and Practice;* AAP: Canada, U.S., 2015; pp 112–130.

CHAPTER 17

MACROMOLECULES VISUALIZATION THROUGH BIOINFORMATICS: AN EMERGING TOOL OF INFORMATICS

HERU SUSANTO[1,2,*] and CHING KANG CHEN[3]

[1]Department of Computer Science & Information Management, Tunghai University, Taichung, Taiwan

[2]Computational Science, Indonesian Institute of Sciences, Serpong, Indonesia

[3]School of Business and Economics, University of Brunei, Bandar Seri Begawan, Brunei

*Corresponding author. E-mail: heru.susanto@lipi.go.id, susanto.net@gmail.com

CONTENTS

ABSTRACT

Bioinformatics supports biologists and scientists to identify more quickly, easily, and most importantly they are able to browse through the web to search appropriate tools or software for their work fields. In a way, it saves times and the process can be made in such a short time without having to wait for years for the analysis to complete. A comprehensive database and software saves time and efforts in searching the appropriate tools to be selected.

17.1 INTRODUCTION

17.1.1 INTRODUCTION TO BIOINFORMATICS

17.1.1.1 DEFINITION OF INFORMATICS

It refers to the study of the science of processing, gathering, storing, retrieving, and organizing the information or data used. In this era of advanced technology, it continually becomes more and more powerful in terms of their input of technology which has enabled data to be processed in just a few seconds. If technology has not been advanced, it may take several years to process a data which is very time consuming to all users.[10] Bioinformatics is basically the compilation and management of sciences, data analysis in genomics, proteomics, and has a wider field area in biological sciences. The tools or the software used in the bioinformatics include the analysis of quality check, trimming, base calling, and assembly as well as gene prediction.[6]

17.1.1.2 COMBINATIONS OF INFORMATICS AND COMPUTER SCIENCE

Computer technology is not only being used by common workers such as accountants, lawyers, secretaries, teachers, or any other private or government firms but it is also being widely used by the researchers especially those who are working in the laboratory fields. There are a few examples of computer science that uses the informatics data such as bioinformatics, biomedical informatics, chemoinformatics, ecoinformatics, geoinformatics, health informatics, neuroinformatics, social informatics, and veterinary informatics.[10]

These lab researchers uses computer informatics to process their lab work or the lab data. They not only work in lab with hands on tube or thermometer doing experiments but they also need computers in order to store their findings as well as to compute their end results. It really helps the lab researchers to store their data in order for the data to be used in the future, understand more of their studies, be able to perform correctly with the correct use of data, and be able to determine the results in a split second.[10]

17.1.1.3   BIOINFORMATICS

Claverie and Notredame[14] have stated that it is commonly known that highly trained professionals used computers to carry out their tasks. However, biologists also use the computers as much as they do. Computers are not only used for writing down some memos and sending out e-mails but biologists also use computers to deal with very specific problems. As a whole, these tasks are called bioinformatics; simply referred as molecular biology's computational branch.[14]

According to Al-Ageel et al., bioinformatics is an interdisciplinary methodology that makes use of gathered information and displayed data to analyze biological data.[15]

The use of computer appliances helps the biologists to analyze more data easily, to store as much as data or findings, do a wide research regarding their tasks as well as to retrieve any of the biological information. The study of bioinformatics basically involves a biologist. In a way, they use computers to analyze a DNA sequence as well as to retrieve advanced study of the molecular genetics.[8,10]

17.1.1.4   EXAMPLES

One of the accomplished users in the bioinformatics field is the human genome. Scientists use the computer to process and to map down the understanding of the aspects of human life. For example, a scientist or even a normal human being is unable to explain why children resemble the face of their parents? In what way the resemblance came from? But now scientists are able to identify and predict how a child will look like by determining the different DNA, height, or even the hair color. Apart from looking at human genes, scientists are able to perform more advanced study in terms of finding a cure to diseases in the human body. Sometimes diseases can be inherited from past generations to their generations. So lab scientists are able to identify the cure to the diseases to prevent it from being inherited to the next generations. In this way, it really helps the generations to stay healthier and to prescribe the medicine as what has been consulted by the doctors.[10]

17.1.2   PAST, PRESENT, AND FUTURE OF BIOINFORMATICS

17.1.2.1   PAST

The Chinese merchants during the 14th century used a form of biometrics technology for identification purposes. They took the children's fingerprints by using ink for identification. This was found by a European explorer named Joao de Barros.

In 1890, the technology of body mechanics and measurements was studied by Alphonse Bertillon. The police used the "Bertillonage method" because it could help to identify criminals. Unfortunately, the method was no longer in use because there had been a false identification of some subjects. The method was removed in favor of fingerprinting and they brought back the method used by Richard Edward Henry of Scotland Yard.

By 20th century, Karl Pearson from the University College of London had made an important finding in the field of biometrics. He studied about the correlation and statistical history which he referred as animal evolution. His works include the method of moments, chi-squared test, correlation, and Pearson system of curves.

In year 1960s and 1970s, signature biometric authentication procedures were advanced. However, the biometric field remains firm until the security agencies and military further developed and investigated the biometric technology more than just the fingerprints recognition.

17.1.2.2   PRESENT

To get the Food and Drug Administration (FDA) into approving a drug requires the protracted process which averagely takes 12 years between the lead identification and the approval of the FDA. The FDA approves the existing drug and uses it to treat different diseases other than what is intended. For example, the original purpose of viagra was to treat the heart disease, whereas it was repurposed to treat the erectile dysfunction. This saves time and money.[20] Different approaches need to be taken into account in order to repurpose the drug. One of them is to compare the signature of molecular in disease and the signatures observed in cells of the animal, or the people that have been tested with different drugs. If there are anticorrelated signatures found, the drug's administration for that disease can at least lessen the symptom or they can even give treatment to the condition. One of the examples of computational approaches is the one that is used by Sirota et al. (2011). They use it to discover the antiulcer drug cimetidine as a means to cure lung adenocarcinoma and approve the off-label usage in vivo through an animal model of lung cancer.[20]

Other than using the computational methods that have been mentioned above, there is also the experimental approach to drug repositioning. One of the examples is when Nygren et al. was screening 1600 known compounds to be compared with two different colon cancer cell lines by using the connectivity map (CMap) data in order to analyze their findings into more details and identify mebendazole that has the ability to have therapeutic effect in colon cancer.[20]

17.1.2.3 FUTURE

The system in biometric technology consists of verification and authentication. Cognitive biometric system was developed to use brain response, stimuli, face perception, and mental performance for search of high-security area.[1] This technology has been advanced so that it has created a variety of recognition systems. For example, biometric strategies such as retinal scan, face location, hand geometry, iris scan, DNA, earprint, and fingerprint are developed. In the coming future, this biometric technology will be the number one choice to solve any threats in the world and able to increase the management of information security.

17.2 HOW BIOLOGISTS USE BIOINFORMATICS

17.2.1 ANALYZING GENE TO PROTEINS

Antonino et al.[4] have mentioned that the working molecules of a cell is symbolized by the protein, yet studying the proteins' functions is not sufficient to comprehend the cell machinery completely. This is because it is more essential to have the interactions among the proteins as the cell's biological activity cannot be identified by the functions of the proteins. The interactions between a group of proteins help to control and to give support among the proteins during the biological activities, and this is called as a protein complex; a functional module of the cell. RNA polymerase and DNA polymerase are the examples of the protein complexes.[4,18]

According to Buehler and Rashidi,[5] the word "bioinformatics" is referring to organizing the work, analyzing them as well as predicting the complex data that have arose from the modern molecular and biochemical techniques as living organisms, and the different human studies life are very complex. However, the meanings tend to vary from one's point of view to another. Some define it depends on the information flow concept in the systems of biological refer to the transmission of genetically encoded information transmitted from genes to proteins, from the blueprint to the machinery of life. The bioinformaticians try to comprehend the saying that the gene codes for the physiological characteristics for the disease. Bioinformaticians can assist in explaining the genetic complexity mechanism and the evolutionary relationships among the organisms via the creation, annotation, and biological databases' mining. Bioinformatics is relying on the assumption that quantifiable relationships are not only present between the genes activity and their existence inside the genome but also between the genes sequence and the proteins structure and role.[16] The techniques of database mining are used by the bioinformaticians in order to study the protein complexes, pathways of metabolic, and the networks of the gene.[5]

Buehler and Rashidi[5] also stated that gene sequence and proteins structures are the important elements of bioinformatics. Genes are the units of life that are heritable and it is history's only molecular source of information, that is, the genomes of contemporary organisms. It is very important for the practicability of the organism to replicate the genetic information from one generation to other generations correctly. It is also crucial for the biological evolution, for example, for the alteration of organisms in adapting the changes in environment. Mutation can cause disease or even death of carriers. New traits are also found in some cases which will provide advantages in the reproduction of the organisms. Some of the mutations are selected to be retained naturally while majority of the mutations are removed, therefore, there are differences in our genomes that form a species. This diversity is characterized by the bioinformatics by comparing the genetic sequences and protein structure directly.[5]

Buehler and Rashidi[5] have mentioned that one of the uses of bioinformatics is to study evolution. As evolutionary changes are difficult to be studied in real time, bioinformatics uses systems/scientific study such as the Linnaean *taxonomic system* to enable to name and rank as well as to classify the organisms *paleontologically* to study the fossil records. Some micro-evolution such as the rapid mutation rate of the *human immunodeficiency virus* (HIV) and *influenza* (flu) can be studied in real time and thus, helps the viruses in escaping the immune detection.[5]

According to Buehler and Rashidi,[5] studying the mutation in living things helps a better understanding of the pathogenic diseases and rising infection. Mutations allow the molecular biologists to track the traits and record the genomes as they play role as the genetic markers. Geneticists use animal models in medicine to analyze and cure the human disease by testing toxicity and effectiveness of the new drugs. However, animal models cannot represent the human subjects in studies such as the human behavior.[5]

Buehler and Rashidi[5] claimed that it is crucial to comprehend the source, quality, and biological significance of the data which relies on the genes and proteins sequences and structures. Bioinformaticians use raw data to discover the connection between the genes, genomes, and proteins. The raw data used are the nucleic acid sequences and protein structures. For correctly interpreting their biological significance, the quality and the precision of the data is essential. For example, when the scientists assess the mutation of genes that is engaged in the cancer, in order to obtain the sequence of information in coding the protein and regulating the component, it is important to clone the genes that are involved in the cancer. The *restriction fragment length polymorphisms* (RFLPs) are used to discover the chromosomal DNA fragment in the individuals that have an effect on. In cloning a gene, a fragment that is not included in the genome will be put into a small customized vector DNA to get its nucleotide sequence, as inserting the secluded fragment into the small customized vector DNA will be resulting in recombinant DNA. Another technique to amplify the DNA is by using the *polymerase chain reaction* (PCR).[5]

Bioinformatics technology has successfully created opportunities for biologists in shaping modern biological research. This is true according to Charles Darwin's Theory of Evolution. The cloning of human DNA can be taken by PCR by the most simplest and effective method, and it also prevents the difficulty of obtaining large amounts of human tissues for the purpose of DNA sequencing. Every sequence of dozens of building blocks has the ability to encode genetic information. The DNA can also be used for studying gene expression and synthesize large amounts of protein for analysis. Another use of bioinformatics technology is that it enables to understand human responses when taking drugs, which is so-called pharmacogenomics. It can be noticeable to those who are susceptible to drugs.[5]

In addition, biometrics technology provides better understanding on the relationships between genes and diseases, such as potassium channel mutations, and heartbeat regulation, for example, long QT syndrome.[2] As more genome projects or the full DNA sequence was completed, more information was obtained. The information then transformed from the molecular biology laboratory into computer labs, which in return, making the information more accessible and free from public databases.[3] This will open up the role of genetic networks such as embryonic development, the structure and the function of proteins, memory, and aging.[5]

17.2.2 SOFTWARE CATEGORY

BioInfoKnowledgeBase (BIKB) is one of the data used to assist the researchers with numerous tools, software, methods, and databases in biological information. It is easily accessible for the biologists to carry out their research since the databases provide a wide range of biological information such as the nucleotide, protein, bacteria, cyanobacteria, fungi, and other biological terms.[12]

A few samples of the tools or databases used at the BIKB under the software category are as follow:

17.2.2.1 GENOME ANNOTATION

Genome annotation refers to the process of locating genes and genome coding to determine what each gene can do. The genome annotation uses tools such as Apollo genome annotation, ASAP, COFECO, GoGene, IBM Genome, TIGR software tools, KAAS, KOBAS, MADAP, MICheck, Sequin, and a lot of tools are used in the genome annotation software. Under the genome annotation, another subcategory can be found such as gene prediction, genome assembly, and ontology analysis.[11]

17.2.2.2 GENOME ANALYSIS

Genome analysis compare features such as the DNA sequence, structural variation, gene expression which uses a few of the sample tools such as bioconductor, bioDAS, bioverse, CARGO, BRIGEP, IslandPath, sockeye, and taverna which can be found under the use of genome analysis. The subcategories under genome analysis are metagenome analysis, genome alignment, and genome browser.[11]

17.2.2.3   GENE ANALYSIS

Gene analysis is the overall study of genetics and molecular biology which involve tools such as WebGestalt, PANTHER tools, GeneTrail, GeneCodis, COFECO, ToppGene Suite, endeavor, babelomics, MirZ, and the SVC. The subcategories are gene regulation, transcription element analysis, splicing, expression analysis, and pathway analysis.[11]

17.2.2.4   NUCLEOTIDE ANALYSIS

Nucleotide analysis is the process of imperiling the DNA and RNA to a wide use of analytical methods in order to understand each of their features. Tools used such as AliasServer, AMOD, BioMart, DINAmelt, feature extract, gendoo, FIE2.0, gene set builder, genoCAD, BCM search launcher sequence utilities, MKT, UGENE, virtual ribosome, and the Pegasys; work-flow management for bioinformatics. The subcategories are motif identification, nucleotide alignment genome alignment, SNP identification, and the structure and sequence detection.[12]

17.2.2.5   PROTEIN ANALYSIS

Protein analysis is the process of determining the different types of acids in the protein structure especially occurring in the human backbone. The biologists use a number of tools such as the AACompldent, InterProScan, and the masssearch. The subcategories are 3D-motif analysis, secondary structure, transmembrane segment, modeling and simulations, structure visualization, and the functional pattern recognition.[12]

17.2.2.6   EVOLUTIONARY ANALYSIS

Evolutionary analysis is basically identifying the gene family, gene discovery as well as the genetic diseases origins. It uses tools such as the Gblocks, MEGA, EEEP, BAMBE, Arlequin, Joes site phylogeny programs,

Weighbor, tree editors, signature, puzzleboot, POWER, orthologue search service, and the OGtree.[11]

17.2.3   DATABASE CATEGORY

The databases categories are as follows:

17.2.3.1   NUCLEOTIDE DATABASES

Nucleotide databases are basically the collection of several sources such as the GenBank, RefSeq, TPA, and PDB which are involved in the gene, genome, and the biomedical research data. It uses tools such as the genome database, genome database for *Rosaceae*, HGVbase, MGIP, projector 2, and the TIGR software. It is divided into subcategories of sequence submission and retrieval system and the genome sequence databases.[12]

17.2.3.2   PROTEIN DATABASES

Protein databases use a three-dimensional structural data of proteins and the nucleic acids. It uses tools such as SWISS-2D PAGE, SCOP, MPDB, HSSP, FSSP, DSSP, and BioMagResBank.[11]

17.2.3.3   BACTERIAL DATABASES

Bacterial databases basically store and analyze the bacterial isolate which is linked to isolate records which uses tools such as database of magnetotactic bacteria, database of natural luminescent bacteria, extra train, ICB database, HOBACGEN; homologous bacterial genes database, and the VFDB.[11]

17.2.3.4   CYANOBACTERIAL DATABASES

Cyanobacterial databases are the database used to collect all cyanobacteria for photosynthesis such as the information, gene information, gene annotations, and the mutant information. It uses tools such as cTFbase, Cyano-Base, CyanoClust database, CyanoData, CyanoMutants, Cyanosite, and the SynechoNET.[11]

17.2.3.5 FUNGAL DATABASES

Fungal databases are where fungal organisms are kept to be used for further research in the medicine, agriculture, and the industry. It uses tools such as the Cortbase, UNITE, phytopathogenic fungi and oomycete database, MycoBank, MycoRec, and the dimorphic fungal database.[11]

The databases and the software as stated above consists of more than 1000 bioinformatics tools which are widely accessible to biologists or the researchers to carry out their extraction and data analysis. Database contains six main sections, "Home," "About Us," "Software," "Databases," "Search," and "Team." These sections show a brief information about the use of database, the role of the institute, search findings, and present the group who developed this database.[12]

17.3 COMPUTATIONAL TOOLS

Computers and computational methods now have a very important device that can be used for scientific applications; thus, these methods are playing a vital role in sciences. The applications are covering the use of computers in molecular biology which is called the classical bioinformatics to complex physiological systems, that is, numerical models. The number of advanced computing tools has been increased therefore this enables the models of physiological system to be more developed with high in details.[13]

Buehler and Rashidi[5] stated that science experiments are involved around test tubes, liquids, and microscope. Experimenting is examining one's idea that involves observing set of samples as well as testing it in a certain period of time. Clearly, planning and organization are needed in testing the idea.[5]

This is where the computers come in handy. It is mentioned by Buehler and Rashidi[5] that the computers are essential in an experiment in terms of designing, executing, and analyzing. Computers help in modernizing almost all activities in the laboratory such as the use for word processing, analyzing with spreadsheet, and accessing the internet. Scientists are able to independently carry out their work with the high in computational power.[5]

Buehler and Rashidi[5] claimed that the computers help the scientists in calculating and measuring the cells as well as recording the cells activities. Sophisticated computers that are equipped with cell-sorting machine, video cameras, and digital oscilloscopes help the scientists to accurately record the data from the experiment. Large number of data can be controlled, performed, manipulated, and stored in the computers.[5]

Processors are needed to be inserted in the machine in this modern science, for example, with the spectrophotometers which are used in assessing the absorption of the light. Immediate experiments are allowed by the computer control in order to perform the experiments, for example, in observing any chemical composition changes in the test solution. The small window controls the microprocessor by putting one or more codes or command text that is allowed to be typed in on view from a short menu. The accurate experiments are not only caused by the computers but also depend on the use of the quality of the instrument.[5]

Before the introduction of computers, the accuracy and the high-quality measurements were already attained in science world. The accurateness of the experiment is determined by the quality of the instrument that is used by the scientist and not just by the computer. For example, high-quality instrument is needed to cut the sample of frozen cell in the electron microscopy or measuring the neuron's electrical activity.[5]

The scientists work differently since the introduction of internet, as it is the vital advanced technique. The communication among the researchers has been improved and the collaboration in the field works has also encouraged. The effectiveness of efforts of research has increased since the existence of data management systems, for example, the National Center for Biotechnology Information (NCBI) and European Bioinformatics Institute (EBI). The specialized and specific management system specifies the biological data. One of the examples is the Protein Data Bank (PDB). The PDB is the storage of the data of the proteins where the relationships among the molecules that are stored are also available. Links of related data should be given by the data management system.[5]

Internet has the interactive mode which is beneficial to science. Remote computers are used to run the internet to access the interactive tasks. Majority of the computing task use PCs, as PCs are less expensive, have big capacity, and they can also interconnect the local and nonlocal networks which enable them to gather the parallel computing processors. The internet needs to be linked to switches, routers, and fiber-optic cables. It also has a massive parallel computing which is similar to the supercomputers.

17.3.1 LIMITATIONS OF COMPUTATIONAL TOOLS

Buehler and Rashidi[5] have claimed that it is not possible for the molecular biology to be without the information storage and recovery, statistical analysis, data fitting, and computer simulation as these are important roles. It is important to use the computers in order to process the large number of data in a particular time frame. Instruction is required for the computers as the analytical process within the system is done by the human operator. Therefore, it is time-consuming and is easy to make an error. Because of this, numbers of modern processes that are conducted by the computers are equipped with a simple recognition to the human regulatory and supervisory. Expert system such as the robots is used to perform the human tasks and this requires huge computational power. Neural networks have the ability to learn the human intervention, but to apply it effectively is very difficult as there are manipulations of symbols that are not in the training data. The computers are going to have changes in the input and output data constantly after setting up and implementing the algorithm.[4, 17] This enables the adjustment to be made by the machine and still is depending on the human input.[5]

One of the limitations of the computers, according to Buehler and Rashidi[5] is the spell checkers. The downside of it is lack of understanding the language. For example, the computers cannot detect if there is a typo that is spelled correctly, however, have misplaced meaning. The spell checking is not good in understanding the context and is less suitable for identifying the correct spelling, style, and grammar. However, computers are suitable for solving the numerical problem, controlling the machine, editing, searching, finding data, and managing databases.[5]

They also mentioned that another limitation is when selecting which data are suitable for interpretation. This is because experimenter with experience is needed in selecting the data. Computer support is becoming critical as biological data is increasing each day. The important decision maker is the scientist's intuition and bias. However, the computers protect against the bias of the experimenter and do not read the bias vision of the scientists.[5]

17.4   THE EXAMPLES OF BIOINFORMATICS TECHNOLOGY

The bioinformatics technology can also be used for information security. It is important because it has the ability to verify and authenticate information effectively. It is a type of software which could detect identification of a person's unique physiological and biological traits. This will open a wide range of opportunities to develop more security and safety from any threats or criminals in the world. This technology has a vital role in preventing any incidents because of its effective tools to detect criminals. The use of biometric technologies provides a beneficial and reliable verification to those who are traveling abroad; the bioinformatics technology will help strengthen the security of passport, visa, and other identifications. The reason behind this action is because of the concerning terrorism in the world lately. Therefore, the main focus of people is based on their security. The next discussion will be elaborated based on its biometrics tools, in addition to its benefits and limitations.

17.4.1 TOOLS OF BIOMETRICS EQUIPMENTS

17.4.1.1 SIGNATURE VERIFICATION TECHNIQUE

The signature recognition is used for tracking people's dynamic signature. It can be measured in terms of the direction, the pressure, acceleration, and the length of the strokes, a number of strokes, and their duration. The advantage of this technology is that the frauder cannot steal any information on how to write and copy the signature similar to the one that writes it previously. Other devices such as traditional tablets capture 2D coordinated and pressure. This technology is mainly used for e-business and other applications where signature is accepted for personal authentication.

The limitation of using signature verification technique is that the signature can look different from the usual user signature. Because of the digitalized signature, it changes slightly compared to how it was written on paper. Moreover, the user does not see what they have already written. They have to look back at the computer monitor to see the signature. This is a disadvantage to those inexperienced users.

17.4.1.2 VOICE RECOGNITION

Voice recognition has the feature of speech that is different between individuals. Mr. Sanjay Kumar and Dr. Ekta Walia argue that there are different styles of spoken input: text-dependent, text-prompted, and text-independent. Most applications use text-dependent input, which involves enrollment of more than one voice passwords. It is used whenever there is concern about imposters. This can avoid the imposters from unlocking the passwords, therefore, security of the user is guaranteed.

The voice recognition, however, can also lead to a disadvantage. This is because of performance degradation. Voice may change due to aging which needs to be inscribed by the recognition system. Furthermore, the results may change from behavioral attributes of the voice, verification, and enrollment on another telephone.

17.4.1.3 FACIAL RECOGNITION

Facial recognition has the ability to detect facial image of a person's individual characteristics. The advantages of biometric identification on photos are: it does not require a direct relation with the subject, it is confidential, it provides unique and stable biological parameters of a person, and its biometric database are complete and widespread. Furthermore, another benefit is, if there is a problem of image processing, it can be transferred to the professionals automatically. It will enable it to process it, specify image control points, which strengthen the accuracy of the system.

The limit of this technology is that the features are resistant to hairstyle changes, people wearing the glasses, and other transformations.

17.4.1.4 PATTERN RECOGNITION

Pattern recognition was traditionally found in the 1950s; the measurement of data measured shows results in a lacking form of formula which therefore takes a lot of time to analyze the results without the use of advanced technology computation. This, therefore, slows down the process and the development is decreased.[9]

With the help of advanced technology, biologists are able to compute data mining, analyze data sets, syntax, statistics structure, recurrent patterns, machine learning, developed effective modeling data, and desired outputs. Pattern recognition basically involves matching and detecting pattern sequences as well as it covers the theory and methods for clustering that refer to how do objects form into natural groups, dimensionality reduction, and classification.[9]

Example of pattern recognition is the diagnosis of breast cancer; pattern recognition is used to identify specific medicines to cure the molecular hidden in the cancer cells. When they can retrieve the cure, they are able to use it for medical purposes, which are more certified after they have used the pattern recognition test. Patient diagnosed with cancer is now able to consult with their doctors on medical instructions. Cancer can be of any type , from brain cancer to bone cancer and basically may be anywhere in the human body; so scientists do play an important role here in finding the right cures, doing screening test, and treatment for patients. As we all know, cancer treatment do uses technology, such as in chemotherapy, radiotherapy, hormone therapy, biological therapy, and transplant.[9]

There are three ways in which pattern recognition works: first, the clustering methods where Gaussians mixture is used in order the data to be fitted well; second, the dimensionality reduction, where the data are in line into a linear analysis; and last, the classification, where boundary is made to separate two different classes of data. The cluster data basically used in performing a large cluster data, for example, experimenting genes which involves different types of genes and cells and this data is used in order to retrieve uncharacterized genes, similarity, patient samples, and most importantly, to discover disease inherited if any. So cure can be identified as early as they can to prevent diseases to spread into a difficult situation. There are two types of clustering the procedures; the first is the partitional and the second is hierarchical. The partitional is used to fit a number of data into one such methods are the k-means and the Gaussian mixture models. The hierarchical clustering is used to group certain objects in the data.[9]

However, the cluster method may not produce a better or an objective result as it may produce confusing results between the clusters and the classes as the groups have the same label, classes may overlap sometimes, and unsupervised classification may occur. By using the pattern recognition, biologists may face challenges in performing the data, such as storage may be limited and may not fit the whole data that they needed to perform. The works need to be packed down which therefore may cause the data to scatter. Some biolaboratories may not have enough budgets to invest advanced software or databases, which is likely to decrease their research activity and may slow down their experiment research fields. Cells of a human body or even organisms may change at times; biologists are depending on advanced computerized system to further their research into their studies. So scientists need to put a lot of investment to make improvements into their computer systems.

As such example, using the pattern recognition, scientists should employ a more skilled biologist that is able to perform the research by using the pattern recognition system or database. In a way, this shows that, computer plays an important role in employing skilled labor who are able to use computer system to perform daily tasks. If biologists are unable to use computer system, this may cause difficulty in the laboratory and therefore, this may slow down the process and may cause error results if computer systems are not used properly.[9]

The data represented are still perform in a very good way although sometimes data may perform into a slightly error probably due to storage error. Sometimes it is impossible for a biologist to retrieve a better result as what they expected. Biologists may use common feature vector such as zero mean, unit standard deviation, log, square root, add, or multiply elements. Sometimes complicated data results may occur such as binary of 0/1, ordinal 1/2/3, qualitative of red–blue–green, or sequential data of ACTGAATA in biologists term which basically shows a loss of information to some part of the research and it needs to convert into reliable information data. These are few examples of a mislead data performed by biologists in the bioinformatics fields.[9]

17.4.1.5   FINGERPRINTING

Fingerprint recognition gives a wide opportunity for security purposes. "A chance of two users having the same identification in the biometrics security technology system is nearly zero" (Tistarelli, 2009). One of the technologies that can be used is Siemens ID Mouse Finger Prints. The results of scanning the fingerprints will produce a very high-quality image while remaining subpixel geometric accuracy. Afterwards, the image will be stored and processed to extract the pixel information. According to Subhra Mazumdar and Venkata Dhulipala, a fingerprint sensor captures the digital image of a fingerprint that is called *live scan*. The live scan is digitally processed to create a biometric template which can be stored and collected for matching. Furthermore, it also has matching algorithms which can be used to solve a problem. Stored templates of fingerprints can be used against candidate's fingerprints for authentication and verification purpose. The process starts either using original image with the candidate image to make comparison. Pattern-based algorithm compares the basic fingerprint, such as its arc, whorl, and loop. The template includes the size, type, and orientation of pattern found in the fingerprint image. This can also be differentiated with the template to determine which matches with the match score generator.

Despite its ability and advantages of this bioinformatic technology, there is a probability that it may lead to drawback, for example, if the user uses fingerprints for security reason, and an incident occurred that could cause loss of his or her fingernails, it will be difficult for the verification process (PBWorks, 2006).

17.4.1.6 IRIS SCAN

Iris scan is also one of bioinformatics technology. It is found to be reliable and accurate for authentication process. It is another biometric that can scan the iris pattern quickly from the left and right. The scan can be used for identification and verification purposes. The highly variance of appearance makes this biometric well organized. According to Chirchi, Waghmare, and Chirchi (2011), the iris scan is an accurate measurement and has following features:

1. It avoids from the difficulty of forging and using as an imposter person
2. It is intrinsic isolation and protection from the external environment
3. It is extremely data-rich physical structure
4. It is a genetic property—no two eyes are the same. The characteristic that is dependent on genetics is the pigmentation of the iris, which defines its gross anatomy and color
5. Its stability overtime: the impracticality of surgically modifying it without risk to physiological and vision responses to light, which gives a natural test against artifice

In terms of its iris detection, the irises are detected even if the image is in constructions, has visual noise, and has different levels of illumination. The lighting reflections, eyelids, and eyelashes barriers are eliminated. Based on security, it is an important technology for security system. In the verification process, it will first verify and check if the user data that was put in is correct or not, for example, username and password. But on the second stage, which is the identification stage, the system tries to find who the subject is without any input information. Therefore, the use of iris scan is also a part of bioinformatics tools which detects out biological data processing and analysis.

17.5 BENEFITS AND LIMITATIONS OF BIOINFORMATICS TECHNOLOGY

Computational approaches have been very successful in facilitating, extending, and complementing experimental investigations including bioinformatics. There are benefits and limitations when it comes to using bioinformatics as they have become an indispensable component of many areas of biology. One of the benefits is that bioinformatics has quick sequencing capabilities which aid in saving time as it would be impossible for projects such as human genome project (HGP) to take place when it requires sequencing and storing of 3 billion nucleotides. If the human genome was to be analyzed using human brain instead of using bioinformatics, it would have taken many generations to complete. The second one is that with the help of advanced technology, it will eventually be possible to sequence DNA of individual patients which would provide physicians with new ways to identify diseases and cure them effectively (personalize medications). Also, bioinformatics can store a massive storage in orderly manner (HDD and indexing) and provide researchers with the storage space they require to store the data they collect. This is very helpful to researchers when they want to compare and analyze the DNA as they are already well organized.

Moreover, bioinformatics techniques such as image and signal processing allow extraction of useful results from a large amount of raw data in experimental molecular biology likewise in structural biology, the use of bioinformatics assists in the simulation and modeling of DNA, RNA, and protein structures as well as molecular interactions. Similarly, in the field of genetics and genomics, bioinformatics help in sequencing and annotating genomes and their observed mutations. Bioinformatics plays a role in the textual mining of biological literature and the development of biological as well as gene ontologies to organize and query biological data. Additionally, they also play an important part in the analysis of gene and protein expression, and regulation.

The other benefit of bioinformatics is that the tools of bioinformatics aid in the correlation of genetic and genomic data and more broadly in the understanding of evolutionary aspects of molecular biology. At a more integrated level, this bioinformatics tools helps to evaluate and catalogue the biological pathways and networks that are an essential part of systems biology.

Computational models of system-wide properties could provide as a basis for experimentation and discovery which will, therefore, result in not only explicit understanding of how organisms are built but also the capability to exhibit specific traits, to discover the casualty of diseases, and also to foresee organisms' responses to changes in the environment. As a result of this, diseases such as HIV, AIDS, malaria, and other diseases can be avoided and treated as well as aid in the preservation of the environment. In the agriculture field, bioinformatics can also be beneficial. Bioinformatics alters the genetic makeup of a certain crop and eventually results in a higher yield and a much more improved quality of food production. (Seung, 2005)

In spite of all the benefits of bioinformatics, there are also limitations to it. The most evident drawback of using bioinformatics is that they are very costly to use as it requires computers and other technology tools. There is also a possibility of losing all the data collected due to a virus. Similarly, there is a chance for the algorithm to make an error in rare instances and a possibility of sequence matching due to a chance. String matching is also a major problem of the computational biology, text processing, and pattern recognition.

The other limitation of using bioinformatics is that biologists will generally have a much larger boundary of their own department of expertise and contribute most of their time on the computer instead of spending their time at the bench. In addition, analyzing other people's data will be much more commonplace since the concept of ownership will also change because of bioinformatics. Because of this change, scientists will be encouraged, if not forced, to concentrate more to the quality of data annotation and actively participate in their improvement. Lastly, the key to successful bioinformatics is close face to face teamwork between the biologist, biostatistician, and bioinformaticist. However, this is not always achievable due to the distances between institutions.

17.6 TRANSLATIONAL BIOINFORMATICS

Translational bioinformatics has become a significant aspect in the area of biological advancements. Thus, it has become the third major domain of informatics.[21] According to the American Medical Informatics Association (AMIA), translational bioinformatics is defined as "… the development of storage, analytic, and interpretive methods to optimize the transformation of increasingly voluminous biomedical data, and genomic data, into proactive, predictive, preventive, and participatory health. Translational bioinformatics includes research on the development of novel techniques for the integration of biological and clinical data and the evolution of clinical informatics methodology to encompass biological observations. The end product of translational bioinformatics is newly found knowledge from these integrative efforts that can be disseminated to a variety of stakeholders, including biomedical scientists, clinicians, and patients."[20]

There are several reasons as to why translational bioinformatics is on reoccurring demand for the last decade.

First, the information technology (IT) infrastructures developed for translational research allow data managers and statisticians to acquire the data and advanced services required for them to accomplish their tasks.[19]

Second, there is an increase in tools that help in the assessment of molecular states. We can now measure a tremendous number of molecules at the same time. For example, the gene expression microarray has enabled us to determine the number of diseases and found novel subgenotypes of diseases in RNA across thousands of genes.[21] Furthermore, the tools are easily available to the public.

Last, the IT required may assist citizens to have a better healthcare approach as there will be access to prevention strategies, diagnosis, and therapy beforehand as a result from the anticipated results of translational research.[19]

As translational bioinformatics consists of evolved and advanced methods, tools, and services that have improved the connection between researchers and healthcare, there are challenges along the way.

One of the challenges is that semantic interoperability is hard to achieve. This is because, within health information systems, clinical data is often changed due to the continuous use of health informatics standards.

Another challenge is that biomolecular databases are becoming progressively large, complicated, and interoperated. This increases the possibility of risks in the growth, placement, and sharing of the database. Data warehousing can be a solution as it supports the reuse of gathered clinical data that is to be utilized for a different purpose and thus, increases the quality of clinical data and restricts repetitions and errors.

17.7 CONCLUSIONS

Advanced technologies have occupied a lot in this modern world and basically have been part of the human life, from education, regular institutes, departments, workplace to business. It really plays an important role in acquiring the technology world so they are able to engage in ambitious projects. It gives an opportunity to people especially the scientists to have such an advancement that will help them to improvise their tasks or studies and has basically made the human life much easier. Technology and computer advancement improves the scientist's world in terms of gaining faster results, providing easy access to internet world, enabling to process reliable data, and gathering as much information as they can. However, in acquiring the technology world, it is also quite expensive to invest such a big amount from thousands to billions or even millions in the investment of computers, update software, and databases. Without technology advancement, people may experience difficulty and may not be able to stay up to date. Basically, we are living in the world of vice versa. So it is important as well to have technology as a part of our lives in order to improve the standard of living of the people which enables rapid and cost-effective generation.

Moreover, the use of biometrics technology can have its own advantages and also consequences. The unique features of bioinformatics technology do provide high accuracy and validity, which even humans are not capable of doing. In the future, as IT will continue to advance, this will provide improvement for a biologist by making his work much convenient and less time-consuming. Not just in biology area, but almost all parts of area, such as businesses, social, and sciences.

17.8 SUMMARY

Scientists have been using a lot of different scientific methods for analyzing their research or experiments, hoping to find a cure to every disease that they find, analyzing organisms, further research on human body, and in a way, developing an advanced medical treatment to people ever since the early 13th century. Gregor Mendel (1822–84) was one of the scientists that prove a successful experiment in his research of a pea plant, where he recorded every single of his observation only through a notebook without any use of modern computers or tablets.

The time scale during the 13th century shows a very long time was consumed in the research activity in manual way without any technology. But now, because of advanced technology, scientists can now record thousands and thousands of analyzed data in a short period of time. As technology kept growing in the 1940s, it was seen as an important deal to scientists; Charles Babbage (1791–1871) invented a new way of performing mathematical calculations automatically and to build a more advanced computing device. After that, computer scientists learnt a lot more about advanced computers, hardware, software, or databases in order for them to be able to perform simple operations in the laboratory.[7]

Along with his partner, Ada Lovelace (1815–52) developed sorts of computer programs to work with numbers as well as programming languages such as Smalltalk, Pascal, and C++ that can sync with Charles' invention.[7]

During the mid-1990s, computational and quantitative analysis in the biological fields exploded where scientists begin to use biological programming to use in for their research. Such example will be the DNA sequence where scientists are able to disentangle the beauty of thousands and millions of genome and nucleotides in a matter of time. As they see the technology is getting more and more important for their lab research, biologists need to cooperate in learning deeper towards the technology world. As they realize that in order to manage, store, analyze the data into a more visualization data, and in order to perform into a statistics results, they are able to produce a bioinformatics results that are seen as reliable data that can be used to other scientists' field. This is where bioinformatics was born.

As technology era improved year by year, biologists now are able to develop algorithms or advanced procedures to further improve their understandings in different types of data collections such as nucleotide, amino acid sequence, protein domains, DNA sequence, pattern recognition as well as structure organisms. It is clear that computer informative system does play an important role in helping scientists or biologists to perform their research in order to help further predictions for the future.[7]

KEYWORDS

- **bioinformatics**
- **information technology**
- **database**
- **informatics**
- **DNA**

REFERENCES

1. Debnath, B.; Rahul, R. A. Farkhod, A.; Choi, M. Biometric Authentication: A Review. *Int. J. u- and e- Serv. Sci. Technol.* **2009,** *2*(3), 13–27.
2. Kumar, S.; Ekta, W. Analysis of Various Biometric Techniques. *Int. J. Comput. Sci. Inf. Technol.* **2011,** *2*(4), 1595–1597.
3. Kazimov, T.; Mahmudova, S. The Role of Biometric Technology in Information Security. *IRJET* **2015,** *2*(3), 1509–1513.
4. Antonino, F.; La Rosa, M.; Alfonso, U.; Ricardo, R.; Salvatore, G. A Knowledge-based Decision Support System in Bioinformatics: An Application to Protein Complex Extraction. *BMC Bioinform.* **2010,** *1,* 51–55.
5. Buehler, K. H.; Rashidi, H. H., Eds. *Bioinformatics Basics: Applications in Biological Science and Medicine;* Taylor & Francis Group: Florida, 2005; pp 11–35.
6. Altman, R. B.; Fernald, G. H.; Capriotti, E.; Daneshjou, R.; Karczewski, K. J. Bioinformatics Challenges for Personalized Medicine. *Bioinformatics* **2011,** *27*(13), 1741–1748.
7. Gopal, S. *Bioinformatics: A Computing Perspective;* The McGraw Hill: 1221 Avenue of the Americas, New York, 2009; p 445.
8. Marguerat, S. RNA-seq: From Technology to Biology. *Cell. Mol. Life Sci.* **2009,** *11,* 569–579.

9. Ridder, D. D. Briefings in Bioinformatics. *Pattern Recogn. Bioinform.* **2013,** *15.*

10. Shelly, G. B. *Discovering Computers Complete: Your Interactive Guide to the Digital World;* Cengage Learning: United States, 2012; pp 140–155.

11. Singh, D. *BioInfoKnowledgeBase,* September 18, 2010 BioInfoKnowledgeBase. http://webapp.cabgrid.res.in/BIKB/edb_home.html (accessed Oct 2017).

12. Singh, D. BioInfoKnowledgeBase: Comprehensive Information Resource for Bioinformatics Tools. *Am. J. Bioinform.* **2015,** *6.*

13. Cannataro, M.; Santosb, R. W.; Sundnesc, J. Biomedical and Bioinformatics Challenges to Computer Science: Bioinformatics, Modeling of Biomedical Systems and Clinical. *Procedia Comput. Sci.* **2011,** *1,* 1058–1061.

14. Claverie, J. M.; Notredame, C. Finding Out What Bioinformatics Can Do for You. In *Bioinformatics—A Beginner's Guide;* Claverie, J. M., Ed.; Wiley India Pvt. Ltd.: New Delhi, 2009; p 9.

15. Al-Ageel, N.; Al-Wabil, A.; Badr, G.; AlOmar, N. Human Factors in the Design and Evaluation of Bioinformatics Tools. *Procedia Manuf.* **2015,** *2,* 2003–2010.

16. Jason, H. M.; Folkert, W. A.; Scott, M. W. Genetics and Population Analysis: A Review. *Bioinform. Chall. Genome-wide Assoc. Stud.* **2010,** *26*(4). http://bioinformatics.oxford-journals.org/content/26/4/445.full#abstract-1 (accessed Oct 2017).

17. Sara, A.; Shehab, A. K.; Hany M. Fast Dynamic Algorithm for Sequence Alignment Based on Bioinformatics. *Int. J. Comp. Appl.* **2012,** *7.* http://research.ijcaonline.org/volume37/number7/pxc3876636.pdf (accessed Oct 2017).

18. San Diego State University. (n.d.) Bioinformatics. *Biological & Medical Informatics.* http://informatics.sdsu.edu/bioinformatics/

19. Daniel, C.; Albuisson, E.; Dart, T.; Avillach, P.; Cuggia, M.; Guo, Y. Translational Bioinformatics and Clinical Research Informatics. In *Medical Informatics, e-Health;* 2013. http://link.springer.com/chapter/10.1007/978-2-8178-0478-1_17 (accessed Oct 2017).

20. Tenenbaum, J. D. Translational Bioinformatics: Past, Present and Future. In *Genomics, Proteomics & Bioinformatics;* 2016. http://www.sciencedirect.com/science/article/pii/S1672022916000401 (accessed Oct 2017).

21. Butte, A. J. Translational Bioinformatics: Coming of Age. *J. Am. Med. Infor. Assoc.* **2008**. http://jamia.oxfordjournals.org/content/15/6/709 (accessed Oct 2017).

22. Sirota, M.; Dudley, J. T.; Kim, J.; Chiang, A. P.; Morgan, A. A.; Sweet-Cordero, A.; Butte, A. J. Discovery and Preclinical Validation of Drug Indications Using Compendia of Public Gene Expression Data. *Sci. Transl. Med.* **2011,** *3*(96), 96ra77.

CHAPTER 18

INFORMATICS APPROACH AND ITS IMPACT FOR BIOSCIENCE: MAKING SENSE OF INNOVATION

HERU SUSANTO[1,2,*] and CHING KANG CHEN[3]

[1]Department of Computer Science & Information Management, Tunghai University, Taichung, Taiwan

[2]Computational Science, Indonesian Institute of Sciences, Serpong, Indonesia

[3]School of Business and Economics, University of Brunei, Bandar Seri Begawan, Brunei

*Corresponding author. E-mail: heru.susanto@lipi.go.id; susanto.net@gmail.com

CONTENTS

ABSTRACT

In today's society, bioinformatics research is putting a great emphasis on answering "when," "what if," and "why" questions with the help of information system and technology and some researchers had argued that it could be the key factor in facilitating and attaining an efficient decision-making in medical research. Hence, the main purpose of this report is to explore the application of information system in biological world and to study the extent to which technology could bring opportunities as well as challenges in science. Also, this report will examine the importance and evolution of bioinformatics over the past decades.

With the current deluge of data, computational methods have become indispensable to biological investigations. Originally developed for the analysis of biological sequences, bioinformatics now encompasses a wide range of subject areas including structural biology, genomics, and gene expression studies. Additionally, nowadays biological data is proliferating rapidly. With the advent of the world wide web and fast internet connections, the data contained in these databases and a considerable amount of special-purpose programs can be accessed quickly and efficiently from any location in the world. As a consequence, computer-based tools now play an increasingly significant role in the advancement and development of biological research. Hence, this report will investigate the relationship between information system and science as well as the consequences and implication of technology in supporting medical research.

18.1 INTRODUCTION

Information system is a part of Information Technology. It has been well defined in terms of two perspectives: one relating to its purpose; the other relating to its structure. From a functional perspective, an information system is a technologically implemented medium for the purpose of recording, storing, and disseminating linguistic expressions as well as for the supporting of inference making. While from a structural perspective, an information system consists of a collection of people, processes, data, models, technology, and partly formalized language forming a cohesive structure which serves some organizational purpose or function. However, they also can be defined as a set of interconnected components that assemble (or retrieve), process, store, and allocate information in order to support decision-making and control in an organization. In addition to supporting decision-making, coordination, and control, information systems may also aid in helping workers in analyzing problems, visualize complex subjects, and create new products.

18.2 GENERAL REVIEW OF INFORMATION SYSTEM IN SCIENCE

Due to the availability of large data sets of digital medical information, the use of informatics to improve health care and medical research has made possible where they provide a new trail for investigation and medical discovery. This is because informatics focuses on developing new and effective methods of using technology to process information. In today's society, informatics is being applied at every stage of health care from basic research to care delivery and includes many specializations such as bioinformatics, medical informatics, and biomedical informatics.

The field of bioinformatics has exploded within the past decade to keep pace with advancements and development in molecular biology and genomics research where researchers could use bioinformatics to obtain an effective understanding of complex biological processes which includes examining DNA sequences or restructuring protein structures.

Furthermore, informatics has also had huge impact on the field of systems biology as systems biology could use computer modeling and mathematical simulations to predict how complex biological systems will behave.. By applying the computer models in the study, researchers can obtain a better and more comprehensive understanding of how diseases affect an entire biological system in addition to the effects on individual component.

18.3 ROLE OF INFORMATION SYSTEM IN INFORMATICS

The use of informatics to improve health care and medical research has become possible due to the availability of large data sets of digital medical information. This is due to the development of a new trail for investigation and medical research. Informatics highlights on improving a new and effective method by using technology to process information. In today's society, informatics is being applied at all health care phases from elementary study to care delivery, including a considerable amount of specializations such as bioinformatics, medical informatics, and biomedical informatics.

Within these past decades, the study of bioinformatics has exploded. This is to keep pace with progression and development in molecular biology and genomics research. Thus, researchers could use bioinformatics to obtain an effective understanding and knowledge of complex biological processes which includes examining DNA sequences or modifying the protein structures.

Furthermore, the prediction on how complex biological systems will behave also influenced by the development of informatics as it had a huge impact on the field of systems biology, and systems biology could use computer modeling and mathematical simulation. National Institute of Health has claimed that researchers have created models to simulate tumor growths. Therefore, researchers can obtain a better and more comprehensive understanding on how diseases may affect an entire biological system in addition to the effects on individual components by applying the computer models in the study.

18.4 THE RELATIONSHIP BETWEEN INFORMATION SYSTEM AND BIOINFORMATICS

According to Cannataro et al.,[3] bioinformatics is the application of computational tools and techniques to the management and analysis of biological data. Over the past few decades, rapid developments in genomic and other molecular research technologies and developments in information technologies have combined to produce a tremendous amount of information related to molecular biology. The primary goal of bioinformatics is to increase the understanding of biological processes.

Bioinformatics develops algorithms and biological software of computer to analyze and record the data related to biology including the data of genes, proteins, drug ingredients, and metabolic pathways. A study has concluded that the creation and analysis, such as; group of sequences, large data sequences and adding new modules for visual representation through Microsoft Windows platform. Bioinformatics is the field of science in which biology, computer science, and information technology merge to form a single discipline.[16] Biological data is in need of certain storage house in which the data can be stored, organized, and manipulated. Thus, biological software and databases provide the scientists and researchers this opportunity, enabling them to extract data from these database efficiently and effectively.

Bioinformatics could be considered to be the combination of several scientific disciplines that include biology, biochemistry, mathematics, and computer science. This is due to the availability of enormous amounts of public and private biological data and compelling need to transform biological data into useful information and knowledge. Additionally, understanding the correlations, structures, and patterns in biological data is the most important task in bioinformatics. Thus, the knowledge and understanding obtained from these disciplines could be sensibly used for applications that cover drug discovery, genome analysis, and biological control. Furthermore, it involves the use of computer technology and statistical methods to manage and analyze a huge volume of biological data regarding DNA, RNA, protein sequences, protein structure, gene expression profiles, and protein interactions.

18.5   AIMS FOR BIOINFORMATICS

There are three aims in using bioinformatics. First, bioinformatics is used as a data organizer in a way that it allows researchers to access existing information as well as making new entries of fresh data. However, information stored in bioinformatics databases will not be useful unless they are analyzed and which extend the purpose of bioinformatics. Second, bioinformatics is also used as developing tools and resources to analyze the information. For an example, sequence of a particular protein needs more than a simple text-based search and program which is needed to be supported by biologically significant match. Additionally, this development process needs an expertise which is not only required in computational expert but also needed in understanding of a medical research. The third aim of bioinformatics is to use the developing tools in analyzing the data and interpreting the result in a biologically meaningful manner. Traditionally, biological studies are conducted on individual system and comparison process. With the help of bioinformatics system, the analysis process can be conducted globally with large range data available and aim of open common principle across many systems.

18.6   THE PROGRESS OF BIOINFORMATICS

Grid infrastructures played an important role in the recent decade as supporting scientific computer-based analysis. However, the increasing complexity of bioinformatics resulted in finding new solutions to speed up computational time. Grid infrastructure is not completely satisfactory in terms of providing services and managing data that are reliable for presenting bioinformatics.

Another key issue is represented by the fact that the grid is offering poor chances to customize the computational environment. In fact, it is quite common in computational biology to make use of relational databases and/ or web-oriented tools to perform analyses, store output files, and visualize results, which are difficult to exploit without having administration rights on the used resources. Another related problem derives from the huge amount of bioinformatics packages available in different programming environment (such as R, Perl, Python, and Ruby) that typically require many dependencies and fine-tuned customizations for the various users.

These are the reasons why the cloud computing is the best solution. Computation is moving from in-house computing infrastructure to cloud computing delivered over the internet. Cloud computing provide cheap, reliable large scale of data where small-size organization can get the same information as well-funded organization. Bioinformatics grew with rising use of the internet which allows creation and sharing of large biological data and offers rapid publication of research results. The internet also provides the researchers with supercomputing system that are complex such as grid infrastructure.

18.7   THE APPLICATION OF BIOINFORMATICS

Some of the grand area of research in bioinformatics includes the following:

18.7.1   SEQUENCE ANALYSIS

It is the most primitive operation in computational biology where the operation includes finding which part of the biological sequences are alike and which part differs during medical analysis and genome mapping processes. Hence, the sequence analysis implies subjecting a DNA or peptide sequence to sequence alignment, sequence databases, repeated sequence searches, or other bioinformatics methods on a computer.

18.7.2   ANALYSIS OF MUTATIONS IN CANCER

The arrangement of the genomes of affected cells is complex where a huge sequencing strength is needed to identify previously unknown point mutations in a variety of genes in cancer. By producing specialized automated systems, a management sheer volume of sequence data could be produce, manage and create new algorithms. This algorithm function to compare the sequencing results with the growing collection of human genome sequences and germ-line polymorphisms. Another type of data that requires novel informatics development is the analysis of lesions found to be recurrent among many tumors.

18.7.3 MODELING BIOLOGICAL SYSTEMS

Modeling biological systems are significant in biology system and mathematical biology, where computational systems biology aims to develop and use efficient algorithms, data structures, and visualization and communication tools for the integration of large quantities of biological data with the goal of computer modeling. It involves the use of computer simulations of biological systems, including cellular subsystems such as the networks of metabolites and enzymes, signal transduction pathways, and gene regulatory networks to both analyze and visualize the complex connections of these cellular processes.

18.7.4 HIGH-THROUGHPUT IMAGE ANALYSIS

Computational technologies are used to accelerate and facilitate the processing, quantification, and analysis of a considerable amount of high-information-content biomedical images. Additionally, modern image analysis systems enhance an observer's ability to make measurements from a large or complex set of images. A fully developed analysis system may completely replace the observer. Biomedical imaging is becoming more important for both diagnostics and research. Some of the examples of research in this area are clinical image analysis and visualization, inferring clone overlaps in DNA mapping, and bioimage informatics.

18.7.5 DRUGS DISCOVERY

Traditionally, pharmaceutical companies may only be attracted to introduce new drugs when any well-known pharmaceutical company had been successful in developing them. However, in today's society, company has invested heavily on approaches that can speed up the development process. The pressure of producing drugs in short period of time with a high standard of safety's concern has resulted in extremely enhancing interest of the researchers in bioinformatics. Bioinformatics algorithm as an identification of biological candidate and could be the storage of information. Drugs only can be produced if the drug target is studied and identified. For an example, human genome sequence information can be found in the system that can help in drug-making process.

18.7.6 PREVENTION AND TREATMENT OF DISEASES

Bioinformatics is a scientific discipline that deals with earning, analyzing, distributing, processing, and storing of biological information. It uses scientific knowledge such as algorithm and computer science in order to understand the biological significance of a wide variety of data. With this, it enables researchers to find new strategies to look for clues in the prevention and treatment of diseases. Bioinformatics has turned into a key ingredient with the alliance of genomics, proteomics, and drugs in today's world.

In fact, bioinformatics owes its creation to the need to handle large amounts of data produced by these "-omic" technologies (genomics, proteomics, and more recently metabolomics). This method of information is generated by high-performance methods such as gene sequencing, DNA microarrays, and mass spectroscopy. For this reason, bioinformatics can be called a transverse activity because it is applicable to all the subsectors of biotechnology and life sciences. However, its main application is biomedicine. Bioinformatics manages and decodes "-omic" data and it facilitates the translational medicine concept by helping to distribute information throughout the entire health care value chain. This covers the discovery and analyzing of genes, the protein structures coded by these genes, and the design of molecules and drugs to counter these proteins up to their clinical application, which is where bioinformatics is executing a leading role in the development of specific medicine.

18.7.7 STUDYING GENETIC DISEASE

There is a growing market in the use of microarrays for studying diseases associated with genetic characteristics. The widespread acceptance of this technique is driving demand for a more user-friendly version of the software and bioinformatics companies are supporting this idea in their latest product developments. The big pharma companies are using systems biology in their drug discovery processes. For example, Novartis has created Novartis Biologics: a new division that incorporates bioinformatics at all levels of the drug creation value chain. Programmers have recently developed an extended markup language exclusively for systems biology, called systems biology markup language (SBML).

This language makes it possible to integrate the software applications of different providers. As a result, bioinformatics is also moving toward standardization of the language used for developing software. This will accelerate the production of new applications and utilities by small (nonindustrial) developers, using an open-source environment. In fact, experts expect that within a few years, all legacy applications in bioinformatics will be available via the internet and will run in ordinary browsers. Consequently, it will be important for bioinformatics companies to adapt their existing products for online use or to develop new applications that are suitable for this purpose.

18.8   THE BENEFITS OF USING INFORMATION SYSTEMS IN BIOINFORMATICS

Bringing together large data sets of medical data and tools to analyze the data offers the potential to enlarge the research capabilities of medical researchers where they could use this vast source of biological and clinical data to discover and develop new treatments and better understand illnesses. Pharmaceutical companies could use the biomedical data to create drugs targeted at specific populations. Furthermore, health care providers can use the data to better inform their treatments and diagnoses.[4]

Etheredge as cited in Ref. [4] claimed that applying informatics to health care creates the possibility of enabling "rapid learning" health applications to aid in biomedical research, effectiveness research, and drug safety studies. For example, using this technology, the side effects from drugs newly introduced to the market can be monitored in real time, and problems, such as those found with the recently withdrawn prescription drug Vioxx, can be identified more quickly. Moreover, the risks and benefits of drugs can be studied for specific populations yielding more effective and safer treatment regimens for patients.

Etheredge had concluded that using rapid learning techniques can not only improve patient safety but also can lead to substantial improvements in the quality and cost of care by turning all of the raw digital data into knowledge where these rapid learning health networks can enable doctors and researchers to better practice evidence-based medicine. Evidence-based medicine is the use of treatments judged to be the best practice for a certain population on the basis of scientific evidence of expected benefits and risks.

18.9   THE LIMITATIONS OF INFORMATION SYSTEM IN BIOINFORMATICS

However, achieving the vision of an intelligent and fully connected health research infrastructure has not yet been realized. While various pilot projects have shown success and have demonstrated the potential benefits that can emerge from a ubiquitous deployment of informatics in health research, many technical obstacles still need to be overcome. Doolan as cited in Ref. [4] believes that these obstacles include making data accessible, connecting existing data sources, and building better tools to analyze medical data and draw meaningful conclusions. Much medical research data is not accessible electronically.

Achieving the widespread use of electronic health records is a necessary requirement for creating the underlying data sets needed for bioinformatics research. Access to the electronic health records of large populations will help researchers apply informatics to various problems including clinical trial research, comparative effectiveness studies, and drug safety monitoring. However, collecting medical data in electronic format is only the first step. Interoperability poses a substantial challenge for biomedical research. This is because the vast amount of electronic medical data cannot fully be utilized by researchers because the data resides in different databases. Even when the organizations that collect and distribute biomedical data are willing to share data, incompatible data formats or data interfaces can create challenges for analyzing data across multiple data sets.

Thus, Stein as cited in Ref. [4] claims that researchers wishing to use multiple data sets must devote significant resources simply to managing the differences between the data and, as a result, have fewer resources available for working with the data.

Highly trained workers that are familiar with life sciences are needed to be able to maintain the system in bioinformatics. In addition, researchers will also need to be trained that may cost the organization.

18.10 THE EVOLUTION OF BIOINFORMATICS

Bioinformatics deals with computer management and analysis of biological information: genes, genomes, proteins, cells, ecological systems, medical information, robots, and artificial intelligence as there are many applications of bioinformatics from the combination of computer and biology. The evolution of technology helps in supporting bioinformatics in discovering diseases and application in forensics using software packages and bioinformatics tools. For example, the evolution of technology by bioinformatics such as IIIumina next-generation sequencing (NGS) to provide accurate sequencing. NGS technology can provide valuable and useful information for a better understanding in health and diseases.

The use of bioinformatics can also be useful in determining the order of the four chemical building blocks called bases that make up the DNA molecule. This is because the sequence provides scientists regarding what kind of genetic information is carried in a particular DNA segment. Moreover, the sequence data can highlight the changes in a gene that may cause disease.

18.10.1 FORENSIC DNA AND BIOINFORMATICS

Bioinformatics and forensic DNA are fundamentally characterized by both studies and draw their techniques from statistics and computer science which facilitate in solving the problems in law and biology. It could be useful to identify victim or suspect with personal relatedness to other individuals that focuses of forensic DNA analysis. It is a common event in forensic analysis especially by crime and investigation unit or CSI[1] by looking at close connections; for example, paternity disputes, suspected incest case, corpse identification, alimentary frauds (e.g., GMO,[2] poisonous food, etc.), semen detection on underwear for suspected infidelity, insurance company fraud investigations when the actual driver in a vehicle accident is in question, criminal matters, and autopsies for human identification following accident investigations. All of these problems may be solved by using bioinformatics methods.[10]

Also, genetic tests have been widely used for major catastrophic events such as terrorist attacks, airplane crash, and tsunami disaster. It can be used for mass fatality identification and forensics evidences. Personal identification

---

[1]Crime Scene Investigation.
[2]genetically modified organism.

relies on identifiable characteristics as the human body has a personal identity that is unique biologically (such as blood, saliva, and DNA) and physiological difference (such as fingerprints, eye irises and retinas, hand palms and geometry, and facial geometry), and also behaviorally different (such as body posture, habits, signature, keystroke dynamics, and lip motion, and on combination of physiological and dynamical characteristics such as the voice).

Hence, genetic testing results are integrated with the information collected by multidisciplinary teams composed of medical examiners, forensic pathologists, anthropologists, forensic dentists, fingerprint specialists, radiologists, and experts in search and recovery of physical evidence. Officers could have access to the personal information where biological data can be obtained from hospital records and behavioral data may be collected from banks or office document such as fingerprint or signature just by looking at the database.

Therefore, the application of genetic testing in large-scale tissue sampling and long-term DNA preservation plays an important role in mass fatalities which have been recently labeled.[10] Thus, DNA has become the most important personal identification characteristic because all genetic differences whether being expressed regions of DNA (genes) or some segments of DNA have characteristic of a person and it poses coding pattern of inheritance that can be monitored and used as markers.

18.10.2 BIOINFORMATICS AND CANCER

According to the Cancer Research United Kingdom,[2] cancer is one of the leading causes of death worldwide where 14.1 million cases of cancer are recorded and about 8.2 million worldwide deaths have been estimated in 2012. A leading cause of cancer is any malignant growth or tumor caused by abnormal and uncontrolled division and the changes of DNA in cell by mutation. Also, errors in the genes may cause this abnormal behavior to be cancerous. These changes develop when exposed to a certain type of cancer-causing substances.

In the era of postgenome, age holds phenomenal promise for identifying the mechanistic bases of organismal development, metabolic processes, and diseases. Bioinformatics research will lead to a wide understanding at the regulation of gene expression, protein structure determination, comparative evolution, and drug discovery. Presently, 2D gel protein pattern can be easily analyzed using bioinformatics technology where these software applications possess user-friendly interfaces that are incorporated with tools for linearization and merging of scanned images.

New techniques and new collaborations between computer scientists, biostatisticians, and biologists are required in today's research. There is a need to develop and integrate database repositories for the various sources of data being collected, to develop tools for transforming raw primary data into forms suitable for public dissemination or formal data analysis, to obtain and develop user interfaces to store, retrieve, and visualize data from databases, and to develop efficient and valid methods of data analysis.[1]

Cancer DNA sequencing using NGS provides better information and is less time-consuming compared to normal gene sequencing using gel structure. With NGS, researchers can perform whole genome studies, targeted gene profiling, and tumor–normal comparisons. Therefore, it is easy to detect tumor and DNA fragments with detailed quantitative measurements from the database.

Furthermore, the prediction of genes is likely to be linked to a newly developed disease or a modified version of old disease that evolve or mutate. The use of bioinformatics can easily recognize and relate genes that as similar to any function or characteristic from an original gene such as the similarity in percentage of DNA sequence. The highest challenge is to identify enormous markers of DNA as the application of molecular links to diseases will continue to face technological challenges as well as biological and algorithm. The human body consists of very complicated and diverse features because it is continually evolving and responding to changes.

As for using bioinformatics to replicate, the structure of a new DNA provides a challenge in technology. This is because other interrelationship such as cells that may not be visible through the microscopic view. Thus, the already designed computer frameworks or databases may not cope with the expanding network-level measurements and information.

18.10.3 ETHICAL ISSUE

According to Johnson (1999), "computer ethics has followed computer technology in its evolution, and for the same reason computer ethics as a separate discipline will disappear in the near future. In fact, when computing becomes a mature technology, the problem of its (urgent) ethical and social impacts due to policy vacuums will diminish, and using computers as a means of achieving some goals will become part of ordinary human action."

What he said was, as time changes from period to period, technology changed. He believes that when the technology changes, the ethics of computer also changes when there is an adaptation of easily accessible technology.

Another citation from Johnson "Once the new instrumentation is incorporated into ethical thinking, it becomes the presumed background condition. What was for a time an issue of computer ethics becomes simply an ethical issue. Copying software becomes simply an issue of intellectual property. Selling software involves certain legal and moral liabilities. Computer professionals understand they have responsibilities. Online privacy violations are simply privacy violations. So, as we come to presume computer technology as part of the world we live in, computer ethics as such is likely to disappear."

18.10.4   INFECTIOUS DISEASES ETHICS

The emphasis of human bioethics in the 1950s and 1960s coincided with a widespread belief in particular area and time (but, with hindsight, unwarranted and dangerous) that the problems of infectious diseases had been solved by sanitation, immunization, and antibiotic therapy. The much quoted pronouncement that "it is time to close the book on infectious disease" is usually attributed to former US Surgeon General William Stewart. Although there appears to be no evidence that he ever actually said this, "the sentiment was certainly widely shared" at the time (Sassetti & Rubin, 2007). This widespread complacency remained largely unchallenged throughout most of the 20th century. It was dispelled by the unfolding of HIV pandemic and the plethora of other emerging and reemerging infectious diseases that followed (or in some cases preceded) it, but it had already contributed to the gross neglect of infectious diseases by bioethicists (Selgelid & Selgelid, 2005).

AIDS was a rare exception, but many of the ethical issues raised—confidentiality, discrimination, patient's rights, and sexual freedom—were not specifically related to its status as an infectious disease. Belatedly, this neglect is now being addressed; infectious diseases have at last come to the attention of bioethicists. During the 21st century, public health ethics has become a rapidly growing subdiscipline of bioethics, and much of the public health ethics literature has focused on infectious disease in particular. In addition to AIDS, attention has especially being focused on severe acquired respiratory syndrome (SARS), pandemic influenza planning and issues related to bioterrorism[14]

There has also been debate about the ethics of issues such as: intellectual property rights, relating to antimicrobial agents and their implications for the access to essential treatment of infectious diseases and the relationship between marketing of antimicrobials and the emergence of antibiotic resistance (Selgelid, 2007).

By the citation of the researchers, it can be argued that using bioinformatics technology can make it more efficient and much more reliable. But it might as well give the people a negative impact such as the confidentiality as well as the consent of the participants.

18.11   EXAMPLES OF BIOINFORMATICS IN INFORMATION CONCEPT SYSTEM

Generally, information system concept in bioinformatics has the same aim, and hence has similar flow of procedure. It organizes data which all users had input and access existing data, analyze each input and interpret the results in a biological manner.

18.11.1   EUROPEAN BIOINFORMATICS INSTITUTE

The European Bioinformatics Institute (EBI) is a research and services center in bioinformatics. This database provides researchers with molecular biology, genetics, medicine, biotechnology, and industries related to pharmaceutical and chemicals.

There are various ways for data entry. Data input can be done via web, accessing their website; or via Sequin, a developed tool accessed via an FTP server; or via e-mail to users whose internet access is through e-mailing services.

Data inputted by users are then analyzed whether there are new data or existing data. For instance, a sample of an unknown virus shows similar signs and symptoms as an existing virus. This data is then compared to each other and produce a result to which it is further elaborated and understood to be interpreted by the user.

18.11.2   PANCANRISK

PanCanRisk is a European project which aims to identify cancer vulnerability and clinical management via bioinformatics. With this, they are able to predict the treatments to cancer as the cancer genome sequencing is very challenging to be understood.

The company intends to give a deliberate, cross-disciplinary structure for a superior comprehension, joining and utilization of tumor clinical information in the assessment of the large number of hereditary variations and changes included in growth vulnerability, for the immediate advantage of disease patients.

Similar to that of EBI, PanCanRisk uses existing data to compare to newly provided sample and looks for varieties of genotypes vulnerable of getting cancer.

18.11.3 GENOGRAPHIC

The project is anonymous, nonmedical, and nonprofit, and all results will be placed in the public domain following scientific peer publication. It is an anonymous, nonmedical, and nonprofitable project, sponsored by National Geographic Society and Waitt Family foundation which helps with migratory history of the human species. This is so the project will be able to compile the data in collaboration with the indigenous and traditional people globally. This will study historical DNA patterns from contributors worldly to better recognize our anthropological genetic heritages. All results are published and accessible to the public.

18.11.4 GEMINI

It is a flexible software for exploring all forms of human genetic variation. It is designed for reproducibility and flexibility for biologists and researchers with a standard framework for medical genomics. Gemini incorporates genetic variation with various and adjustable set of genome annotations into a unified database to enable interpretation and data exploration.

Among many bioinformatics service providers and softwares, these are the four which stands out and yet still have improvement that can be done. The development not only will take time but also will be very costly. Most of the concerns raised along with the advancement of the system is the purchase of the machines required to run such experiments for researchers and students majoring in bioinformatics.

18.12 OPINION

We believe that bioinformatics deals with computer management and analysis of biological information such as genes, genomes, proteins, cells, ecological systems, medical information, robots, and artificial intelligence since there are many applications of bioinformatics from the combination of computer and biology. The evolution of technology helps in supporting bioinformatics in discovering diseases and application in forensics using software packages and bioinformatics tools. Also, we believe that information system could be the key factor in facilitating and attaining an efficient decision-making in medical research as the knowledge and understanding obtained from bioinformatics could be sensibly used for applications that cover drug discovery, genome analysis, and biological control.

18.13 CONCLUSION

In conclusion, with the current deluge of data, computational methods have become indispensable to biological investigations. Thus, bioinformatics tools hold a huge potential for use in medical research and clinical practice as the analysis of genetic information offered by bioinformatics and the study of systems behavior with detailed mathematical models may lead to huge benefits for drug development and personalized health care. Also, the research and education in life sciences are increasingly dependent on bioinformatics and advanced information system to support the evidence using large set of data.

KEYWORDS

- **bioinformatics**
- **information system**
- **technology**
- **cancer**
- **DNA**

REFERENCES

1. Bensmail, H.; Haoudi, A. Postgenomics: Proteomics and Bioinformatics in Cancer Research. *BioMed Res. Int.* **2003**, *4*, 217–230.
2. Cancer Research UK. Worldwide Cancer Statistics, 2014. http://www.cancerresearchuk. org/health-professional/worldwide-cancer-statistics (accessed Oct 2017).

3. Cannataro, M.; Santos, R. W.; Sundnes, J. In *Bioinformatics' Challenges to Computer Science: Bioinformatics Tools and Biomedical Modeling, Part I*, International Conference on Computational Science, May 25, 2009; Springer: Berlin, Heidelberg, 2009; pp 807–809 (LNCS 5544).

4. Paila, U.; Chapman, B. A.; Kirchner, R.; Quinlan, A. R. GEMINI: Integrative Exploration of Genetic Variation and Genome Annotations. *PLoS Comput. Biol.* **2013,** *9*(7), e1003153.

5. Mandoiu, I.; Narasiman, G.; Zhang, Y. In *Bioinformatics Research and Applications*, 5th International Symposium, Fort Lauderale, FL, USA, May 13–16, 2009; Springer Science & Business Media: Berlin, 2009.

6. Merelli, I.; Cozzi, P.; Ronchieri, E.; Cesini, D.; D'Agostino, D. Porting Bioinformatics Applications from Grid to Cloud: A Macromolecular Surface Analysis Application Case Study. *Int. J. High Perform. Comput. Appl.* **2017,** *31*(3), 182–195 (Institute of Biomedical Technologies, National Research Council of Italy).

7. Katara, P. Role of Bioinformatics and Pharmacogenomics in Drug Discovery and Development Process. *Netw. Model. Anal. Health Inform. Bioinform.* **2013,** *2*(4), 225–230 (Center for Bioinformatics, IIDS, University of Allahabad).

8. Lucia, B.; Pietro, L. Forensic DNA and Bioinformatics. *Brief. Bioinform.* **2007,** *8*(2), 117–128.

9. Luscombe, N. M.; Greenbaum, D.; Gerstein, M. What is Bioinformatics? An Introduction and Overview. *Yearb. Med. Inform.* **2001,** (1), 83–99 (Department of Molecular Biophysics and Biochemistry, Yale University, New Haven, USA).

10. Reid, L. Diminishing Returns? Risk and the Duty to Care in the SARS Epidemic. *Bioethics* **2005,** *19*(4), 348–361.

11. Rodriguez-tomé, P. EBI Databases and Services. *Mol. Biotechnol.* **2001,** *18*(3). http://link.springer.com/article/10.1385%2FMB%3A18%3A3%3A199 (accessed Feb 20, 2016).

12. Santhaiah, C.; Reddy, R. M. Role of Computers in Bioinformatics by Using Different Biological Datasets. *J. Comput. Eng. (IOSR-JCE)* **2014,** *16*(2), 80–83 (2278–8727).

13. Yao, J.; Zhang, J.; Chen, S.; Wang, C.; Levy, D.; Liu, C. A Mobile Cloud with Trusted Data Provenance Services for Bioinformatics Research. In *Data Provenance and Data Management in eScience;* Springer: Berlin Heidelberg, 2013; Vol. 426, pp 109–128 (Information Engineering Laboratory, CSIRO ICT Centre, Australia).

14. Selgelid, M. J. Ethics and Infectious Disease. *Bioethics* **2005,** *19*(3), 272–289.

15. Johnson, D. G. *Computer Ethics;* Englewood Cliffs, NJ, 1985.

INDEX

Printed and bound by CPI Group (UK) Ltd, Croydon, CR0 4YY

23/10/2024

01777701-0016